SOLVING FERMI'S PARADOX

The Search for Extraterrestrial Intelligence (SETI) has for 60 years attempted to solve Fermi's paradox: if intelligent life is relatively common in the universe, where is everybody? Examining SETI through this lens, this volume summarises current thinking on the prevalence of intelligent life in the universe and discusses 66 distinct solutions to the so-called paradox. It describes the methodology of SETI and how many disciplines feed into the debate, from physics and biology, to philosophy and anthropology. The presented solutions are organised into three key groups: rare-Earth solutions, suggesting that planetary habitability, life and intelligence are uncommon; catastrophist solutions, arguing that civilisations do not survive long enough to make contact; and non-empirical solutions, which take theoretical approaches, such as that our methodology is flawed. This comprehensive introduction to SETI concludes by looking at the future of the field and speculating on humanity's potential fate.

DUNCAN H. FORGAN is Associate Lecturer at the Centre for Exoplanet Science at the University of St Andrews. He is a founding member of the UK SETI research network and leads UK research efforts into the Search for Extraterrestrial Intelligence. His work includes simulations of civilisation evolution, developing observables of intelligence and policy for post-detection scenarios.

CAMBRIDGE ASTROBIOLOGY

Series Editors
Bruce Jakosky, Alan Boss, Frances Westall, Daniel Prieur, and Charles Cockell

Books in the Series:

1. *Planet Formation: Theory, Observations, and Experiments*
   Edited by Hubert Klahr and Wolfgang Brandner
   ISBN 978-0-521-18074-0

2. *Fitness of the Cosmos for Life: Biochemistry and Fine-Tuning*
   Edited by John D. Barrow, Simon Conway Morris, Stephen J. Freeland, and Charles L. Harper, Jr.
   ISBN 978-0-521-87102-0

3. *Planetary Systems and the Origin of Life*
   Edited by Ralph Pudritz, Paul Higgs, and Jonathan Stone
   ISBN 978-0-521-87548-6

4. *Exploring the Origin, Extent, and Future of Life: Philosophical, Ethical, and Theological Perspectives*
   Edited by Constance M. Bertka
   ISBN 978-0-521-86363-6

5. *Life in Antarctic Deserts and Other Cold Dry Environments*
   Edited by Peter T. Doran, W. Berry Lyons, and Diane M. McKnight
   ISBN 978-0521-88919-3

6. *Origins and Evolution of Life: An Astrobiological Perspective*
   Edited by Muriel Gargaud, Purificación Lopez-Garcia, and Hervé Martin
   ISBN 978 0521-76131-4

7. *The Astrobiogical Landscape: Philosophical Foundations of the Study of Cosmic Life*
   Milan M. Ćirković
   ISBN 978 0521-19775-5

8. *The Drake Equation: Estimating the Prevalence of Extraterrestrial Life through the Ages*
   Edited by Douglas A. Vakoch and Matthew F. Dowd
   ISBN 978-1-107-07365-4

9. *Astrobiology, Discovery, and Societal Impact*
   Steven J. Dick
   ISBN 978-1-108-42676-3

10. *Solving Fermi's Paradox*
    Duncan H. Forgan
    ISBN 978-1-107-16365-2

SOLVING FERMI'S PARADOX

DUNCAN H. FORGAN
University of St Andrews, Scotland

CAMBRIDGE
UNIVERSITY PRESS

University Printing House, Cambridge CB2 8BS, United Kingdom

One Liberty Plaza, 20th Floor, New York, NY 10006, USA

477 Williamstown Road, Port Melbourne, VIC 3207, Australia

314–321, 3rd Floor, Plot 3, Splendor Forum, Jasola District Centre,
New Delhi – 110025, India

79 Anson Road, #06–04/06, Singapore 079906

Cambridge University Press is part of the University of Cambridge.

It furthers the University's mission by disseminating knowledge in the pursuit of education, learning, and research at the highest international levels of excellence.

www.cambridge.org
Information on this title: www.cambridge.org/9781107163652
DOI: 10.1017/9781316681510

© Duncan Forgan 2019

This publication is in copyright. Subject to statutory exception and to the provisions of relevant collective licensing agreements, no reproduction of any part may take place without the written permission of Cambridge University Press.

First published 2019

Printed in the United Kingdom by TJ International Ltd. Padstow Cornwall

A catalogue record for this publication is available from the British Library.

Library of Congress Cataloging-in-Publication Data
Names: Forgan, Duncan, 1984– author.
Title: Solving Fermi's paradox / Duncan Forgan
(University of St Andrews, Scotland).
Description: Cambridge ; New York, NY : Cambridge University Press, 2019. |
Includes bibliographical references and index.
Identifiers: LCCN 2018052247 | ISBN 9781107163652 (alk. paper)
Subjects: LCSH: Fermi's paradox. | Life on other planets.
Classification: LCC QB54.F77945 2018 | DDC 999–dc23
LC record available at https://lccn.loc.gov/2018052247

ISBN 978-1-107-16365-2 Hardback

Cambridge University Press has no responsibility for the persistence or accuracy of URLs for external or third-party internet websites referred to in this publication and does not guarantee that any content on such websites is, or will remain, accurate or appropriate.

For Rosa

Contents

Preface			*page* xvii
Part I	**Introduction**		1
1	Introducing the Paradox		3
	1.1	Fermi and His Paradox	3
		1.1.1 The Paradox, in Brief	3
		1.1.2 Fermi's Paradox Is Not His, and It Isn't a Paradox	4
	1.2	Drake and His Equation	7
	1.3	The Great Filter	11
2	Fact A – The Great Silence		15
	2.1	Radio SETI	21
		2.1.1 The Water Hole	21
		2.1.2 Doppler Drift	24
		2.1.3 Radio Frequency Interference	24
		2.1.4 Single-Dish Radio Astronomy	25
		2.1.5 Multi-Dish Radio Interferometry	27
		2.1.6 A History of Radio SETI observations	30
	2.2	Optical SETI	30
		2.2.1 Artefact SETI	35
		2.2.2 A History of Optical SETI Observations	37
3	Classifying Scenarios and Solutions to the Paradox		38
	3.1	Contact Scenarios and the Rio Scale	38
	3.2	The Three Classes of Civilisation – the Kardashev Scale	40
	3.3	The Three Classes of Solutions to Fermi's Paradox	43
		3.3.1 Rare Earth (R) Solutions	43
		3.3.2 Uncommunicative (U) Solutions	44

		3.3.3	Catastrophic (C) Solutions	44
		3.3.4	Hard and Soft Solutions	44

Part II Rare Earth Solutions   45

4   Habitable Worlds Are Rare   47
- 4.1   Worlds Are Not Rare   47
  - 4.1.1   The First Exoplanets – Pulsar Timing   47
  - 4.1.2   The Radial Velocity Method   48
  - 4.1.3   Transit Detection   53
  - 4.1.4   Gravitational Microlensing   56
  - 4.1.5   Direct Imaging   61
  - 4.1.6   Detecting Life on Exoplanets   64
  - 4.1.7   The Exoplanet Population   68
- 4.2   Habitable Zones   71
  - 4.2.1   Circumstellar Habitable Zones   71
  - 4.2.2   Galactic Habitable Zones   75
- 4.3   The Earth as a Habitable Planet   84
- 4.4   Good and Bad Planetary Neighbours   87
  - 4.4.1   Milankovitch Cycles   87
  - 4.4.2   Planetary Neighbours as Asteroid Shields   89
- 4.5   Does the Earth Need the Moon To Be Habitable?   91
  - 4.5.1   What Influence Does the Moon Have on Earth's Habitability?   91
  - 4.5.2   What If the Moon Never Existed?   93
  - 4.5.3   Are Large Moons Common in Terrestrial Planet Formation?   95

5   Life Is Rare   98
- 5.1   What Is Life?   98
- 5.2   The Origins of Life   100
  - 5.2.1   Oparin-Haldane Theory and the Miller-Urey Experiments   101
  - 5.2.2   'Vents' Theories   103
  - 5.2.3   'Pools' Theories   104
  - 5.2.4   Multiple Origins of Life?   104
- 5.3   The Major Evolutionary Transitions in Terrestrial Organisms   104
  - 5.3.1   Free-Floating Self-Replicating Molecules to Molecules in Compartments   105
  - 5.3.2   Independent Replicators to Chromosomes   105

		5.3.3 Cyanobacteria to Photosynthesis	109
		5.3.4 Prokaryotes to Eukaryotes	111
		5.3.5 Single-Cell to Multicellular Organisms	112
		5.3.6 Asexual Reproduction to Sexual Reproduction	115
		5.3.7 The Brain	119
		5.3.8 Solitary Individuals to Social Units	121
		5.3.9 Social Units to Civilisation	125
		5.3.10 Biology to Post-Biology?	125
6	Intelligence Is Rare		127
	6.1	The Evolutionist Argument against Intelligence	127
	6.2	What Is Intelligence?	129
	6.3	How Did Humans Become Intelligent?	130
		6.3.1 Why Be Intelligent at All?	135
	6.4	What Is the Role of Language?	136
		6.4.1 The Origin of Human Language	136
		6.4.2 Possible Biological Origins of Language	139
		6.4.3 What Can We Say About the Language of ETIs?	141
		6.4.4 Non-Human Communication	141
	6.5	What Is the Role of Art?	145
	6.6	What Is the Role of Science and Mathematics?	149
	6.7	What Is the Role of Political Structures?	150
Part III	**Catastrophist Solutions**		**153**
7	Doomsday Arguments		155
	7.1	The Carter-Leslie Argument	155
	7.2	Gott's Argument	157
8	Death by Impact		159
	8.1	Asteroids	159
	8.2	Comets	160
	8.3	The Risk of Impact on Earth	161
	8.4	The Consequences of Impact	164
	8.5	Impact as an Existential Risk to ETIs	166
9	Death by Terrestrial Disaster		169
	9.1	Supervolcanism	169
	9.2	Magnetic Field Collapse	173

Contents

10	Death by Star	179
	10.1 Stellar Activity	179
	10.1.1 Solar Winds	181
	10.1.2 Flares and Coronal Mass Ejections	181
	10.1.3 Earth's Protection from Stellar Activity	184
	10.2 Stellar Evolution	184
	10.3 Death by Stellar Evolution	190
	10.4 Death by Red Giant	191
	10.5 Death by Supernova	194
	10.5.1 Type Ia (White Dwarf Nova)	194
	10.5.2 Type II Supernovae (Core-Collapse)	196
	10.6 Close Stellar Encounters	198
11	Death on a Galactic Scale?	201
	11.1 Active Galactic Nuclei	201
	11.2 Gamma Ray Bursts	204
	11.3 Terrestrial Atmospheric Responses to Very High-Energy Events	206
	11.3.1 UV Radiation	206
	11.3.2 Gamma and X-Rays	207
	11.3.3 Cosmic Rays	207
	11.3.4 Muons and Other Secondary Particles	208
	11.3.5 The Role of Heliospheric Screening	208
	11.3.6 Mitigation against High-Energy Radiation Events	209
12	Death by Unsustainable Growth	212
13	Death by Self-Induced Environmental Change	218
	13.1 Atmospheric Damage	218
	13.2 Destruction of the Land Environment	221
	13.3 Destruction of the Ocean Environment	222
	13.4 Destruction of the Biosphere	223
	13.5 Destruction of the Space Environment (Kessler Syndrome)	224
14	Self-Destruction at the Nanoscale	228
	14.1 Genetic Engineering	228
	14.2 Nanotechnology, Nanomachines and 'Grey Goo'	230
	14.3 Estimates of Existential Risk from Nanoscale Manipulation	231
15	Artificial Intelligence and the Singularity	234
	15.1 The Turing Machine	235
	15.2 A Brief History of AI Techniques	236

		15.2.1 Logic-Driven AI: Symbolic Approaches	236
		15.2.2 Biologically Motivated Approaches	239
		15.2.3 Data-Driven Approaches	241
	15.3	The Turing Test	242
	15.4	Is the Singularity Inevitable?	243
	15.5	Controls on AI	245
	15.6	Is Strong AI Likely?	246
16	War		248
	16.1	The Origins of Human Warfare	249
	16.2	Is Violence Innate to Humans?	250
	16.3	Nuclear War and Weapons of Mass Destruction	251
	16.4	Is War Inevitable amongst Intelligent Civilisations?	253
17	Societal Collapse		254
	17.1	Common Causes of Societal Collapse	255
	17.2	Intellectual Collapse	258

Part IV Uncommunicative Solutions — 259

18	Intelligent Life Is New		261
	18.1	Carter's Argument against SETI	261
	18.2	Global Regulation Mechanisms and the Phase Transition Solution	262
19	Exploration Is Imperfect		265
	19.1	Percolation Theory Models	265
	19.2	Exploration of Realistic Star Fields	267
	19.3	Human Designs for Interstellar Probes	270
		19.3.1 Chemical Rockets	271
		19.3.2 External Nuclear Pulse Propulsion	271
		19.3.3 Internal Nuclear Pulse Propulsion	272
		19.3.4 Antimatter Drive	273
		19.3.5 The Interstellar Ramjet	274
		19.3.6 Light Sails	275
		19.3.7 Gravitational Assists	278
		19.3.8 Relativistic Effects	279
20	Probe Exploration Is Dangerous		282
	20.1	The Self-Replicating Probe	282
	20.2	Can Probes Self-Replicate without Evolving?	284
	20.3	Predators and Prey	285

21	The Aliens Are Quiet	291
	21.1 They Communicate at a Different Frequency	291
	21.2 They Communicate Using a Different Method	296
	21.2.1 Neutrinos	296
	21.2.2 Gravitational Waves	298
	21.3 We Do Not Understand Them	304
	21.4 They Do Not Communicate	306
	21.4.1 Arguments in Favour of METI	306
	21.4.2 Arguments against METI	313
	21.4.3 Can the METI Debate Be Resolved?	316
22	They Live Too Far Away	317
	22.1 The Feasibility of Crewed Interstellar Travel	318
	22.2 The Feasibility of Reliable Interstellar Communication Networks	322
23	The Zoo/Interdict Hypothesis	326
24	The Simulation Hypothesis	330
	24.1 How Much Computing Power Is Needed?	332
	24.2 Where Could We Get It From?	334
	24.2.1 The Matrioshka Brain	335
	24.3 Arguments against the Simulation Hypothesis	336
25	They Are Already Here	338
	25.1 Bracewell or 'Lurker' Probes	338
	25.2 Unidentified Flying Objects and Accounts of Alien Abduction	342
26	They Were Here Long Ago	346
	26.1 The 'Ancient Astronaut' Hypothesis	347

Part V Conclusions   351

27	Solving Fermi's Paradox	353
	27.1 The Future of Fact A	353
	27.1.1 Current and Future Radio/Optical SETI Missions	353
	27.1.2 Gravitational Waves and SETI	355
	27.1.3 Neutrino SETI	355
	27.1.4 Transit SETI	356
	27.2 Future Attempts at Interstellar Missions	358
	27.3 Theoretical and Numerical Advances	360

		27.3.1 Numerical Modelling of Civilisation Growth, Evolution and Interaction	361
		27.3.2 Characterising Observables of Intelligence in Astrophysical Data	364
	27.4	Can the Paradox Be Solved?	367
Appendix A		A Database of Solutions to Fermi's Paradox	370
	A.1	Rare Earth Solutions	370
	A.2	Catastrophist Solutions	372
	A.3	Uncommunicative Solutions	373
References			375
Index			411

Preface

This textbook attempts to tackle the many academic concepts and theories that guide our thinking about the search for extraterrestrial intelligence (SETI), using Fermi's Paradox as a lens. While writing it, I have aimed at two objectives:

1. Provide an exhaustive list of solutions to Fermi's Paradox as of the time of publication, and
2. Provide context to this list of solutions, by describing and explaining their theoretical/numerical/observational underpinnings.

This is no mean feat, and the lack of graduate-level textbooks which attempt to achieve both objectives is probably an indication. We will see that the question raised by Fermi's Paradox requires an extremely broad swathe of academic disciplines to formulate a complete set of solutions. In fact, it could be argued that *all* academic disciplines have contributions to make, and that SETI is a study of the human condition, generalised to include all sentient beings.

Obviously, a textbook that requires contributions from the entirety of academe will be necessarily abbreviated, and in the interests of focus I have been forced to edit out or ignore aspects that would otherwise merit more attention. As such, this book should be seen as an introduction to the most pertinent facts and concepts. For example, in the Introduction I give a brief account of the observational aspects of SETI, but the field is worthy of far more discussion. This can be found in a range of excellent textbooks on observational SETI and radio astronomy in general (see References).

This book is dominated by astronomy in its make-up, principally because SETI remains an astronomy-dominated discipline. What should hopefully be clear upon reading this textbook is that SETI's accounting of other disciplines is quite uneven, and there are many topics highly deserving of further study. I will do my best to highlight areas where I believe that further work is necessary, to encourage the reader to help fill in these gaps, so that subsequent editions of this book can display our more developed understanding of life and intelligence in the Universe.

Part I
Introduction

1

Introducing the Paradox

1.1 Fermi and His Paradox

1.1.1 The Paradox, in Brief

Consider the Milky Way galaxy. It is a barred spiral galaxy, corresponding to the Hubble class Sbc, indicating its spirals are relatively loosely wound. It can be separated into several components (by mass): overlapping thick and thin discs of stars, gas and dust with spiral structure, an inner bulge (and bar), a spheroidal halo of relatively old stars and a far greater dark matter halo, which is the most massive component and encompasses the entire galaxy, extending far beyond its visible component.

The Milky Way's stellar disc is approximately 160,000 light-years (about 50 kiloparsecs or kpc) in diameter (Xu et al., 2015). If we consider an intelligent civilisation that wishes to explore the entire galaxy, it is estimated that it could do so in a timescale, $\tau_{explore}$ at most one hundred million years, and most likely much less than that (we will explore the rationale behind these estimates later).

Now consider the Earth. Its parent star, the Sun, orbits the Galactic Centre at a distance of around 26,000 light-years (around 8 kpc). The Earth's age is established to be at least 4.4 Gyr from radiometric dating of zircons – age estimates of $4.54^{+0.1}_{-0.03}$ Gyr is determined from radiometric dating of meteorites (see, e.g., Dalrymple, 2001). Fermi's Paradox then becomes clear if we express the Earth's age in units of $\tau_{explore}$:

$$\tau_\oplus > 45\tau_{explore}. \quad (1.1)$$

The Paradox is best described in the language of a logical argument. Formally, logical arguments consist of:

- a set of *premises* (sometimes *premisses*), statements or propositions assumed to be true at the beginning of the argument;

- a set of *inferences* from those premises, derived using (for example) rules of propositional logic, and
- a *conclusion*, a statement or proposition derived from the premises and inferential steps above.

A *valid* argument guarantees the conclusion is true if the premises are true. An argument is *sound* if it is valid and the premises are indeed true. We can therefore be in a situation where an argument is valid but unsound – the logical argument is internally consistent but built on premises that are demonstrably false.

Now consider the following premises:

Premise 1. Humans are not the only intelligent species in the Milky Way.
Premise 2. The Earth is not the oldest terrestrial planet in the Milky Way.
Premise 3. Other intelligent civilisations are capable of making themselves known to us.

Some inferential steps might be:

Therefore. Other civilisations currently exist in the Milky Way, and have done so for longer than humanity.
Therefore. Other civilisations have had more than enough time to make themselves known to us.

With a final conclusion:

Therefore. Humans must have evidence for other civilisations in the Milky Way.

The 'Paradox' is the fact that the conclusion is false – humans do *not* have evidence for other intelligent civilisations, despite the validity (and apparent soundness) of the above argument. As will be clear to those familiar with formal logic, this argument does not in fact constitute a paradox, as we will discuss below. It is a valid argument that arrives at a conclusion that is contradicted by astronomical observations and other scientific evidence. We can resolve this by either showing one or more of the premises to be false, making the argument unsound, or showing one of the inferential steps to either be false or not proceed logically from the premises.

1.1.2 Fermi's Paradox Is Not His, and It Isn't a Paradox

There is an ugly truth to Fermi's Paradox – at its base, it does not belong to Fermi, and as we have seen from the previous section, it is not formally a paradox.

It might be better described as the Fermi Question, the Fermi-Hart Paradox, the Hart-Tipler Argument or even, as suggested by Webb (2002), the 'Tsiolkovsky-Fermi-Viewing-Hart Paradox', giving Tsiolkovsky the original credit for the concept in the early 1930s. However, the moniker 'Fermi's Paradox' has taken hold in the literature and the Search for Extraterrestrial Intelligence (SETI) community, so we will persist with the name that is something of a misnomer.

The origins of what is now called Fermi's Paradox are typically traced back to a now-legendary lunchtime conversation at Los Alamos National Laboratory in 1950, between Enrico Fermi, Emil Konopinski, Edward Teller and Herbert York, all four of whom were deeply involved in the development of nuclear technology.

Fermi's own memories of the conversation are not a matter of record (Fermi died in 1954). Later conversations with the other participants are our only source (Jones, 1985; Gray, 2015). The timing of the conversation is confirmed by testimony that indicates it preceded the 'George shot' of 8–9 May 1951, a nuclear bomb test that proved to be a crucial milestone in the development of thermonuclear weapons. Konopinski recalled joining the others as they walked to lunch:

When I joined the party I found being discussed evidence about flying saucers. That immediately brought to my mind a cartoon I had recently seen in the New Yorker, explaining why public trash cans were disappearing from the streets of New York City. The cartoon showed what was evidently a flying saucer sitting in the background and, streaming toward it, 'little green men' (endowed with antennas) carrying trash cans. There ensued a discussion as to whether the saucers could somehow exceed the speed of light.

(Gray, 2015, p. 196)

The cartoon also helps us to pin down the date further, to the summer of 1950. We can already see here the beginnings of the logical arguments that form the foundations of Fermi's Paradox. Fermi is then said to have asked Teller:

Edward, what do you think? How probable is it that within the next ten years we shall have material evidence of objects travelling faster than light?

(Jones, 1985, p. 2)

Teller's estimate of the probability was pessimistic (10^{-6}, i.e., one in a million). Fermi ascribed a 10% probability, and this became a topic of debate until the group sat down at lunch (accounts differ on the total number of people present). The conversation moved on to other topics. Accounts reach consensus on the next moment, where Fermi pronounced, apropos of nothing:

Where is everybody?

(Jones, 1985, p. 3)

The precise wording of the question is also in doubt, but all agree on its substance. Where the discussion pivoted to following this remark is also unclear. Teller dismissed the significance of what was next said:

I do not believe that much came from this conversation, except perhaps a statement that the distances to the next location of living beings may be very great and that, indeed, as far as our galaxy is concerned, we are living somewhere in the sticks, far removed from the metropolitan area of the galactic center.

(Jones, 1985, p. 3)

York suggested that Fermi went on to estimate (in his now famous style) the probability of Earthlike planets, the probability of life appearing on these planets, the probability of intelligent creatures and so on. These factors are direct ancestors of the terms in Drake's equation, as we will see shortly. York's own admission that his memory of this moment was 'hazy' may lead one to the conclusion that York's memories of 1950 are confabulated with subsequent events in the 1960s, which is wholly plausible.

So where do Hart and Tipler fit into this story? Michael Hart is widely credited with formulating Fermi's Paradox as it is understood today, and yet does not receive the appropriate credit. In his paper *Explanation for the Absence of Extraterrestrials on Earth*, Hart (1975) declares 'Fact A' – the lack of evidence for intelligent beings other than humans on Earth. His argument is simple – if intelligent beings exist, and they are capable of space travel, then the Milky Way can be colonised on a timescale short relative to the Earth's age, and hence it seems that intelligent beings should already be here, but are not according to Fact A. Hart begins listing several possible solutions to this incongruity, and is one of the first to publish work on what we now refer to as Fermi's Paradox.

Tipler is usually credited alongside Hart for formulating perhaps the most binding form of the Paradox. He invokes the universal constructor or von Neumann machine as the perfect device for galactic exploration. Essentially computers capable of programmed self-replication, a single probe can grow exponentially into a fleet of probes to rapidly explore the entire Milky Way (we will delve into the details in section 20.1).

Tipler argues that any intelligent species that is capable of interstellar communication will eventually construct von Neumann devices, and as such if other intelligent beings exist, it is highly likely that the Milky Way has already been explored. Tipler then invokes Hart's Fact A to insist that there are no extraterrestrial intelligences, and consequently the SETI endeavour is fruitless (see, e.g., Tipler, 1993).

This formulation is largely considered to be the most stringent form of Fermi's Paradox, as it is independent of civilisation lifetime. Once an initial probe is launched, the exploration continues even after a civilisation dies. It also considers technological trajectories that humans are pursuing, through greater automation and self-sufficiency of computational processes. It seems clear, then, that Fermi's Paradox is strong, and any solution to it must also be strong.

But it is not a Paradox, at least as it is defined by formal logic. In this case, a Paradox is a set of propositions that results in an ambivalent or contradictory conclusion (i.e., the final statement can be either true or false, or both). A classic example of a paradox is the Liar's Paradox, summarised in a single sentence:

This sentence is a lie.

If the above sentence is true, then it is also a lie. If it is a lie, then it is also true.

Fermi's Paradox doesn't satisfy this definition of a paradox, or any of its other, more colloquial definitions (such as 'a concept containing mutually exclusive or contradictory aspects'). At best, it describes a logical argument that is valid but unsound. The phrase 'Fermi's Paradox' appears to have been coined by Stephenson (1977). The celebrity status of Fermi, plus the dramatic allure of the word 'Paradox', has successfully implanted itself into the imagination of both academics and the general public, even if it is not a strictly correct description of the concept. For convenience, we will continue to describe the problem at hand as 'Fermi's Paradox'.

1.2 Drake and His Equation

In 1958, Frank Drake, a recent PhD graduate from Harvard, accepted a role as staff scientist at the National Radio Astronomy Observatory (NRAO) at Green Bank, West Virginia. Radio astronomy was in its infancy in the United States – Drake was only the third person to have held his role at Green Bank, which itself was only founded two years previously. During his graduate studies, Drake had calculated that military radar with a power of 1 megawatt could communicate with a counterpart system at a distance of approximately 10 light-years.

Drake convinced the interim director of NRAO, Lloyd Berkner, to use Green Bank's 26 m radio dish to search for artificially produced signals on the condition that it was done quietly, and that the observations would yield data useful for conventional astronomy. Beginning what he would call Project Ozma, Drake targeted two stars, Tau Ceti and Epsilon Eridani, for artificial transmissions.

His preparations coincided with the publication of Cocconi and Morrison's (1959) seminal paper *Searching for Interstellar Communications*. His observations in 1960 yielded no signs of extraterrestrial intelligence, but his publication of the attempt began to nucleate a variety of thinkers on the concept of intelligent life. An informal meeting was convened at Green Bank in 1961 to discuss the possibility of detecting intelligent life.

In an attempt to focus the discussion, Drake would write his now famous equation on the board:

$$N = R_* f_p n_e f_i f_l f_c L \tag{1.2}$$

The number of civilisations emitting signals at any time, N, is related to:

R_* – The rate of star formation, in stars per year
f_p – The fraction of those stars which have planetary systems
n_e – The average number of habitable planets per planetary system
f_l – The fraction of habitable planets that are inhabited
f_i – The fraction of inhabited planets that possess intelligent technological civilisations
f_c – The fraction of intelligent technological civilisations that choose to emit detectable signals
L – The average lifetime of a civilisation's signal

Each of these seven parameters has assumed a range of values over the years, depending on contemporary knowledge of star and planet formation, as well as rather subjective opinions regarding the evolution of life, intelligence and the longevity of civilisations. Optimists will typically enter relatively large values for the initial parameters, resulting in $N \approx L$, i.e., that the number of signals extant is only sensitive to their longevity. A more restrictive argument can easily yield $N < 10^{-10}$, indicating that we are likely to be alone in the Galaxy.

Because of the current uncertainties in estimating its parameters, it is worth noting that Drake, and indeed no SETI scientist, regards Drake's equation as a useful tool in actually calculating the number of intelligent civilisations we should expect to detect, for a variety of reasons.

There are two principal issues with the equation, which are its fixed nature in space and time. It implicitly assumes a steady state of civilisation birth and death persists across the entire Milky Way. By extension, it assumes that the astrophysical constraints on civilisation birth and death have also reached a steady state at all locations of interest.

It is immediately clear to anyone with a grounding in astronomy that with the exception of some physical constants, the Universe cannot be described as constant in either space or time. The constituents of the Universe are perpetually changing, and this is clear at the level of the Milky Way. The propensity of the Milky Way to produce habitable planets has certainly changed a great deal since the Galaxy's initial disk structure was formed around 9 billion years ago (Haywood et al., 2016), and there are likely to be regions more hospitable to life (and intelligent life) than others (see section 4.2.2). It is possible to allow the seven factors to evolve with time, producing a number of communicating civilisations that also varies with time, $N(t)$, but constructing functions to describe the time dependence of each parameter is challenging. Many authors have discussed how folding in our understanding of star formation history in the Milky Way (i.e., the form of $R_*(t)$), and the properties of the *initial mass function* (broadly, the probability distribution of stellar mass) can make N vary greatly with time.

It is also clear that the availability of chemical elements with atomic numbers larger than helium (what astronomers call *metallicity*) changes greatly with time, as several generations of star formation create heavier elements via nuclear fusion, which are then expelled in stellar death. Sufficiently rich chemistry is a prerequisite for prebiotic molecules (and subsequently life), so a function that describes f_l will contain a term that grows in tandem with growing metallicity. These evolutionary trends have led many to propose that Fermi's Paradox results from a quirk of timing, as we will see in section 18.2.

Even if suitable functions are found for each variable, Drake's equation still struggles to inform us *where* to look for technological civilisations. Manipulating Drake's equation may not yield a more useful expression, but it can deliver insight. Maccone (2012) notes that we could consider an extended form of the equation, where the seven factors are re-expressed as a large (effectively infinite) number of sub-factors X_i. The equation would now consist of the following product:

$$N = \Pi_i X_i \tag{1.3}$$

Let us now assume each factor X_i is instead a random variable. If we take the logarithm of this equation:

$$\log N = \sum_i \log X_i \tag{1.4}$$

then we are afforded insight to the properties of N via the Central Limit Theorem. Consider a set of independent random variables $[X_i]$, which may have arbitrary probability distributions with mean μ_i and variance σ_i^2. The Central Limit Theorem guarantees that the sampling mean of these random variables is itself a random variable with a Gaussian distribution (provided that the variances are finite). This implies that $\log N$ is normally distributed, or equivalently N is lognormally distributed. What is more, the parameters that define the distribution of N are determined by the parameters that define X_i:

$$\mu_N = \frac{1}{N} \sum_i \mu_{X_i} \tag{1.5}$$

$$\sigma_N^2 = \frac{1}{N} \sum_i \sigma_{X_i}^2 \tag{1.6}$$

And the mean number of civilisations is hence

$$\bar{N} = e^{\mu_N} e^{\sigma_N^2/2} \tag{1.7}$$

In a similar vein, Glade et al. (2012) proposed modelling the arrival of civilisations as a Poisson point process, where the time interval between civilisations appearing is Poisson-distributed. Like Maccone's approach, this has the advantage of describing a level of uncertainty in the resulting parameters. It also results in a probability distribution for N at any given time t that is Poissonian. Equivalently:

$$P((N(t + \Delta t) - N(t)) = k) = \frac{(\lambda t)^k e^{-\lambda t}}{k!} \qquad (1.8)$$

Prantzos (2013) undertook a combined analysis of the Drake Equation and Fermi's Paradox, where both were described rather simply (Drake's equation being used to determine the probability of radio contact, and Fermi's Paradox being applied to galactic colonisation). By modelling both the colonisation front and the number of viable two-way conversations as a function of $f_{biotech} = f_l f_i f_c$

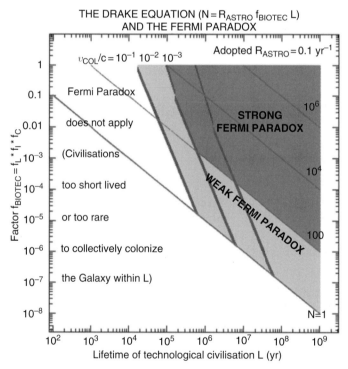

Figure 1.1 Prantzos's (2013) joint analysis of the Drake Equation and Fermi's Paradox. Given the condensed parameters of the Drake Equation $f_{biotech}$ (see text) and L, one can determine regions of parameter space that result in a completely full Galaxy, where we should expect to see evidence of ETI (upper right). In the bottom-left section, Fact A can be explained simply due to either the rarity of civilisations or the inability to colonise on short timescales.

(where the astrophysical parameters are held fixed), Prantzos identifies an important parameter space in $f_{biotech}$ and L where Fermi's Paradox is particularly strong (Figure 1.1). Briefly, if N civilisations are present, and can colonise spherical volumes at a given velocity (i.e., each civilisation occupies a sphere of radius $v_{col}L$), then there exists a typical L for which the entire Galactic volume is occupied, and Fermi's Paradox is strong (upper right portion of Figure 1.1).

The Drake Equation's primary function is as a roadmap. It identifies obstacles to our knowledge, and sheds light on our own ignorance. Indeed, when Drake first wrote this equation, only the first term was understood to any level of certainty. The next two terms have only become accessible to us in the last twenty years. The optimist will state that the march of knowledge continues to move from left to right, and that in time all seven terms of Drake's equation will be understood to within some uncertainty.

What both optimists and pessimists agree on is the deep implications of the structure of Drake's equation. At the leftmost, its terms are dictated by the physical sciences – initially physics and astronomy, proceeding to chemistry, geology and biology. The terms describing intelligent life then require the input of social sciences and humanities – anthropology, psychology, linguistics, sociology and so on.

Drake's equation reminds us that SETI demands all of human knowledge and academe, if the phenomenon of intelligent life in the Universe is to be understood. The science of SETI is the science of the human condition, generalised to encompass all forms of intelligent life.

1.3 The Great Filter

We have so far focused on one particular expression of the problem encapsulated by Fermi's Paradox. It is of course not the only way it can be described. The Paradox's argument is motivated largely by physics and astronomy, but other discussions of what Brin refers to as 'the Great Silence' are based in other disciplines.

The Great Filter is one of these alternatives, and it is motivated by the concept of *existential risk*. Following Bostrom (2002), we can define a risk connected to a specific event in terms of three properties: scope, intensity and probability. In other words, we want to know how likely the event is to happen, how widespread its effects might be and the level of hazard or damage associated with them.

An existential risk affects all of humanity, and threatens its continued existence, i.e., it has a wide scope and a high intensity. Of course, the probability of some risks is higher than others. Without such risks occurring, it is apparently likely that humanity will expand beyond the Earth, assuring its continued survival. If beings such as ourselves come to dominate the Milky Way on timescales as short as a few million years, how does one resolve their absence using the concept of existential risk?

Hanson proposed the existence of the Great Filter as a solution. If we consider some of the crucial milestones that life achieves on the route to intelligent life, and if the probability that any one of these milestones occurring is small, then that milestone is the Great Filter – a barrier that worlds fail to surpass on the way to producing intelligent life. We can see that this concept is very similar to those proposed by Fermi, Hart, Tipler and Drake, but viewed through the lens of existential risk.

Hanson lists nine milestones, but there is no consensus on what should appear in this list, and any such list is unlikely to be exhaustive. The first eight milestones carry an existential risk of **not** occurring. For example, if what Hanson calls 'Reproductive something', by which they mean self-replicating molecules like RNA and the ability to inherit traits, does not occur, then the future existence of intelligent life is in jeopardy.

The last milestone Hanson describes carries an existential risk of occurring, and indeed is a catch-all term for what most existential risk experts study: the risk of humans self-extinguishing their civilisation. Crammed into this milestone are dozens of solutions to Fermi's Paradox, all relying on technological civilisations ending themselves and producing a very low mean civilisation lifetime.

An important result of this argumentation is the inferred consequences of discovering primitive life. Traditionally this would have been considered a very positive indication for humanity, but Hanson demonstrates that this is a mixed blessing. To understand why, we must ask the question: is the Great Filter in our past, or in our future? If the Great Silence continues to hold, then we can arguably place the Great Filter in our past, and we can breathe easy in our unintentional excellence at survival.

If the Great Silence is broken, the Great Filter lies in our future. Currently, we lack the data to confirm the position of the Great Filter, so we must instead consider the probability that it lies in our future. It is worth noting that from a Bayesian viewpoint, our choice of priors becomes important (Verendel and Häggström, 2015).

Briefly, the two principal approaches to estimating the probability of events are frequentism and Bayesianism. In a frequentist approach, the probability of an event depends on how frequently that event is expected to occur, if we subjected the world to a large number of trials.

In the Bayesian approach, the probability of an event A occurring given B has already occurred, $P(A|B)$, can be determined by Bayes's Theorem:

$$P(A|B) = \frac{P(B|A)P(A)}{P(B)} \qquad (1.9)$$

1.3 The Great Filter

This is actually a rather simple rearrangement of a simple rule of probability:

$$P(A|B)P(B) = P(B|A)P(A) (= P(A \cap B)) \qquad (1.10)$$

where $P(A \cap B)$ represents the probability that A and B both occur. The reason that Bayes's Theorem is so powerful is that it can be used to describe the ability of a model M – defined by parameters θ – to fit the data from a current experiment or observation, D:

$$P(\theta|D) = \frac{P(D|\theta)P(\theta)}{P(D)} \qquad (1.11)$$

$P(D|\theta)$ is the *likelihood* of the data occurring, given the model is correct. $P(\theta)$ is the *prior* (an estimate of one's degree of belief in this parameter set θ, perhaps given previous experiments).

Bayes's Theorem then becomes a reliable tool to update the *posterior* $P(\theta|D)$, given previous data (the prior), and the current data (through the likelihood). The denominator is often referred to as the *evidence*, and can be interpreted as a normalisation constant for the likelihood:

$$P(D) = \int_M P(D|\theta)d\theta \qquad (1.12)$$

Applying Bayes's Theorem to a prior/likelihood/evidence combination gives a posterior, which itself can become the prior for the next use of Bayes's Theorem, resulting in a new, more refined posterior distribution. For example, estimating the probability of a die rolling a 2, when it has been loaded to roll a 6, would proceed as follows. A sensible first prior is that all the sides of the die are equally likely ($P(\theta) = 1/6$ for any θ). This is known as a uniform or flat prior. As the die is repeatedly rolled, and it becomes clear that all values are not equally likely, that prior becomes gradually modified until the model predicts that the probability that a 6 is rolled is higher than the other five options.

Now consider a simplified rephrasing of the Great Filter:

$$N f_l f_i f_{sup} \leq 1 \qquad (1.13)$$

where N is the total number of habitable planets, f_l is the fraction that develops primitive life, f_i the fraction of those that become intelligent and f_{sup} the fraction that go on to be highly visible technological civilisations. N is thought to be large, so we must now assign values to the remaining f parameters in the range $[0,1]$, effectively assigning probabilities that a given habitable planet will be alive, intelligent and visibly intelligent respectively.

An initial Bayesian approach to this problem would be to assign uniform priors to all three f terms, i.e., these parameters have equal probability of taking any value in the range $[0,1]$. Applying Bayes's Theorem to obtain the posterior distribution for f_{sup} is essentially our measure of the Great Filter's strength, and that measure changes depending on our information about the Universe. If the Great Silence holds, as N increases, naturally the posterior distribution for f_{sup} declines sharply. If we begin to receive data on f_l (i.e., we detect primitive life on another planet), the resulting effect is that the posterior distribution for f_{sup} declines even more sharply. It is for this reason that some philosophers would be deeply dismayed if we discovered a second origin of life on Mars.

This picture can change dramatically if we select a different starting prior. An example is if the f factors are not independent of each other. A discovery of another intelligent civilisation can then propagate through all three factors as we update the posterior, potentially providing us with hope if we discover signs of life on other worlds.

2
Fact A – The Great Silence

It should now be apparent that the existence of the Paradox is predicated entirely on Fact A – that we have no evidence for intelligent life beyond the Earth (what Brin 1983 calls the Great Silence). But how is this evidence obtained? Fact A has two facets: the lack of evidence for alien intelligence on Earth, and the lack of alien intelligence off Earth.

The first facet is the study of geologists and archaeologists. We will return to this somewhat later when we discuss the more outlandish (and factually erroneous) 'ancient aliens' claims in Chapter 26. At this moment, we should consider the second facet, which is the preserve of astronomers and space scientists.

Observational Searches for Extraterrestrial Intelligence (SETI) have been conducted primarily in the radio and visible parts of the electromagnetic spectrum. We will discuss future surveys which extend into other parts of the spectrum in the final chapters of this book. For now, we shall focus on the history of observational SETI.

Tarter (2007) defines a nine-dimensional search space for artificial signals:

- **Direction –** A signal will originate from some point on the sky, which we represent in two dimensions. Typically, the celestial sphere is described via the equatorial co-ordinate system, which uses *Right Ascension* (sometimes RA, or α) and *Declination* (Dec, or δ).[1] These are somewhat equivalent to longitude and latitude respectively, but their definition is designed to ensure that a star's position (α, δ) does not change if the observer's position on the Earth changes, or as Earth moves in its orbit around the Sun.

  Let us first define the *celestial equator* as the plane that contains the Earth's equator, and the *celestial north pole* as a vector pointing northwards that is perpendicular to this plane (and similarly for the *celestial south pole*). An object's declination (δ) is simply the angle made between the celestial equator and a

---

[1] Other branches of astronomy may elect to use different co-ordinate systems depending on convenience, for example galactocentric co-ordinates.

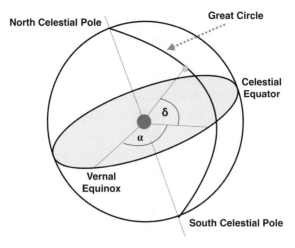

Figure 2.1 The equatorial co-ordinate system. The celestial sphere is defined by the celestial equator (a plane containing the Earth's equator), the celestial north and south poles perpendicular to this plane, and a reference location on the equator (the vernal equinox). The right ascension α and declination δ can then be calculated using this system, with appropriate calculation of the hour angle, which accounts for the Earth's rotation.

vector that connects the Earth to the object on the celestial sphere. The right ascension (α) requires us to draw a *great circle* on the celestial sphere that connects the object and the celestial north pole (known as an *hour circle*), and then calculate the angle between the plane this hour circle lies on, and a reference position along the celestial equator.

The standard reference position is known as the *vernal equinox*, and is defined using the plane of the Solar System (see Figure 2.1). More correctly, it is defined using the apparent path that the Sun follows over the course of the year, known as the *ecliptic*. The Earth's rotational axis is not aligned with its orbital axis (i.e., it has a non-zero *obliquity*). As a result, the ecliptic moves north and south on the sky over the course of a year. The vernal equinox refers to the moment (and location in Earth's orbit) where the Sun's motion crosses the celestial equator (going southward). As a result, it also refers to the day of the year where the day and night components of the 24-hour time interval are equal (hence the name 'equinox', which roughly translates from its Latin root *aequinoctum* as 'equal-night').

The right ascension of the vernal equinox is defined to be zero. However, the rotation of the Earth on its axis requires us to also identify the time of observation for any celestial object. The hour angle of an object on the sky is defined

$$H = UTC - \lambda_{observer} - \alpha \tag{2.1}$$

where UTC is the co-ordinated universal time (more or less equivalent to Greenwich Mean Time, although measured differently), $\lambda_{observer}$ is the longitude of the observer on the Earth's surface.

- **Distance** – This has two crucial impacts on any signal: firstly, electromagnetic signals decay with distance D as $1/D^2$, and hence a weak signal will eventually become invisible to our detectors, and secondly, electromagnetic signals are constrained to travel no faster than the speed of light c. Consequently, a signal received from a distance D will require a travel time of $\Delta t = D/c$, and hence any signal received will contain information from a time Δt in the past.
- **Signal Strength** – Regardless of how we make a measurement, our instrumentation will have a minimum sensitivity limit. A signal is emitted with some power P at a distance D, and received at Earth. The signal strength measured at Earth P_{Earth} is

$$P_{\text{Earth}} = \frac{PGA}{4\pi D^2} \tag{2.2}$$

The *effective isotropic radiated power* PG is composed of the power and the gain G of the transmitting antenna. The gain describes how well focused the transmission is. In terms of solid angle Ω:

$$G = \frac{4\pi}{\Omega} \tag{2.3}$$

Measurement systems with a large collecting area A can detect much weaker signals. Note the degeneracy between P and D – weak transmissions at close range can produce a value of P_{Earth} equal to that of a strong transmission at much larger distances.

- **Time** – We must be searching at the time the transmission arrives at Earth. Even for a transmission that is continuous over a significant time interval (say 300 years), if that interval happens to begin arriving at Earth in the year 1300, humans will not have been able to receive it. Transmissions that arrive at Earth in the modern era may still suffer from poor timing. Energy-efficient transmission schemes prefer pulses rather than continuous signals (Benford et al., 2010). This adds another factor to timing arguments – listening efforts that take place entirely in the interval between pulses will fail to detect the signal.
- **Polarisation** – A quantum of electromagnetic radiation is composed of coupled oscillating electric and magnetic fields (see Figure 2.2). In free space, the electric and magnetic fields are constrained to be perpendicular to each other. Polarisation refers to the orientation of the electric field. While many detectors are insensitive to this orientation, polarisation can be used to encode information, and indeed encodes information regarding scattering events photons may have received on its way to the detector. Determining photon polarisation

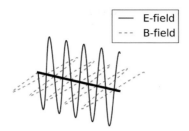

Figure 2.2 The electromagnetic radiation field, composed of perpendicular oscillating electric and magnetic fields.

(*polarimetry*) may be crucial for observers to determine if a signal is natural or artificial.

Consider a photon travelling in the z-direction, with an electric field given by

$$\mathbf{E} = \begin{pmatrix} E_x \\ E_y \\ 0 \end{pmatrix} e^{i(kz-\omega t)} \qquad (2.4)$$

And a magnetic field given by

$$\mathbf{B} = \begin{pmatrix} B_x \\ B_y \\ 0 \end{pmatrix} e^{i(kz-\omega t)} \qquad (2.5)$$

Note that the E_x, E_y, B_x, B_y components can be complex. The orientation of the electric field as a function of time can be encoded in the Stokes parameters (I, Q, U, V):

$$I = \left|E_x^2\right| + \left|E_y^2\right| \qquad (2.6)$$
$$Q = \left|E_x^2\right| - \left|E_y^2\right| \qquad (2.7)$$
$$U = 2\mathbb{R}\mathrm{e}(E_x E*_y) \qquad (2.8)$$
$$V = -2\mathbb{I}\mathrm{m}(E_x E*_y) \qquad (2.9)$$

I is the intensity of the radiation; Q and U represent the degree of linear polarisation (in planes at 45° to each other); and V represents the degree of circular polarisation (see Figure 2.3). We use $\mathbb{R}\mathrm{e}$ and $\mathbb{I}\mathrm{m}$ to denote the real and imaginary parts of the above quantities. In general

$$I^2 \geq Q^2 + U^2 + V^2 \qquad (2.10)$$

If the light is linearly polarised (in this case $U = 0$), we can represent the two electric field components thus:

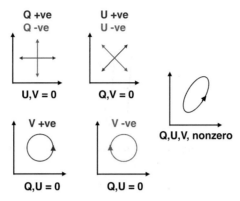

Figure 2.3 Schema of the various forms of polarisation, with the accompanying values of the Stokes parameters (I,Q,U,V).

$$E_x = \sqrt{\frac{1+Q}{2}} \tag{2.11}$$

$$E_y = \sqrt{\frac{1-Q}{2}} e^{i\phi} \tag{2.12}$$

where ϕ is a phase difference. If the phase difference is constant, then the electric field is constrained to oscillate in a plane, and hence the electric field is linearly polarised (either parallel to the x-axis for $Q = 1$, or perpendicular if $Q = -1$). If the light is unpolarised, the electric field vector is randomly oriented (or equivalently the phase difference ϕ varies randomly). If the phase difference varies smoothly in time, then the light is circularly polarised (with left and right circular polarisation denoting whether the phase difference is smoothly increasing or decreasing).

Astronomical radiation can be both polarised and unpolarised depending on the source. Generally, stellar radiation is unpolarised, although reflected starlight (say off a planet's atmosphere) is typically polarised. Synchrotron radiation (which occurs when electrons conduct helical orbits around a magnetic field line) can either be linearly polarised or circularly polarised depending on the observer viewpoint.

Detectors can be insensitive to polarisation, and receive light of all polarisations without discrimination. Others may only be sensitive to one polarisation (or one polarisation at any given time), or act as polarimeters, allowing the Stokes parameters of incoming photons to be determined.

- **Frequency** – The electromagnetic spectrum spans many orders of magnitude. The interstellar medium, a rarefied dusty plasma that exists between the stars, places constraints on the frequencies of light that can be used as communication. This constraint is hardly helpful – the allowed frequency range is approximately $10^5 - 10^{29}$ Hz (or equivalently, wavelengths of approximately 10^{-10} nm–3 km).

As we will see, specific bands in this vast range offer considerable advantages for transmissions, and as a result SETI scientists tend to observe in these bands.
- **Modulation** – This is in fact a second, separate parameter space, constrained more by the ingenuity of the transmitter than the other factors. The modulation of the signal is the principal means by which information is encoded into it. A 'beacon' can be as simple as a powerful narrowband transmission, but contain very little information content. Modulation of the signal allows information to be encoded. A successful encoding scheme requires the receiver to be able to decode the signal – more importantly, it requires the receiver to recognise the signal as a signal. As more information is encoded into a signal, it begins to resemble noise, and becomes challenging to separate from the background.

For our observations to be complete, we must sweep all nine dimensions of this parameter space effectively and sensitively. The magnitude of this endeavour suggests that a truly complete SETI sweep lies deep in our future, and it might suggest that any attempt is indeed fruitless. A common refrain from contact pessimists (those who do not consider detection of ETI to be likely) is that SETI is a project doomed to fail, and hence it is a waste of time and money.

I would regard myself as a contact pessimist; however, I see clear reasons for conducting thorough, detailed surveys for ETIs in the Milky Way. Firstly, we must remember the scientific method. It is insufficient to propose a hypothesis and not to test it with either observation or experiment. There are those that would argue that our observations to date are complete enough to reject the hypothesis that human beings are not alone in the Milky Way, but to me it seems clear that this is not the case.

An incomplete search still provides us with data to constrain our theories. An excellent example is the most recent constraints placed on Kardashev Type III civilisations by infrared studies of distant galaxies (Garrett, 2015). This survey removes a considerable region of search space from our calculations, and it forces us to rethink possible behaviours of advanced super-civilisations.

Like most modern scientific experiments, the approach towards an 'answer' is gradual. Eureka moments are few and far between in large observational surveys or giant experimental apparatus. Rather, the probability of the null hypothesis being correct (i.e., that there are no ETIs) increases with time. Of course, probability is not equal to certainty, and a single counter-example is sufficient to reject the null hypothesis. As we will see towards the end of this book, the current era of survey astronomy can and will provide SETI with significant opportunities to sweep large sections of the nine-dimensional parameter space.

The second reason for SETI is the implications for humanity if we fail to detect ETI. If the null hypothesis probability continues to grow unchecked, this is no less an important result than if ETIs were detected. The growing likelihood that we are alone sheds light on our civilisations' likely trajectory and the risks ahead, which will inform our subsequent actions as a species (see section 1.3).

2.1 Radio SETI

Radio SETI is the oldest of all SETI disciplines. Its formal beginnings lie with Drake's Project Ozma at Green Bank, but there are a significant number of surveys (albeit with very incomplete spatial and temporal coverage). Radio astronomy has many benefits over astronomy in the optical. The Earth's atmosphere is very transparent at such long wavelengths, and hence ground-based observations are easily achievable with relatively cheap instrumentation.

The radio brightness of the Sun is also relatively small, so radio astronomers routinely operate in the daytime. Equally, objects that do radiate strongly in the radio typically exhibit *non-thermal* spectra. Put simply, we expect thermal spectra to decrease in energy as the wavelength is increased. Non-thermal spectra increase in energy as the wavelength is increased, as a result of various high-energy astrophysical processes.

2.1.1 The Water Hole

While the relative noisiness of the long-wavelength universe is extremely useful for radio astronomers, this actually presents a problem for interstellar communicators. Thankfully, this is a mixed blessing, as it limits the range of useful wavelengths for sending transmissions. There are some wavelength ranges where the atmosphere is transparent – the visible spectrum resides in one of these frequency windows. This transparency to optical light is a proximate cause of our eyes being adapted to work in this wavelength range (Figure 2.4).

Another window exists at much lower frequencies (longer wavelengths). Known as the microwave window, it stretches from approximately 1 GHz to 10 GHz (approximately 3–30 cm in wavelength).

At frequencies higher than 10 GHz, transmissions are absorbed by terrestrial planet atmospheres, particularly water (H_2O) and oxygen (O_2). At low frequencies, the Milky Way emits a significant quantity of synchrotron radiation (produced when electrons orbit the Galactic magnetic field). This can be parametrised as follows (Mozdzen et al., 2017):

$$T_{\text{synchrotron}} = 300 \left(\frac{\nu}{150\,\text{MHz}}\right)^{-2.6}, \qquad (2.13)$$

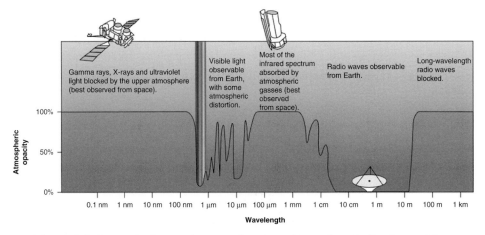

Figure 2.4 Atmospheric opacity as a function of wavelength. Regions of low opacity are suitable for ground-based astronomy. Radio emission between 1 cm and 10 m is clearly observable from the Earth's surface. Original Image Credit: NASA

where we consider the intensity of radiation in terms of its *brightness temperature* (see section 2.1.4). Further constraints come from the *cosmic microwave background*, emitted by the Universe when it first became transparent to its own radiation, resulting in an (effectively constant) brightness temperature of $T_{\text{CMB}} = 2.725$ K (Fixsen, 2009). At very high frequencies, the quantum nature of light results in *shot noise* at the detector:

$$T_{\text{shot}} = \frac{h\nu}{k_B}, \qquad (2.14)$$

where h is Planck's constant, and k_B is the Boltzmann constant. In the 1–2 GHz range (where the window is most transparent), two prominent emission lines reside. The first is one of the most pervasive emission lines in astronomy, the HI line (at a wavelength of 21 cm or approximately 1420 MHz). This emission line is produced by hydrogen atoms, and corresponds to a change in spin alignment between the proton and the electron. As hydrogen is the most common element in the Universe, and 21 cm emission is used to map atomic hydrogen on vast scales, it is commonly assumed by SETI scientists that intelligent civilisations engaging in astronomy would eventually discover radio astronomy and the pervasive 21 cm line.

The second line, at 1.66 GHz (and a third at 1.712 GHz), comes from the hydroxyl radical OH. The presence of H and OH lines immediately invokes the concept of water, as photodissociation of water gives:

$$H_2O + \gamma \rightarrow H + OH \qquad (2.15)$$

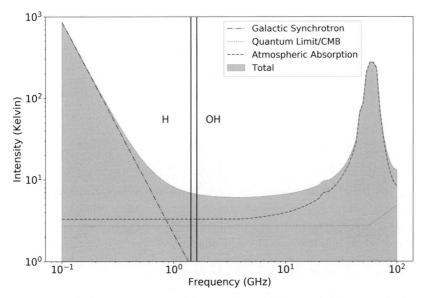

Figure 2.5 The Terrestrial Microwave Window (commonly known as the Water Hole). At low frequencies, galactic synchrotron radiation dominates the measured intensity. At high frequencies, atmospheric absorption dominates. The locations of the hydrogen (H) and hydroxyl (OH) lines are also indicated.

Water is perhaps the most bioactive chemical species on Earth, being a crucial solvent for chemical reactions as well as a key piece of cellular infrastructure. If life on Earth is a typical example of life in the Universe generally, then it is likely that alien astronomers are interested in signs of water and its constituents.

The combination of these facts suggests that human-like astronomers will be naturally drawn to the 1–10 GHz range to study the Universe. In 1971, during Project Cyclops, Bernard Oliver, then Vice-President of Research and Development at the Hewlett Packard Corporation, proposed that the H and OH lines represented clear markers to any alien civilisation. This could be an elegant solution to the waveband selection problem – choosing a frequency related to these lines (say $\pi \times \nu_{HI}$) would be easier for aliens (and us) to guess.

It was Oliver who coined the term the 'Water Hole' (Morrison et al., 1977, Colloquy 4), with its double meaning of the frequency space where water-relevant lines are emitted, and a place where multiple species may gather.

Transmitting in the Water Hole (see Figure 2.5) does reduce the search space significantly, but it does not completely solve the problem, as we still have a huge number of wavelength choices. Blair and Zadnik (1993) identified at least 55 dimensionless physical constants and arithmetic combinations of H and OH frequencies that could be used to identify a frequency ($C \times \nu_{HI}$).

If these assumptions are correct, potential transmitters and receivers will be focusing on a rather narrow band of the electromagnetic spectrum. As a result, most SETI searches (and interstellar messages transmitted from Earth) have focused on the Water Hole, and specifically frequencies around the HI line.

2.1.2 Doppler Drift

It is almost certain that relative to our receivers, any alien radio transmitter will be moving. If that is the case, then we must account for the Doppler effect in the radio signal. If the receiver moves with speed v_r towards the transmitter, and the transmitter or source moves towards the receiver at speed v_s, then the frequency of the signal v is shifted away from its emitted frequency v_0 by

$$v = \frac{c + v_r}{c - v_s} v_0 \tag{2.16}$$

If the signal is being transmitted from the surface of a planet (and received from the surface of our planet), then the *radial velocity* (i.e., the speed of transmitter and receiver towards each other) will vary smoothly according to the period of each planet. The frequency of the signal will be expected to drift over a range related to the orbital speed of the bodies.

Consider a stationary transmitter ($v_s = 0$), transmitting towards the Earth in its orbital plane. The radial velocity of the Earth towards the target will vary over the course of a year between its minimum and maximum of

$$v_r = \pm \frac{2\pi a_\oplus}{P_\oplus} \approx 29,840 \text{m s}^{-1} \tag{2.17}$$

And hence the transmitter frequency will vary between

$$v = [0.9999, 1.0001] v_0 \tag{2.18}$$

For a transmission at 1.4 GHz, this results in drifts of hundreds of kHz over the Earth's orbital period of 1 year. This is a relatively slow drift – we can, of course, imagine a transmitter on a planet with a much shorter orbital period (perhaps only a few weeks), or a transmitter orbiting a planet every few hours, adding a small but rapid variation in frequency.

2.1.3 Radio Frequency Interference

Ironically, the greatest barrier to detecting artificial signals is a constant background of artificial signals, arriving at the telescope from terrestrial sources.

As we will see, almost every signal detected by a radio SETI mission is eventually identified as radio frequency interference (RFI). This can be radio signals that

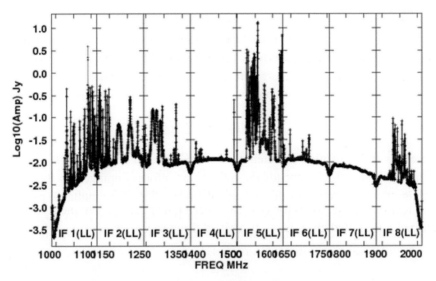

Figure 2.6 Radio Frequency Interference (RFI) measured at the Very Large Array in New Mexico in the L-band (1–2 GHz). Spikes in the above spectrum indicate signals from sources such as GPS and weather satellites, radar pulses from nearby installations, and modem transmissions at the VLA site. Data courtesy of NRAO.

have reflected off the ionosphere and arrived in the telescope's sidelobes (see next section), or radio emission that has bounced off a celestial object (either the Moon or a satellite) and returned to Earth.

As a result, many key bands for SETI searches are heavily contaminated (see Figure 2.6). A critical component of data reduction/processing of potential SETI signals requires ruling out RFI as a source.

2.1.4 Single-Dish Radio Astronomy

The need for effective radar dishes during the Second World War had the happy fortune of spurring on advances in radio astronomy (culminating with the detection of interstellar HI by van de Hulst in 1944). This detection (and the understanding that its mapping would yield significant advances in our understanding of galactic structure) spawned the building of larger and larger radio dishes, such as the Green Bank dish, and Jodrell Bank in the UK.

The intensity of radiation I received by radio transmitters can be easily related to a temperature using the blackbody relation in the Rayleigh-Jeans limit:

$$I = \frac{2k_b \nu^2 T_b}{c^2} \qquad (2.19)$$

where k_b is Boltzmann's constant, ν is the frequency and c is the speed of light. It is common for radio astronomers to discuss the brightness temperature T_b of the source being measured.[2] In practice, the raw data received from a radio telescope is referred to as the 'antenna temperature'. This is a convolution of the brightness temperature with a beam pattern, which is produced as the photons strike various regions of the antenna before coming together at the focus. This beam pattern determines the *angular resolution* θ:

$$\theta = \frac{k\lambda}{D} \qquad (2.20)$$

where λ is the observing wavelength, and D is the dish diameter. k is an order of unity constant which depends on how the waveform impinges on the receiver, which is a strong function of the antenna geometry. Circular symmetric paraboloid dishes produce a power response which is also symmetric relative to the angular distance of the target from the line of sight vector, u. In fact, this response is given by a Bessel function of order zero (Figure 2.7). In this case $k = 1.22$, as

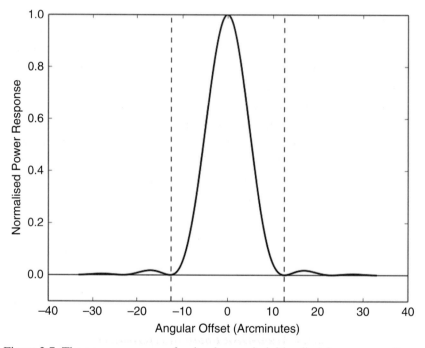

Figure 2.7 The power response of a circular paraboloid radio telescope, as a function of offset from the centre of the field of view. The primary beam is formed within $1.22\lambda/D$ of the centre, where the power response is null (indicated by dashed lines). Note the sidelobes (normalised peak response of 0.0174) at either side of the primary beam.

---

[2] Even if the source does not emit as a blackbody, it is still common practice to discuss a brightness temperature, derived using equation 2.19.

this indicates the position of the first null of the response (dashed lines in the figure). It is important to note that antenna response grows again after the first null. These extra maxima are referred to as *sidelobes*, and typically result in signals being received from off-axis, which must be removed during the deconvolution process.

For observations at 21 cm, with a dish diameter of 100 m, this corresponds to an angular resolution of around 0.15 degrees. Astronomers typically use *arcseconds* ($''$) to describe very small angles, where an arcsecond is 1/3,600 of a degree (an *arcminute* ($'$), which is 1/60 of a degree, is 60 arcseconds), and hence in this scenario $\theta \approx 528'' \approx 8'$. This is much higher than the typical blurring caused by atmospheric motions (referred to as *seeing*), but equally it is not sufficient for resolving objects at great interstellar distance.

The angular resolution limit of single-dish radio telescopes (for a fixed wavelength) is simply the size of the dish that can be created. As D increases, this becomes an engineering challenge.

The largest single-dish telescopes are not dishes in the traditional sense, but rather bowls that are installed inside pre-existing holes or craters. For example, the Arecibo telescope is a 305 m diameter spherical dish, occupying a karst sinkhole (produced by the acidic erosion of limestone). As the primary dish cannot be steered, the secondary is fitted to a series of gantries that allow it to move. Arecibo is famous for its use both as a radio receiver and a radar transmitter.

Arecibo's place as the largest single-dish radio instrument in the world was recently subsumed by the Five hundred meter Aperture Spherical Telescope (FAST). FAST also occupies a karst sinkhole, but its primary reflector surface is composed of several thousand panels with accompanying actuators to convert the spherical dish into a parabolic reflector.

The sensitivity of radio observations depends crucially on the noise present in the telescope itself. The *system temperature* defines this noise, including background and atmospheric radiation, radiation from the ground (which enters via the sidelobes), losses in the system's feed and receiver system, and calibrator signals that may need to be injected.

2.1.5 Multi-Dish Radio Interferometry

An alternative to building single large dishes is to build large quantities of smaller dishes, and use *interferometry* to correlate the signal. Broadly speaking, this is equivalent to building a single dish with maximum angular resolution equal to the maximum distance between dishes (the *baseline*, B):

$$\theta_{max} = \frac{1.22\lambda}{B}. \tag{2.21}$$

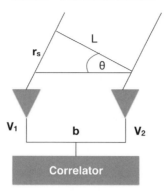

Figure 2.8 Schematic of a two-element interferometer. Signals impinge on both detectors with a time delay given by the path difference L. The voltages induced in both detectors are then correlated to produce a single Fourier component of the sky brightness distribution.

Interferometry relies on the time delay δt between two detectors as emission impinges on them (see Figure 2.8). If we imagine radio waves at a single frequency arriving at the detectors, the two signals received will be in phase if the difference in *path length* the light must travel, L, is equal to an integer number of wavelengths ($L = n\lambda$, where $n = 0, 1, 2$) and similarly if $L = (n - \frac{1}{2})\lambda$, then the two signals received will be out of phase.

Consider a source with direction (unit) vector $\hat{\mathbf{r}}_s$ away from the interferometer, and the baseline vector $\mathbf{b}$ connecting the two dishes. The path length difference is then

$$L = \hat{\mathbf{r}}_s . \mathbf{b} = |b| \cos\theta. \tag{2.22}$$

Each antenna records an output voltage V. If the signal is of a frequency $\nu = \omega/2\pi$, the signal oscillates at this frequency, and the recording of both voltages will register a geometric time delay $t_g = \hat{\mathbf{r}}_s . \mathbf{b}/c$:

$$V_1 = V \cos(\omega t) \tag{2.23}$$
$$V_2 = V \cos\left(\omega\left(t - t_g\right)\right). \tag{2.24}$$

The output voltage from both antennae are then *correlated*. Firstly, the two outputs are multiplied together:

$$V_1 V_2 = V^2 \cos(\omega t) \cos\left(\omega\left(t - t_g\right)\right) = \frac{V^2}{2} \cos\left(2\omega t - t_g\right) \cos\left(\omega t_g\right) \tag{2.25}$$

The correlation process is completed by averaging the signal over a sufficiently long time interval to remove the high-frequency oscillations produced by the $\cos\left(2\omega t - t_g\right)$ term ($\Delta t \gg 2\omega^{-1}$):

$$R = \langle V_1 V_2 \rangle = \frac{V^2}{2} \cos(\omega t_g) \quad (2.26)$$

If the plane wave emanates from a point source, the voltage is proportional to the flux density multiplied by $\sqrt{A_1 A_2}$, where A_i represents the receiving area of antenna i. The correlation process effectively removes uncorrelated noise from the retrieved signal, adding a further advantage to interferometry over single-dish observations.

The response R varies sinusoidally with the source position on the sky. The sinusoids are known as *fringes*, with fringe phase

$$\phi = \omega t_g = \frac{\omega b}{c} \cos \theta \quad (2.27)$$

As the source moves in the sky, the phase varies as

$$\frac{d\phi}{d\theta} = 2\pi \frac{b \sin \theta}{\lambda} \quad (2.28)$$

and hence a radio source on the sky can be located with high accuracy (provided that $b \sin \theta$, the projection of the baseline on the sky, is much larger than the wavelength).

This two-dish example is only sensitive to signals with an angular period on the sky $\frac{\lambda}{b \sin \theta}$ – equivalently, the interferometer can only detect one Fourier component of the sky brightness distribution. If we wish to have access to the other Fourier components, then more baselines are required, with a variety of values of b and $\sin \theta$. This can be achieved by adding more dishes, or if the source is relatively steady, by carrying out multiple observations and allowing the motion of the object on the sky to rotate the object relative to the baseline.

In practice, sources are extended, not points. If the sky brightness distribution is $I_\nu(\hat{\mathbf{r}}_s)$, then the response is

$$R_c = \int I_\nu(\hat{\mathbf{r}}_s) \cos\left(\frac{2\pi \nu b}{c} \hat{\mathbf{r}}_s . \mathbf{b}\right) d\Omega \quad (2.29)$$

This 'cosine response' is an incomplete measure of I_ν, as only even parts of I_ν will be recovered. To retrieve the odd parts, a 'sine response'

$$R_s = \int I_\nu(\hat{\mathbf{r}}_s) \sin\left(\frac{2\pi \nu b}{c} \hat{\mathbf{r}}_s . \mathbf{b}\right) d\Omega \quad (2.30)$$

is also computed (using a second correlator with a 90 degree phase delay). The combination of both parts is the *complex visibility*:

$$V_\nu = R_c - i R_s \tag{2.31}$$

which is more succinctly described as

$$V_\nu = \int I_\nu(\hat{s}) \exp\left(i \frac{2\pi \nu b}{c} \hat{r}_s . \mathbf{b}\right) d\Omega \tag{2.32}$$

This is a one-dimensional visibility, measured along the axis of the available baselines. Interferometric imaging requires a 2D generalisation of the above. We define the *uv*-plane as that containing vectors along the east-west baseline (u) and north-south baseline (v). If we also define l and m as the projections of r_s onto u and v, then

$$V_\nu = \int I_\nu(\hat{s}) \exp\left(-2\pi i \left(ul + vm\right)\right) dl dm \tag{2.33}$$

which is amenable to a 2D Fourier transform to obtain the sky brightness distribution I_ν.

2.1.6 A History of Radio SETI observations

Table 2.1 gives a potted history of historic and recent radio surveys, from its beginnings in Project Ozma to contemporary efforts. Most have focused their efforts at frequencies around 1420 MHz, with a variety of target stars, and in some cases blind or all-sky surveys with no specific targets. This list is far from exhaustive, but gives a sense of the key players in radio SETI. The history is dominated by US and USSR teams (with some contributions from radio astronomy observatories in the UK and Western Europe).

2.2 Optical SETI

As Drake and others were beginning radio SETI observations, the concept of optical SETI was already in its early stages of formulation.

During and following the Second World War, the Allies were thoroughly convinced of the need and value of radar, and general manipulation of the radio bands of the EM spectrum. The development of quantum theory between World War I and World War II (most notably Einstein's insights into stimulated emission of photons) encouraged physicists to manipulate light at increasingly shorter wavelengths, in an attempt to probe molecular and atomic structure.

At the time, radar systems relied on cavity magnetrons, vacuum tubes which passed a stream of electrons past a series of open cavities. Applying a magnetic field to the apparatus excites radio waves inside the cavities. Such systems were suitable for centimetre band radar, but were not viable at shorter wavelengths.

Table 2.1 Selected historic and recent radio SETI surveys. This table does not include an extensive list of one-off observations and opportunistic targeting of potential sources.

Survey Name/Observers	Site/Instrument	Size	Search Frequencies (MHz)	Spectral Resolution (Hz)	Flux Limit ($W\,m^{-2}$)
Ozma	Green Bank Telescope	26 m	1420 ± 0.4	10^2	4×10^{-22}
Kardashev & Sholomitskii	Crimea Deep Space Station	920	10^7	N/A	
META II	Institute for Argentine Radioastronomy	30 dishes	1420, 1667, 3300	0.05 (33)	$1 \times 10^{-23} - 7 \times 10^{-25}$
SERENDIP IV	Arecibo	305 m	1420 ± 50	5×10^{-24}	
Southern SERENDIP	Parkes Telescope	64 m	1420.4 ± 8.82	0.07-1200	4×10^{-24}
Troitskii, Bondar & Starodubtsev	Gorky, Crimea, Murmansk, Ussuri	53 m	1420	10	1.5×10^{-21}
Wielebinski & Seiradakis	Max Planck Institute for Radioastronomie (Germany)	100 m	1420	2×10^6	4×10^{-23}
ARGUS	Multiple (amateur-driven)	1	1420–1470	$\sim 1 \times 10^{-21}$	
SIGNAL	Westerbork Synthesis Radio Telescope (Netherlands)	3000 m baseline	1420.4	1200	10^{-24}
Phoenix	Arecibo (Lovell, Jodrell Bank)	305 m (76 m)	1200–3600 (dual polarisation)	0.67	1×10^{-26}
- *Kepler* Targets, Siemion et al. 2013	Green Bank Telescope		1500 ± 800	0.3	2×10^{-18}
VLBI SETI Rampadarath et al. 2012	Australian Long Baseline Array	1230–1544	1.9×10^3	$\sim 5 \times 10^{-26}$	

Charles Townes, working at the time at Columbia University, recognised that microwave radiation could be used to stimulate further emission from a population of molecules. This led in 1954 to the invention of the *maser* (Microwave Amplification by Stimulated Emission of Radiation). This used excited ammonia molecules inside a cavity to emit an initial burst of microwave radiation, which went on to stimulate further emission from molecules that had absorbed the initial burst. A continuous supply of excited molecules ensured a self-sustaining chain reaction of stimulated emission (Gordon et al., 1954). Schawlow and Townes (1958) were quick to suggest the technology could be extended to much shorter wavelengths, especially the infrared, where the benefits for high-resolution spectroscopy of molecules was obvious. Their vision – the laser – was quickly realised (Maiman, 1960).

The laser concept was soon understood to be an advantageous means by which one might communicate over large distances. The *collimation* of the laser ensures that the received flux from a laser remains strong at large distances. As optical SETI is interested in the detection of relatively low photon counts, we should rewrite equation (2.3) in terms of photon flux F:

$$F = \frac{PGA}{4\pi h \nu D^2} \, \text{s}^{-1} \qquad (2.34)$$

where ν is the photon frequency, and h is Planck's constant. As we are dealing with laser communications, the gain G is relatively high. The laser wavelength is typically much shorter than the radio (either the optical or infrared), and as such a beam can encode a great deal more bits per second (see section 21.1). Going to shorter wavelengths (or higher frequencies) reduces the numbers of photons sent for a given transmitter power and time interval. As a result, it is assumed that optical SETI communications will use higher transmitter power at short time intervals (pulses) for manageable total energy costs.

Disadvantages of Optical SETI

There are of course clear disadvantages to attempting to communicate in the optical. Firstly, the Universe is spectacularly loud at these frequencies, and the background signal can be extremely large in some configurations, reducing the signal-to-noise of the transmission significantly.

Secondly, optical photons are easily absorbed by interstellar dust (Figure 2.9), and re-emitted typically in the infrared. This is especially true of shorter-wavelength (blue) light, which in other circumstances might have been preferred for its greater ability to concentrate information in a pulse than red lasers. Photons emitted in the Lyman continuum range (~ 50–90 nm) are preferentially absorbed to ionise hydrogen. An optical pulse from a nearby galaxy such as M74 will be

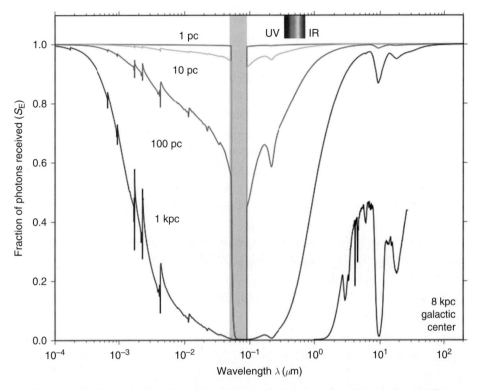

Figure 2.9 Transmission fraction of optical photons through the interstellar medium (see Draine 2003). The shaded region represents the Lyman continuum, which is opaque for even nearby targets due to its ability to ionise neutral hydrogen. Data kindly provided by Michael Hippke (private communication).

reduced in signal strength by approximately a factor of two due to absorption by interstellar dust.

These problems are far from insurmountable. For the sender, they can typically be overcome by selecting as long a wavelength as possible (typically above 10 μm) to reduce dust absorption by a factor of five, and emitting a highly collimated pulse to maximise the transmitted power towards the receiver (and hence the received flux).

Timing is perhaps the biggest concern in OSETI. It is clear that pulses are the most energy-efficient way to transmit information (Benford et al., 2010), and it is quite clear that without good fortune (and ETI's intent to signal Earth), most observations will coincide with ETI signalling each other with highly collimating pulses which are effectively invisible to us (Forgan, 2014), and not transmitting towards us. Serendipity is key, and is not something that can be agreed upon prior to initial contact.

Detection of Optical SETI signals

There are two complementary OSETI strategies. Searches consider either (a) narrowband signals that can be continuous or pulsed, in much the same vein as radio SETI, or (b) broadband pulses that are significantly more powerful than the background emission generated by the Universe.

Narrowband optical pulses are accessible to virtually any astronomical telescope above 1 m in diameter without requiring significantly modified instruments, with the very small bandwidth of the signal being sufficient to confirm its artificial nature. On the other hand, broadband pulses are expected to have very short pulse durations (for energy efficiency reasons). A pulse that lasts only a few nanoseconds or less, over a range of bands where celestial objects emit large numbers of photons, is not detectable without specific instrumentation.

Optical SETI typically adopts nanosecond photon counters when conducting a SETI survey in this mode. These instruments observe the Universe on nanosecond exposure times, which dramatically reduces the number of photons received by natural sources. A nanosecond photon counter observing the Sun would measure approximately one photon per exposure. An optical SETI signal produced using human-like technologies would produce a photon count hundreds of times higher!

This estimate is based on what are now quite old studies of twentieth-century pulsed laser technology in use at laboratories studying nuclear fusion (Howard et al., 2004). These inertial confinement fusion experiments rely on powerful laser pulses incident on small targets. Several lasers are positioned around the target, with synchronised pulses resulting in extremely high pressure and temperature in the outer layers of the target. The shock wave induced in the target compresses and heats the inner layers further, in the hope that the target reaches the conditions necessary for ignition.

The standard figure of 100 photons per nanosecond is derived from a pulsing laser built at the LLNL, which has a power rating of 10^{15} watts in picosecond pulses (with a pulse rate of one per hour). Modern pulsing lasers, such as those at the National Ignition Facility in the LLNL, still achieve petawatt power, but now focus on maintaining the pulse for longer durations, and improving energy absorption by the target.

Optical SETI is therefore a relatively straightforward observing strategy. One does not require significant amounts of data processing, as is needed for radio interferometry. One does not even need a particularly large telescope. If we were to observe a star while ETI was transmitting a pulse from an orbiting planet, we can expect the broadband emission from the star in a given filter to increase by up to six orders of magnitude! Observing unfiltered stellar emission would still yield a boost of 30 to 100 x (hence the 100 photons per nanosecond figure). These numbers can obviously be boosted further by increasing the beam energy and reducing the solid angle of the beam.

In practice, OSETI observations require two independent photon counters, to measure counts both on-source and off-source. As electronics have decreased in price, this has allowed OSETI to operate effectively on telescopes in the 1 to 2 m class.

2.2.1 Artefact SETI

A notable offshoot of optical/infrared SETI is the search for physical artefacts that either emit or reflect light in these bands. There are two principal artefact types that account for the majority of attempts at artefact SETI (sometimes referred to as the Search for ExtraTerrestrial Artefacts or SETA).

The most famous of these is the putative Dyson sphere. Freeman Dyson's concept builds on the more general concept of *macro-engineering* or *astro-engineering*, where advanced civilisations construct structures with scales comparable to those of celestial bodies, and owes its genesis to earlier ideas from Stapledon, Tsiolkovsky and Haldane (Ćirković, 2006).

A Dyson sphere is a sphere, of order an AU in diameter and a few kilometres thick, that completely encompasses a star. In most conceptualisations of Dyson spheres, the sphere diameter is selected such that its inner surface would receive radiative fluxes similar to that of the Earth. If this sphere was built as a single solid construction around a Sunlike star, then the interior of the sphere would provide a habitable surface A_{sphere} many, many times the total surface area of the Earth $A_\oplus$:

$$\frac{A_{\text{sphere}}}{A_\oplus} = \left(\frac{1\text{AU}}{1 R_\oplus}\right)^2 \approx 5.5 \times 10^8 \quad (2.35)$$

It is also the most efficient way to harvest stellar radiation, as the star's entire output can be absorbed by the sphere, and not just the fraction that is emitted in the solid angle that contains an individual planet such as the Earth. An alternative construction (sometimes referred to as a 'Dyson swarm') relies on a large number of small collectors, arranged in the same spherical configuration, with the sole purpose of harvesting stellar energy.

As a Dyson sphere/swarm's equilibrium temperature is expected to be tuned to that suitable for a biosphere (around 300 K), it should therefore radiate approximately as a blackbody with a similar temperature (with a peak emission of order 10 μm). The precise temperature of the body depends on, of course, the preferred temperature of its inhabitants, and the thermodynamic efficiency of the sphere as it re-radiates into the wider Universe.

Table 2.2 *Selected historic and recent optical SETI surveys. This table does not include an extensive list of one-off observations and opportunistic targeting of potential sources.*

Survey Name/Observers	Site/Instrument	Size	Search Wavelengths (μm)	Targets
Wischnia (1973)	Copernicus Observatory, Poland	1 m	2.8	3 stars
Witteborn (1980)	Mount Lemon		8.5–13.5	Search for Infrared Excess from Dyson Spheres
Kingsley et al. (1995)	Columbus Optical SETI Observatory	0.25 m	0.55	Nearby Solar-type stars (nanosecond pulses)
Blair and Zadnik (2002)	Perth-Lowell Telescope, Australia	0.61 m	Optical	Search for Optical Beacons (> 1 Hz repetition rate)
Beskin et al. (1997)	Special Astrophysical Observatory, Russia	6 m	N/A	Wide-field, short-duration pulses
Timofeev et al. (2000), Carrigan (2009)	IRAS	0.57 m	12–100	> 250, 000 sources (search for Dyson spheres)
Stone et al. (2005)	Lick Observatory	1 m	55–70	4,605 stars for laser pulses
Howard et al. (2007)	Oak Ridge Observatory	1.8 m	Optical	80% of the sky (150 hours total)

Dyson spheres would be extremely odd celestial objects, as their effective temperatures, and their resulting spectra, are similar to those of *brown dwarfs* (low-mass substellar objects that bridge the gap between planet and star), but with a radius several thousand times larger.

Ergo, Dyson spheres would be reasonably obvious in any large-scale survey of the Milky Way at infrared wavelengths, especially surveys looking for objects like brown dwarfs. The straightforward nature of this search has meant that artefact SETI also commonly goes by the name 'Dysonian SETI'.

The other principal searches for artefacts occur in our own Solar System, with various observers attempting to detect visiting interstellar probes in distant orbits around our own Sun (we will address this in more detail in section 25.1).

2.2.2 A History of Optical SETI Observations

Table 2.2 gives a selection of past and present optical SETI surveys. Like Table 2.1, this is not intended to be an exhaustive list, but again gives a sense of the key players in optical SETI.

3

Classifying Scenarios and Solutions to the Paradox

3.1 Contact Scenarios and the Rio Scale

There are clearly many different means by which alien intelligences could reveal themselves to us. We may receive a transmission or message from a nearby star system (or even another galaxy, a possibility not outside the laws of physics, yet still highly improbable). We may see unintentional signals of intelligence, such as astroengineering or changes to a planetary atmosphere due to industrial activity. An uncrewed interstellar probe may enter the System (or reveal itself having been present for millions of years). Finally, in perhaps the least likely of all scenarios, an alien spacecraft may emulate the ultimate trope of landing on the White House lawn.

We should note that the last scenario is what originally interested Fermi, and what motivated those diverting discussions at Los Alamos. Over the years, Fermi's Paradox has grown to encompass any and all scenarios under which humanity learns of the existence of other intelligent civilisations. The risk or probability of these events occurring, plus the magnitude of such an event in terms of societal response, has led to some discussion of how to classify these situations.

Almàr and Tarter defined the Rio Scale in response to this discussion (Shostak and Almar, 2002; Almár, 2011), where the Rio Scale Index (RSI) for a given event is:

$$RSI = Qd \qquad (3.1)$$

where Q is an estimator for the consequences of a discovery, and d is the assessed credibility of a discovery. Clearly, aliens landing on Earth would possess a high Q, and many describe such events vividly, but these discoveries without exception possess a low d, and hence a low RSI.

The estimator Q is determined by the sum of three factors:

1. Class of Phenomenon
2. Discovery Type
3. Distance

Table 3.1 *The Rio Scale 1.0: some examples of detection scenarios, and their resulting value of Q (where $RSI = Qd$, and d is the credibility of the detection)*

Phenomenon	Discovery Type	Distance	Q
Traces of astroengineering	From archive data, unverifiable	1000 light-years	3
Leakage radiation	Amateur observation, unverified, not repeated	10 light-years	5
Earth-specific message	Verified observation that cannot be repeated	30 light-years	12
Artefact with general message	Verified observation that can be repeated	Within the Solar System	13

The full rubric for each of these factors can be found in Almár (2011). We reproduce a few representative scenarios (and a corresponding Q value) in Table 3.1. The value for Q can range between 1 and 10, with the credibility d ranging between 0 and 1.

Recently, this scale was revised (Rio 2.0) to reflect the new realities of public news consumption and social media (Forgan et al., 2018c). The Q/d structure of the scale is maintained, and RSI still ranges between 0 and 10. The principal changes have been to the computation of d, which is the main concern of most SETI practitioners.

The new scale places a great deal more scrutiny on how a signal has been observed; how many times it has been observed, by multiple groups and instruments; and the likelihood of human error, anthropogenic effects or simply the signal being a hoax.

More explicitly, d is now computed by

$$d = 10^{\frac{A+B+C-30}{2}} \tag{3.2}$$

where A, B and C are ratings of the signal's veracity, the likelihood it is instrumental, and the likelihood it is not a natural or anthropogenic signal. The values for A, B and C (and Q) are obtained by filling in a short online quiz, and the resulting values are then displayed with some appropriate context.

Usefully, we can also use the logarithmic credibility

$$J = A + B + C - 20 \tag{3.3}$$

As a means of comparing signals with better dynamic range than a linear scale:

$J < 1$ No interest warranted
$J = 1$–4: SETI interest potentially warranted; no press interest warranted

$J = 5–6$: SETI interest probably warranted; technical popular press interest potentially warranted

$J = 7–8$: SETI interest definitely warranted; technical popular press interest probably warranted; possible off-beat news item for general press, if expressed with appropriate caveats. If not aliens, still very interesting.

$J = 9$: Significant mainstream press interest warranted, heavy coverage by technical popular press. Broad agreement that the signal could be due to aliens.

$J = 10$: Aliens. Front page of every major newspaper.

This discussion might feel a little tangential from this textbook's purpose, but it is important to keep in mind that the mere detection of intelligent life has implications for Fermi's Paradox, but the detection must be credible if we are to apply it to the problem. And, what is more important, the scale of a detection's impact on society varies greatly according to its nature.

The International Academy of Astronautics (IAA) hosts a Permanent Committee on SETI, which has adopted two protocols for SETI scientists in the event of a successful detection. The first, 'a Declaration of Principles Concerning Activities Following the Detection of Extraterrestrial Intelligence', is known colloquially as the First Protocol, or the Post-Detection Protocol, and gives scientists a guide as to how to act in the event of a detection. This includes appropriate steps for confirming the detection, e.g., by having other teams attempt to measure the signal with different instrumentation and methodologies as well. Once a detection of ETI is confirmed, the protocol then describes how this should be announced to the world.

The Second Protocol, 'a Draft Declaration of Principles Concerning Sending Communications with Extraterrestrial Intelligences', is known colloquially as the *reply protocol*. This deals with the appropriate steps to take in returning communications (more on this in section 21.4.1).

3.2 The Three Classes of Civilisation – the Kardashev Scale

If we assume that human civilisation will survive and prosper for even another thousand years, what will that civilisation look like? Given the enormous technological progress made since the year 1000, how can we begin to consider how a civilisation much older than ours might think or act?

Kardashev (1964) provides us with a route grounded in physical law. The laws of thermodynamics place strong constraints on the manipulation of energy, which as far as we can see must apply to all intelligent species. We can define the three laws of thermodynamics thus:

3.2 The Three Classes of Civilisation – the Kardashev Scale

1. For a closed system with internal energy U, a differential change

$$dU = dQ + dW \tag{3.4}$$

where dQ is the differential change in heat, and dW the differential work done on the system. This statement is equivalent to that of the conservation of energy.

2. The entropy S of a closed system cannot decrease:

$$dS = \frac{dQ}{T} \geq 0 \tag{3.5}$$

Entropy has multiple definitions, depending on whether one approaches it from a thermodynamics or statistical mechanics viewpoint. In thermodynamics, entropy is related to the dissipation of heat in a system ($dQ = TdS$). In statistical mechanics, the entropy measures the number of possible states of a system (microstates), given its macroscopic properties (macrostate). Systems with high entropy can be arranged into many microstates for a given microstate, showing entropy as a measure of uncertainty or disorder in the system. For example, the Gibbs entropy

$$S = -k_B \sum_i P(i) \log P(i) \tag{3.6}$$

where k_B is Boltzmann's constant, i refers to the possible micro states of the system, and $P(i)$ is the probability that the system is in state i.

The Second Law of Thermodynamics is hence equivalent to the statement 'disorder in closed systems tends to increase'.

3. As the temperature of a closed system approaches absolute zero, the entropy change of a reversible process also tends to zero. Equivalently, a system cannot reach a temperature of absolute zero in a finite number of steps.

These laws are sometimes summarised rather colloquially using gambling parlance:

1. You can't win, i.e., you can't get something for nothing, as energy is always conserved.
2. You can't break even – you cannot return to a previous energy state, as entropy has increased, depleting available *free energy* (for example, the Gibbs free energy is defined as $G = U + PV - TS$).
3. You can't leave the table – you cannot remove all energy from a system.

In short, thermodynamics is a set of economic principles, where the currency in question is free energy. We see the applications of these principles throughout physical systems, from the Earth's atmosphere to the internal machinations of organisms.

Kardashev's suggestion was that we consider the energy consumption of civilisations as a measure of their development. Energy consumption can be considered a proxy for both their ability to send powerful communication signals, and their potential detectability through unintentional energy leakage. His classification system gives three civilisation types:

Type I: a civilisation that consumes energy at a rate equivalent to the radiative energy deposited at their home planet per unit time. In other words, a civilisation with power demands equal to that generated at their planet's surface by the Sun.

Type II: a civilisation that consumes the entire power output of their star, i.e., their energy demands per unit time are equal to the Sun's luminosity.

Type III: a super-civilisation that consumes the entire power output of a galaxy, i.e., the luminosity of several hundred billion stars.

Sagan (1973) defined a simple expression that interpolates between the types, giving a non-integer Kardashev number K:

$$K = \frac{log_{10}P - 6}{10} \quad (3.7)$$

where the power consumption P is measured in watts. A Type I civilisation in this scheme consumes power at a rate $P = 10^{16}$ W, and Type II and III civilisations possess $P = 10^{26}$ and 10^{36} W respectively. The Earth consumed approximately 100,000 terawatt hours of energy in 2012 (out of a total of approximately 150,000 terawatt hours generated). This gives an average power consumption of approximately $P = 10^{13}$ W, giving humanity a Kardashev number of $K = 0.7$.

Several attempts have been made to expand on the Kardashev scale. Barrow (1998) considered our increasing ability to manipulate increasingly smaller spatial domains as an indicator of development (the so-called Barrow scale, intended to complement Kardashev), but this is less likely to yield insights into observations of phenomena at interstellar distances. Sagan extended the definition into two dimensions, by considering information as well as energy. The Sagan scale is defined

$$S = K\{I\} \quad (3.8)$$

where I is one of the 26 letters of the Roman alphabet, with each letter being assigned a number of bits according to its position, e.g., $A = 10^6$ bits, $B = 10^7$ bits, $C = 10^8$ bits and so on.

It has been estimated that as of 2007, humanity's data storage is approximately 10^{21} bits (Hilbert and López, 2011). Combined with our energy consumption, humanity would be classified as $0.7\{P\}$.

As should be clear from the laws of thermodynamics, information and energy are linked through the concept of entropy. It can be shown by comparing the thermodynamic entropy and the information entropy, defined by Shannon as

$$S_{\text{inf}} = -\sum_i P(i) \log P(i), \tag{3.9}$$

that changing the information content of a system will always incur a greater increase in entropy (Brillouin, 1953). However, this linkage does not enforce a one-to-one dependence of information on energy, and hence civilisations should be able to attain high levels of information classification without necessarily being high energy consumers.

3.3 The Three Classes of Solutions to Fermi's Paradox

If we consider Drake's equation, we can use it to derive a simple mathematical expression for the Great Filter:

$$f_l f_i f_c L \ll 1 \tag{3.10}$$

The exact position of the Filter is equivalent to asking which of the variables on the left-hand side is small. After all, any one of these variables being small is sufficient for the above equation to be true. We can therefore produce three classes of solution, which I will use throughout this textbook.

As the Great Filter is a product of three terms, it need not be the case that only one term is small. We could conceive of circumstances where a combination of terms is small. This makes an attempt to classify solutions in this manner less than simple. However, it is usually the case that even if multiple terms are small, one is significantly smaller, and as a result is the governing term in the Great Filter. I will therefore persevere with this classification scheme, but readers should note this caveat, and be aware that this is not the only classification scheme that is possible. Other authors in this area have selected different schemata by which solutions are classified, in a tradition that extends all the way back to Hart, in his classification of temporal, social and physical constraints.

3.3.1 Rare Earth (R) Solutions

In this paradigm, f_i (and perhaps even f_l) is small. As a result, the number of planets in the Milky Way conducive to intelligent life is small. Note that this makes no statement about the number of habitable worlds, but it does make claims about the number of *inhabited* worlds.

3.3.2 Uncommunicative (U) Solutions

If f_c is small, then the number of intelligent civilisations could be large, but their visibility to us is extremely poor. As we will see, this classification is epistemologically challenging, as it requires us to make statements about phenomena which by definition we cannot measure.

3.3.3 Catastrophic (C) Solutions

Perhaps most depressingly, it may well be the case that communicating civilisations have a short lifespan, and hence L is small. From a consideration of the various existential risks to humans, we will see that the list of catastrophic solutions is worryingly large, leaving many in the SETI community to place the Great Filter in our not-too-distant future.

3.3.4 Hard and Soft Solutions

From its earliest exposition, it has been clear that not all solutions to Fermi's Paradox are equal. Hart was among the first to note that solutions could be categorised according to how they relate to physical, temporal or sociological factors, and that these solutions have varying levels of constraining power on civilisations.

In modern terminology, solutions are described as either 'hard' or 'soft'. Hard solutions are applicable to (or prescribe the behaviour of) all possible intelligent civilisations. Typically, hard solutions are motivated by fundamental physical or biological principles that prevent the arisal of intelligent life, or the invention of means to communicate.

'Soft' solutions are only applicable to a limited subset of intelligent civilisations. They often attempt to prescribe behaviour or motive, relying on political or sociological arguments. This approach may represent part of the reason why the Great Silence exists, but soft solutions cannot be solely responsible. Soft solutions permit a single civilisation to deviate from the pattern: a single 'rogue' civilisation which does not conform to the template imposed by a soft solution is sufficient to break the Silence. Brin (1983) refers to these as 'non-exclusive' solutions. The fraction of 'rogues' in the population of Galactic civilisations may be low, but in most cases it is likely to be non-zero, and hence these solutions are regarded as 'soft'.

The collated list of solutions to Fermi's Paradox can be found in Appendix A. I classify them into three as described above: Rare Earth (R), Uncommunicative (U) and Catastrophic (C). Where appropriate, they will be referred to in the main text (e.g., Solution R.x, the xth solution in the Rare Earth category in Appendix A).

Part II
Rare Earth Solutions

Perhaps the most prosaic solution to Fermi's Paradox is that we are rare, possibly unique in the Universe. In other words, we should ascribe low values to the f_p, n_e and f_l terms in Drake's equation. We will begin by exploring our understanding of these terms of Drake's equation, from our growing knowledge of the Galactic population of planetary bodies to our understanding of how our technologically advanced, intelligent species flourished on the Earth. Along the way, we will consider the factors that could frustrate planetary habitability, and place the Great Filter at various points along these first few terms.

4
Habitable Worlds Are Rare

The most obvious Rare Earth solution is that the supply of habitats – warm rocky surfaces with moderate gravity, liquid water and appropriate chemical/electron gradients – is rare. Until the late twentieth century, this solution was left wildly unconstrained. It is only in the last few decades that we now possess the data to begin addressing this question.

4.1 Worlds Are Not Rare

The advent of extrasolar planet (exoplanet) detection put us on the path to understanding the distribution of niches for Earth-esque lifeforms. Using a variety of observational techniques, it has become increasingly clear that the number of planetary systems in the Milky Way is very large indeed. Given our remarkable ability to detect substantial quantities of extrasolar planets (or exoplanets) in the local Solar Neighbourhood, we must conclude that the fraction of stars with planetary systems, f_p, is extremely close to 1!

But how do we detect these planets?

4.1.1 The First Exoplanets – Pulsar Timing

The first exoplanets to be detected orbited a *pulsar*, PSR 1257+12. Pulsars are a special class of neutron star, the remnants of massive stars which have ceased fusing hydrogen in their cores. Once hydrogen fusion has ended, the star attempts to fuse other elements to provide thermal support against gravitational collapse. Once the star attempts to fuse iron, fusion becomes energetically disfavourable, and a supernova explosion typically occurs, leaving a neutron star behind (see section 10.5).

Angular momentum conservation ensures that as the star contracts from some 10^5 km down to ~ 10 km, the rotation rate increases significantly. The first pulsar to be discovered (by Jocelyn Bell Burnell and Antony Hewish, then of the University of Cambridge in 1967) had a measured rotation period of 1.33 seconds. Pulsars are highly magnetised, and therefore they emit light preferentially along the axis of their magnetic field. To an observer at an appropriate orientation to the pulsar, this rapidly rotating object emits a pulse of light every rotation towards the observer, much as a lighthouse does.

The extremely rapid pulsation rate – along with its astonishingly precise period – led some to initially dub it 'the LGM star', with LGM standing for Little Green Men. While this explanation was quickly quashed in favour of a natural explanation (Penny, 2013), pulsars remain a useful tool in the search for exoplanets.

Indeed, it is the pulsar's precision that is of use to astronomers attempting to detect exoplanets. Wolszczan and Frail (1992) investigated the pulsar PSR 1257+12, measuring the time of arrival (TOA) of the pulses. Pulsars typically exhibit highly regular pulses, and as such a Fourier transform of the TOA data should reflect a single period – the pulsar's rotation period, in this case 6 milliseconds. Wolszczan and Frail were able to show that the TOA of 1257+12 varied smoothly over time. These variations were fitted as the pulsar's period plus a combination of two periods, one at 66.6 days and one at 98.2 days.

These periods were inferred to be due to two planets of a few Earth masses ($M_\oplus$) orbiting the pulsar. The pulsar's TOAs are affected by the gravitational field of the planets, as all the bodies in the system – including the pulsar – orbit the centre of mass (see more in the next section). This implies that the pulsar is continually moving towards and away from the observer according to gravitational interactions with the other bodies. This motion causes periodic changes in the distance between the pulsar and the observer, and hence changes in the TOA. These changes are small, but not compared to the pulsar's rotation period, so they can be distinguished with ease.

While pulsar timing yields fascinating, exotic planetary systems, it is yet to yield statistically significant quantities of exoplanets. Also, pulsars are stellar corpses – the end of a star's nuclear fusion, and to some degree the end of its ability to provide biologically friendly radiation (although see Patruno and Kama, 2017).

On the search for habitable planets, we are more disposed to search for exoplanets around *main sequence* stars, the initial stage of nuclear fusion, where hydrogen is fused into helium in the star's core. This means we must turn to other methods.

4.1.2 The Radial Velocity Method

The first exoplanet to be detected around a main sequence star was detected using the *radial velocity* method. This relies on the fact that not only does the gravitational field of the star exert a force on the planet, but vice versa – the planet's gravitational field exerts a force on the star. As a result, both objects orbit the system's centre of mass or barycentre.

Consider astronomers observing a star-planet system from the Earth. The bodies have masses M_* and M_p respectively. The planet executes a Keplerian orbit about the centre of mass. Keplerian orbits are elliptical, and hence described by six parameters. The orbit's shape is given by its *semi-major axis*, a, and its *eccentricity e*. The period of the orbits of both bodies is given by Kepler's Third Law:

$$P^2 = \frac{4\pi^2}{G(M_* + M_p)} a^3 \quad (4.1)$$

The orientation of the orbit is described by three angles, the definition of which depends on our co-ordinate system. Let us define a reference plane, with normal vector equal to the Cartesian $\mathbf{z}$ vector (in other words, the $x - y$ plane). The planet's orbit defines an *orbital plane*, which in general does not coincide with the $x - y$ plane.

The (specific) angular momentum of the orbit is defined by the body's position and velocity vectors $\mathbf{r}$ and $\mathbf{v}$, defined relative to the orbital focus.

$$\mathbf{h} = \mathbf{r} \times \mathbf{v} \quad (4.2)$$

where $\times$ indicates the usual vector product. As a result, $\mathbf{h}$ is perpendicular to the planet's orbital plane. We can therefore define the *inclination*, i, as the angle between our reference plane and the orbital plane:

$$\cos i = \hat{\mathbf{h}}.\hat{\mathbf{z}} \quad (4.3)$$

We use the $\hat{}$ convention to indicate that we are dealing with unit vectors. The *longitude of the ascending node*, Ω, denotes the angular location of the ascending node (where the orbital plane intersects the system reference plane). Ω is determined by:[1]

$$\cos \Omega = \left(\hat{\mathbf{z}} \times \hat{\mathbf{h}}\right).\hat{\mathbf{x}} \quad (4.4)$$

The vector product $\hat{\mathbf{z}} \times \hat{\mathbf{h}} \equiv \mathbf{q}$ gives a vector that points at the ascending node, allowing us to calculate the angle via a scalar product with the reference direction $\mathbf{x}$.

The third orientation angle is the *argument of periapsis*, ω, the angular distance between the point of the planet's closest approach to the centre of mass, and the

---

[1] As is usually the case with scalar products, we must be careful to define the desired angle, as the scalar product will always return the minimum angle between two vectors. This can be determined by taking the magnitude of the vector product, and checking its sign. If the magnitude is positive, the scalar product returns the 'correct' angle. Otherwise, the appropriate angle is the complementary angle, in this case $2\pi - \arccos\left(\hat{\mathbf{z}} \times \hat{\mathbf{h}}\right).\hat{\mathbf{x}}$. The only exception in this context is the inclination, which is usually defined as the smallest angle, appropriately retrieved by the scalar product.

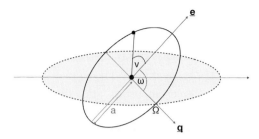

Figure 4.1 A Keplerian orbit. The orbital ellipse is determined by the semi-major axis a, and the eccentricity e. The eccentricity vector **e** points to the periapse point, and the node vector **q** points to the ascending node.

longitude of the ascending node. To calculate this, we must construct a vector which points at the periapsis point. The vector in question is the *eccentricity vector* (sometimes known as the Laplace-Runge-Lenz vector):

$$\mathbf{e} = \frac{\mathbf{v} \times \mathbf{h}}{G(M_* + M_p)} - \hat{\mathbf{r}} \qquad (4.5)$$

This also provides us with the eccentricity, which we obtain from the magnitude of **e**. The argument of periapsis is

$$\cos \omega = \mathbf{q} \cdot \mathbf{e} \qquad (4.6)$$

Finally, an angle is required to place the planet at a specific location on this orbit. As Keplerian orbits are elliptical in general, the angular velocity of the planet changes with time. Indeed, Kepler's Second Law dictates that bodies on Keplerian orbits sweep out equal areas of the ellipse in equal time. In polar co-ordinates, this is equivalent to

$$\frac{1}{2} r^2 \frac{d\phi}{dt} = |\mathbf{h}| = \text{const.} \qquad (4.7)$$

which informs us that Keplerian orbits conserve angular momentum, as expected. Kepler's Second Law gives the motivation for defining the *mean anomaly*:

$$M = nt = \frac{2\pi}{P} t \qquad (4.8)$$

n is usually referred to as the *mean motion*. t describes the time since the planet last passed its periapse. This angle does not represent a position *per se*. To obtain a position angle, we must compute the *true* anomaly v. This is best calculated via the *eccentric anomaly* E:

$$M = E - e \sin E \qquad (4.9)$$

The solution for E is non-trivial as it does not possess a closed form. In practice E is usually calculated through an iterative numerical procedure. The eccentric

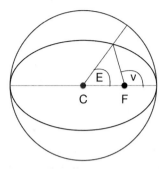

Figure 4.2 The eccentric and true anomaly of a Keplerian orbit. Bodies orbit in ellipses around the focus (F). The eccentric anomaly describes the angle that the orbiting body would make with the centre of the ellipse (C) if the body executed a circular orbit. Hence, the eccentric and true anomaly are equivalent when the orbit is circular.

anomaly is the angle between the body's position vector, the *centre* of the ellipse and the argument of periapsis (as shown in Figure 4.2, where the periapsis point is aligned with the x direction). Note that the star does not occupy the ellipse centre, but instead the ellipse *focus*. The *true anomaly* is the angle subtended by the body's orbit and the reference direction, centred at the system focus. This can be derived from the eccentric anomaly:

$$\nu = \frac{\cos E - e}{1 - e \cos E} \qquad (4.10)$$

or from the eccentricity vector:

$$\nu = \arccos \frac{\mathbf{e}.\mathbf{r}}{|\mathbf{e}||\mathbf{r}|} \qquad (4.11)$$

However, the second approach is invalid if the orbit is circular, as the argument of periapsis is undefined. The first equation is therefore a generally safer approach. With the true anomaly known, the system is closed. Indeed, we can determine r from these elements using the ellipse equation:

$$r = \frac{a(1 - e^2)}{1 + e \cos \nu} \qquad (4.12)$$

As we have discussed already, the star is not fixed in position at the ellipse focus, but instead moves around the focus on its own elliptical orbit. In fact, we should define two semi-major axes:

$$a_* = \frac{M_p}{M_* + M_p} a \qquad (4.13)$$

$$a_p = \frac{M_*}{M_* + M_p} a \qquad (4.14)$$

As we can see, the distance of the star's centre from the system barycentre is typically very small. For the case of the Sun and Jupiter, where $a = 5.2$ AU, this gives $a_* = 0.0045 AU$, or about 95% of the solar radius.

The motion of the star around the barycentre results in a changing radial velocity – that is, the velocity along the line of sight of the observer. If the planet's orbit is circular, and aligned with the observer's line of sight, the radial velocity semi-amplitude of the star, K, has a particularly simple form:

$$K = \frac{2\pi a_*}{P} \qquad (4.15)$$

which can be deduced rather intuitively. As the orbit is circular, the angular speed of the planet and star are constant in time, and K is simply the circumference of the star's orbit divided by the duration of an orbit. The radial velocity of the star follows a simple sine curve with semi-amplitude given by K:

$$v_r = K \cos \nu \qquad (4.16)$$

where ν is the true anomaly as defined previously.[2]

If the planet's orbit is inclined and elliptical, then

$$K = \frac{2\pi a_* \sin i}{P\sqrt{1-e^2}} \qquad (4.17)$$

and the changing angular speed of both star and planet introduces new complexity into the radial velocity:

$$v_r = K \left(\cos(\nu + \omega) + e \cos \omega \right) \qquad (4.18)$$

Observers measure the radial velocity via Doppler shifts in the star's spectrum. Given a sample of several robust absorption or emission lines, radial velocity curves are subsequently fitted to determine the planet's orbital properties. A more explicit definition of K can be made by invoking Kepler's Third Law on a_*:

$$K = \left(\frac{2\pi G}{P} \right) \frac{M_p \sin i}{(M_* + M_p)^{2/3}} \frac{1}{\sqrt{1-e^2}} \qquad (4.19)$$

Note that measurements of the radial velocity (and ultimately K) are degenerate in $M_p \sin i$. This means that radial velocities measure the 'minimum mass' of exoplanets, as $|\sin i| \leq 1$ always.

The presence of multiple planets in a system will result in a superposition of radial velocities. For relatively low-mass planets at moderate distances, this can be modelled by simply adding multiple versions of equation (4.18) together.

---

[2] Strictly, the true anomaly is not defined for circular systems, and is usually replaced with a much simpler angular parameter that is equivalent to the mean anomaly if the inclination is zero.

When attempting to detect a planetary signal in radial velocity data, periodograms are constructed. However, this approach requires continuous monitoring of the target star, which is not possible from the ground if planets of relatively long periods are sought. Gaps in the radial velocity data can produce spurious signals when periodograms are constructed. Lomb-Scargle periodograms are typically employed to combat this problem (Lomb, 1976; Mortier et al., 2015).

For the Sun, Jupiter induces a K of around 12 m s^{-1}. The Earth produces a K of approximately 0.1 m s^{-1}. The first planets to be detected with this technique were 'Hot Jupiters', orbiting close to their stars, with strong radial velocity signals, i.e., $K \sim 50$ m s^{-1}.

These Hot Jupiters are hardly our most promising candidates for habitable worlds. Gas giants with surface temperatures of several thousand Kelvin, orbiting so close to their parent star that tidal forces have sapped the planet's rotational angular momentum, until it is pseudo-synchronous with its orbit, i.e., its orbital period and rotation period are close to equal.

Habitable worlds will typically be less massive, and at greater distance from their star. Current attempts to detect Earth analogues require extraordinary precision, and these levels are being achieved with the latest instrumentation. The newest obstacle is *stellar noise*. The motions of plasma in the Sun create radial velocity signals of order that of the Earth's. Characterisation and subtraction of this stellar activity is one of the most important and intensely studied aspects of radial velocity surveys (see, e.g., Haywood, 2016, for a review).

4.1.3 Transit Detection

The *transit method* has become extremely popular for exoplanet detection thanks to the provenance of extremely stable photometry – that is, the ability to measure the total electromagnetic flux from a source with precisions as high as a few hundred parts per million (ppm). Transit detection relies on the exoplanet passing between the host star and the observer (sometimes referred to as primary eclipse), causing a small dimming of the starlight (Figure 4.3).

Consider a planet of radius R_p, orbiting a star of radius R_*, and mass M_*. If this planet transits its star, relative to some point of observation, the relative flux level will change as the planet moves across the stellar disc. The maximum relative change in flux received by observers as the planet passes between them and the star is:

$$\Delta I \propto \left(\frac{R_p^2}{R_*^2} \right) \quad (4.20)$$

If the stellar radius is well characterised, this *transit depth* will provide observers with the planet radius. As this transit will be repeated once every orbital period of the planet, measuring the interval between transits gives P.

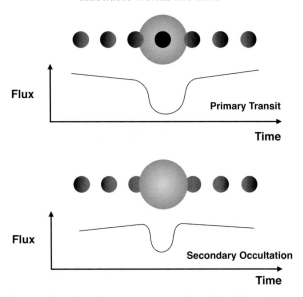

Figure 4.3 The transit of a planet across a stellar disc. The primary transit (top row) blocks out a small fraction of the star's light, while the occultation or secondary eclipse (bottom row) blocks out the planetary light. In between transits and occultations, small portions of the planet's surface become visible, giving rise to variations in the out-of-transit flux (the so-called phase curve).

However, transits require strongly constrained inclinations to be observed at all. Broadly speaking, the probability of observing a transit is

$$P_{transit} \approx \frac{R_*}{a_p}\left(\frac{1}{1-e^2}\right) \tag{4.21}$$

where we have assumed that all inclinations relative to the observer line of sight are equally likely, and that $R_* \gg R_p$. For the case of the Earth, this equation yields a transit probability of ~ 0.005. Even if the planet transits, it is unlikely to transit the centre of the stellar disc. The impact parameter b indicates the closest approach the planet makes to the stellar disc centre (in units of the stellar radius):

$$b = \frac{a_p \cos i}{R_*}\left(\frac{1-e^2}{1+e\sin\omega}\right) \tag{4.22}$$

More information can be gleaned by determining the epoch of four different *contact times*:

1. First contact of the planet and stellar disc (beginning of ingress)
2. End of ingress, where the transit depth reaches its minimum
3. Beginning of egress, where the planet begins to leave the stellar disc
4. End of egress, when the planet has completely left the stellar disc and the transit is complete.

4.1 Worlds Are Not Rare

These are usually labelled $t_I - t_{IV}$ respectively. If the orbital elements of the planet are completely known, then the total transit duration is given by integrating the following:

$$t_{IV} - t_I = \frac{P}{2\pi(1-e^2)} \int_{v_I}^{v_{IV}} \left(\frac{r(v)}{a}\right)^2 dv \quad (4.23)$$

and so on for other combinations of the four contact times. In the case of circular orbits, the solution is:

$$t_{IV} - t_I = \frac{P}{\pi} \sin^{-1}\left[\frac{R_*\sqrt{1 + \left(\frac{R_p}{R_*}\right)^2 - b^2}}{a \sin i}\right] \quad (4.24)$$

and the length of the full transit

$$t_{III} - t_{II} = \frac{P}{\pi} \sin^{-1}\left[\frac{R_*\sqrt{1 - \left(\frac{R_p}{R_*}\right)^2 - b^2}}{a \sin i}\right] \quad (4.25)$$

If the orbits are eccentric, an approximate solution can be obtained by multiplying the above by $\frac{\sqrt{1-e^2}}{1+e\sin\omega}$.

Occultations, or secondary eclipses, occur where the star passes between the planet and the observer. The impact parameter is then

$$b_{occ} = \frac{a_p \cos i}{R_*}\left(\frac{1-e^2}{1-e\sin\omega}\right) \quad (4.26)$$

and the occultation depth becomes

$$\Delta I_{occ} = \frac{I_{planet}}{I_{planet} + I_{star}} \quad (4.27)$$

where we now use disc-averaged intensities I for the star and planet. Measuring the combined intensity of the system throughout an orbital period gives us access to the planet's emission, which varies as the observer sees increasing and decreasing fractions of the illuminated surface (astronomers usually refer to this as the *phase curve*, as illustrated in Figure 4.3).

Spectroscopic measurements of transits measure different transit depths at different wavelengths of light. This is essentially measuring the planet atmosphere's ability to absorb or transmit different wavelengths of light. With careful modelling, this can yield chemical composition information about the planet's atmosphere. Combined with a radial velocity measurement to obtain the planet's mass, this can yield a great deal of information on the planet's properties.

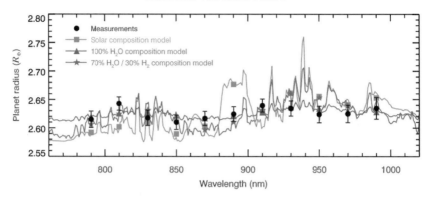

Figure 4.4 The spectrum of the super-Earth GJ1214b, as determined by measuring the planetary radius from transit data at a number of wavelengths. The lack of features in this spectrum indicates that cloud cover is playing an obscuring role. Figure taken from Bean et al. (2010).

A common feature of exoplanet spectra is that they are relatively flat. This lack of change in received flux with wavelength is best modelled by the presence of *clouds* in the upper atmosphere. Figure 4.4 shows this for the exoplanet GJ1214b, a super-Earth detected both via radial velocity and transit. Its mass is estimated at 6.3 $M_\oplus$, and its radius at around 2.8 $R_\oplus$. Various attempts to characterise the measured radius as a function of wavelength have shown a flat, featureless spectrum. The relatively low density of GJ1214b has led many to dub it a 'water world', and early modelling of its spectrum suggested this could be the case, but until observations can penetrate these clouds, we will remain uncertain.

4.1.4 Gravitational Microlensing

At present, general relativity (GR) is the best description of how the gravitational force operates in nature. It represents a fundamental shift from classical Newtonian gravity, as it demands that space and time are not separate entities, but are instead aspects of a single entity, *space-time*, which is distorted by the presence of matter or energy. This distortion then affects the motion of objects in this curved space. As we cannot perceive the curvature of space-time, it appears to us as if a force acts on the body, which we label 'gravity'.

If GR can be condensed to a single equation, it is Einstein's field equation:

$$G_{\mu\nu} = R_{\mu\nu} - \frac{1}{2} g_{\mu\nu} R = \frac{8\pi G}{c^4} T_{\mu\nu} - g_{\mu\nu} \Lambda \qquad (4.28)$$

where $G_{\mu\nu}$ is the *Einstein tensor*, and $R_{\mu\nu}$ is the Ricci curvature tensor which describes how the shape and geometry of space-time deviates from flat or *Euclidean space*. Einstein's field equation is soluble given judicious choices of the *metric tensor*, $g_{\mu\nu}$.[3] The metric essentially determines how distances are measured in space-time. For example, if we wish to model empty space in Cartesian co-ordinates, we choose the metric

$$ds^2 = c^2 dt^2 - dx^2 - dy^2 - dz^2 \tag{4.29}$$

The equivalent metric tensor is

$$g_{\mu\nu} = \begin{pmatrix} -c^2 & 0 & 0 & 0 \\ 0 & 1 & 0 & 0 \\ 0 & 0 & 1 & 0 \\ 0 & 0 & 0 & 1 \end{pmatrix} \tag{4.30}$$

$T_{\mu\nu}$ is the *momentum-energy* tensor, which amounts to a census of all sources of matter and energy in the region. This gives rise to the classic aphorism:

Matter tells space how to curve – space tells matter how to move.[4]

In flat space-time, light rays or photons will follow straight-line trajectories. In curved space-time, light rays continue to follow straight-line trajectories as projected onto the curved surface (*geodesics*). In the presence of strong curvature then, non-converging or divergent light rays can be focused at a focal point in a manner exactly analogous to a lens.

This *strong lensing* effect can be seen around clusters of galaxies. Objects at great distance from the cluster can have their emitted photons lensed around the cluster and received at the Earth, in some cases producing multiple images of the object. At cosmological scales, the images of galaxies are weakly distorted by dark matter between the galaxy and Earth. This *weak lensing* only mildly affects the galaxy's magnification or *convergence* and the ellipticity, expressed as the 2-component *shear*. Measured on the sky, the convergence and shear fields encode information about the lenses at play – in effect, the entire intervening dark matter distribution.

Given some assumptions about the typical ellipticities of galaxies (and their alignments to each other), cosmologists can deduce important statistical properties of the intervening dark matter, and other properties of the Universe such as

---

[3] Whether one writes the cosmological constant Λ on the left- or right-hand side of the field equations largely depends on (a) one's reading of history, and (b) one's belief as to what Λ represents. On the right-hand side, Λ represents an additional source of energy-momentum in the Universe. On the left, it represents an additional source of curvature. Einstein originally proposed that Λ should be inserted *ad hoc* into the field equation to ensure a steady-state universe. He is famously said to have described it as 'my biggest blunder', but whether he actually did is at best unclear. Λ is now used by cosmologists to describe the apparent repulsive force seen throughout the Universe, known as *dark energy*.

[4] This is attributed to John Wheeler, a Manhattan Project scientist and giant in the field of relativity. He would supervise record numbers of PhD students at Princeton that would go on to define the field of GR.

dark energy (more can be found on this in Bartelmann and Schneider, 2001, and Schneider, 2005).

In the case of exoplanet detection, *microlensing* is the phenomenon of relevance. This refers to lenses of much smaller mass, such as individual stars. While microlensing shares many features with strong lensing events, the multiple images it produces are not resolved, as their separations on the sky are small.

Consider a star which emits photons in the approximate direction of the observer (the *source* star). If the line of sight between source and observer contains a sufficiently high stellar density, a *lens* star will pass accross the line of sight. As the multiple images produced by the lens are unresolved, the source will appear magnified, brightening and darkening on a characteristic timescale of order a day. If the lens star hosts a planet, then this planet may act as a second lens, producing another magnification event as the star system crosses the line of sight. This secondary event can be as brief as an hour, but encoded within the light curve of this microlensing event is both the mass and semi-major axis of the planet. To see how, let us consider the problem formally (Figure 4.5). It will suffice for us to consider the problem in 1D, but in practice lensing is a 2D phenomenon on the sky.

The source lies at a distance D_S from the observer, the lens at a distance D_L.[5] We will define P_S and P_L as the *source plane* and *lens plane* respectively. Light from the source is deflected by the lens at an angle of $\hat{\alpha}$ towards the observer. This deflection causes images of the source to appear at an angular separation of θ from the lens. The *lens equation* describes how this apparent angular separation is related to the true separation β, which would be observed if no lensing occurred.

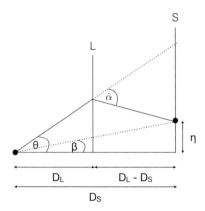

Figure 4.5 The geometry of a lensing event. A source (in the source plane denoted S) emits light which is then lensed by mass in the lens plane L. The lens equation is then derived from this geometry to solve for the mass of the lens.

---

[5] Note these are *angular diameter distances*. In GR, space-time curvature ensures that different methods of measuring distance, such as using apparent angular size given a known physical size, or apparent flux given a known luminosity, do not agree on their result!

We can derive this lens equation through some simple geometry and the small-angle approximation. The physical separation between the observer's sight line and the source location (in the source plane), η, is

$$\eta = D_S \tan \beta \approx D_S \beta, \tag{4.31}$$

where we have used the small angle approximation. The physical separation between the lens and the image in the lens plane, ξ, is

$$\xi = D_L \tan \theta \approx D_L \theta \tag{4.32}$$

We can re-express η as

$$\eta = \frac{D_S}{D_L}\xi - (D_L - D_S)\hat{\alpha} = \frac{D_S}{D_L}\xi - D_{LS}\hat{\alpha}, \tag{4.33}$$

where we have defined $D_{LS} \equiv D_L - D_S$ for convenience. Dividing throughout by D_S gives us the lens equation:

$$\beta = \theta - \frac{D_{LS}}{D_S}\hat{\alpha} \tag{4.34}$$

It now merely remains to calculate the deflection angle α. Newtonian gravity gives us a prediction for the deflection, which is:

$$\hat{\alpha}_{Newton} = \frac{2GM}{c^2 \xi} = \frac{2GM}{c^2 D_L \theta} \tag{4.35}$$

However, to obtain the deflection angle from Einstein's field equations we must select a metric. We will use the Schwarzschild metric, which describes the curvature of space-time – in spherical co-ordinates (r, θ, ϕ) – of a single non-rotating point particle with mass M:

$$ds^2 = -\left(1 - \frac{2GM}{rc^2}\right)c^2 dt^2 + \left(1 - \frac{2GM}{rc^2}\right)^{-1} dr^2 + r^2 d\theta^2 + r^2 \sin^2\theta d\phi^2. \tag{4.36}$$

This yields an extra factor of two, and hence

$$\hat{\alpha}(\xi) = \frac{4GM}{c^2 D_L \theta} \tag{4.37}$$

This yields the lens equation as:

$$\beta = \theta - \frac{D_{LS}}{D_S D_L}\frac{4GM}{c^2 \theta} \tag{4.38}$$

If $\beta = 0$ (i.e., the lens and source are in perfect alignment), then the image forms an *Einstein ring* with angular radius equal to the *Einstein radius*, θ_E:

$$\theta_E = \sqrt{\frac{4GM}{c^2}\frac{D_{LS}}{D_S D_L}} \qquad (4.39)$$

The lens equation can be divided by the above to make it dimensionless:

$$u = y - y^{-1} \qquad (4.40)$$

which is a simple quadratic expression, and therefore typically has two solutions and hence two images are seen (if $\beta \neq 0$):

$$y_\pm = \pm\frac{1}{2}\left(\sqrt{u^2+4} \pm u\right) \qquad (4.41)$$

Lensing produces a magnification of the source star in both images:

$$A_\pm = \left|\frac{y_\pm}{u}\frac{dy_\pm}{du}\right| = \frac{1}{2}\left|\frac{u^2+2}{u\sqrt{u^2+4}} \pm 1\right| \qquad (4.42)$$

with a total magnification

$$A(u) = \frac{u^2+2}{u\sqrt{u^2+4}} \qquad (4.43)$$

Note that at the Einstein radius, $u = 0$ and the magnification is undefined. The motion of the source and lens is most simply modelled as a smooth change in source-lens angular separation. If we know the *proper motion* of the source relative to the lens, i.e., the rate of change of angular separation $\mu_{rel} \equiv \frac{du}{dt}$, we can define the timescale $t_E = \theta_E/\mu_{rel}$, and hence

$$u(t) = \left(u_0^2 + \left(\frac{t-t_0}{t_E}\right)^2\right)^{1/2} \qquad (4.44)$$

where u_0 is the minimum angular separation, which occurs at time t_0. Plugging this into equation (4.43) gives the classic 'Paczynski curve', as plotted in Figure 4.6.

If a planet is present, as is the case here, a perturbation to the Paczynzki curve will be seen which indicates its presence. Typically, what occurs is that one of the images produced by the lens star will be lensed a second time by the planet. This extra lensing event has a typical duration of

$$t_{E,p} = \sqrt{\frac{M_p}{M_*}} t_{E,*} \qquad (4.45)$$

Typically, the primary lensing event has a duration of $t_{E,*} = 10\text{--}100$ hours. This means that for Earth-mass planets, the secondary lensing event can be only a few hours. Also, because microlensing events require precise alignments of lens, source

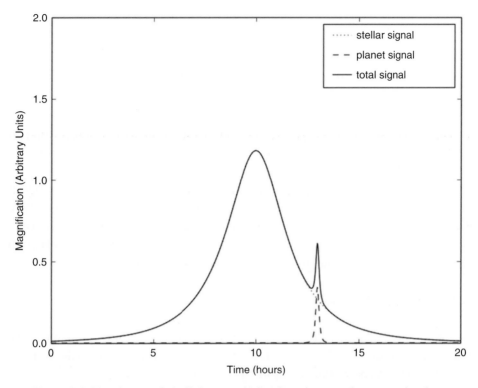

Figure 4.6 The characteristic light curve indicative of an exoplanetary microlensing event. The lens star produces the principal curve, with the planet's addition relatively small in amplitude, and exceptionally brief.

and observer,[6] microlensing systems can only be observed **once**. On the other hand, microlensing events are surprisingly insensitive to the planet's mass, and are particularly sensitive to planets orbiting at the local Einstein radius $r_E = D_L \theta_E$, as images tend to be formed around or upon the Einstein ring. It complements radial velocity and transit searches extremely well, as it performs well at detecting planets at large separations from their star, as does our final detection technique, which we discuss next.

4.1.5 Direct Imaging

While the above methods are relatively indirect, recent imaging advances have allowed exoplanet observers to use perhaps the most intuitive detection method –

---

[6] Indeed, it was this low probability of microlensing that discouraged Einstein from considering them further. As early as 1936, and possibly as early as 1912, he had derived the above equations for star-star microlensing, and had then dismissed them as a practical observational tool, as 'there is no great chance of observing this phenomenon'.

detecting planets by directly measuring the light they emit or reflect, in the same manner that we detect almost all other astronomical phenomena.

While it is the most conceptually simple, it is extremely challenging. The observer must capture photons from the planet, while a much brighter object (the parent star) sits very close to it on the sky. We can define the *contrast* between a planet and its host star (at a given wavelength λ) as:

$$C_\lambda = \frac{F_{\lambda,p}}{F_{\lambda,*}} \qquad (4.46)$$

where F is the total flux received by each object, p and $*$ designate the planet and star respectively. The higher the contrast, the better the odds of detecting the planet in question. Typically, very young planets are preferentially detected using this technique, as they produce substantially larger radiative flux as they relax into their equilibrium shape (and hence their equilibrium temperature).

The wavelength of observation, λ, will prove important, as the *spectral energy distribution* (SED) of the star and planet will be significantly different, making certain wavelengths preferred for planets of different ages around stars of different types. The scale of the challenge is already evident: at visible wavelengths, the Earth has a contrast of 10^{-10}! As the wavelength is increased into the infrared ($\lambda = 10\mu m$), the contrast increases to 10^{-8}.

These challenges ensured that the first direct detections of exoplanets would yield young, hot planets, orbiting at large separations from their host star. The maximum angular separation of a star and planet on the sky can be estimated using the small angle approximation:

$$\theta = \frac{a_p(1+e_p)}{D} \qquad (4.47)$$

where a_p and e_p are the planet semi-major axis and eccentricity respectively, and D is the distance from the observer to the planetary system. Astronomers define the *parsec* as a helpful unit of distance for calculations such as this. If the orbit is circular, with semi-major axis is equal to the Earth's ($a_p = 1AU$) and the distance to the system is 1 parsec (pc), then $\theta = 1$ arcsec, which is 1/3600 of a degree.

The nearest main sequence star system to the Earth, the Alpha Centauri system, by happy coincidence is approximately 1 pc away. Astronomers typically use a distance of 10 pc as a reference length, and therefore our reference θ (again assuming $a_p = 1$ AU) is 0.1 arcsec, or 100 mas (milli-arcseconds).

The above underlines the need to somehow 'screen' our telescopes from starlight, leaving only the planetary light (both thermally generated and starlight reflected by the planet's upper atmosphere). Broadly speaking, there are two classes of observation available to observers:

1. *coronagraph* techniques, which block out the star using an opaque mask internal to the telescope optics, and
2. *interferometry* techniques, which combine data from multiple telescopes with differing path lengths to the source, to cancel out contributions from the star.

The above descriptions are necessarily glib, and the optics of both techniques deserve a textbook of their own. Indeed, such a textbook would likely state that both techniques are two sides of the same coin: both involve combining the phase and amplitude of the incoming photons of light to produce destructive interference and suppress the stellar signal. I will avoid diving into a deep tangent on optics, but I should mention in brief some important concepts that affect our ability to directly image exoplanets.

The *point spread function* (PSF) of a telescope is an extended pattern produced when a point source is imaged. When imaging a star system, the PSF 'blob' produced by the star is sufficient to completely smother an exoplanet signal if the planet semi-major axis is sufficiently small.

In a perfect optical system, the PSF is an Airy function:

$$I(u) = \frac{1}{(1-\epsilon^2)^2}\left(\frac{2J_1(u)}{u} - \epsilon^2 \frac{2J_1(\epsilon u)}{\epsilon u}\right) \qquad (4.48)$$

I is the surface brightness as a function of distance from the maximum (which is measured by the dimensionless quantity u, which in turn depends on the angular separation and the diameter of the primary aperture). ϵ is the radius of the central obscurating mask, if present. J_1 is a Bessel function of the first kind. This function results in a central core of emission around the maximum (the 'Airy disk'), followed by 'Airy rings' of decreasing intensity.

The PSF is unique to each telescope/instrument. Astronomers and engineers take great pains to characterise the PSF of an optical system, so that it may be subtracted from the raw image.

Imperfections in the optics of a telescope produce *speckle* in the resulting image. A perfect system will focus rays of photons with precise *amplitude* and *phase*, resulting in a perfect *wavefront* received by the detector chip. Imperfections in the system produce errors in the wavefront, resulting in the image artefacts we refer to as speckles. These are the nemesis of direct imaging: a speckle is extremely difficult to distinguish from an exoplanet.

Speckle has a variety of sources, and is not limited to imperfections in the optics. For ground-based telescopes, the turbulence in the Earth's atmosphere can also distort the incoming wavefronts and produce speckles of its own.

Atmospheric speckle can be corrected using *adaptive optics* techniques. Wavefront errors are diagnosed by observing a guide star in tandem (either real or artificially generated by laser excitation of sodium atoms in the upper atmosphere). The real-time errors generated by atmospheric turbulence are then communicated to actuators in the telescope mirror, which deforms to 'correct' for the turbulence and reduce atmospheric speckle.

However, speckles can never be fully removed, as they can result from any unwanted scattering of light in the telescope optics. Single grains of dust on the telescope's mirrors are sufficient to generate speckle! Instead, the astronomical images are processed after the fact to identify sources of speckle that can subsequently be removed.

Angular Differentiation Imaging (ADI) utilises the fact that if the telescope is rotated along the axis parallel to the observation line of sight, the speckle remaining after adaptive optics, etc. has been applied will rotate with it. If there truly are any exoplanets in the image, they will not rotate with the telescope, remaining fixed on the sky. By appropriately subtracting the rotated images from the non-rotated image, speckle patterns can be removed, and the exoplanet will remain.

Simultaneous Spectral Differentiation Imaging (SSDI) takes advantage of the wavelength dependence of the speckle images. The distance of the speckle from the central star changes as the wavelength of the image is changed, and the planet's position will remain fixed. Appropriate rescaling of the image allows the identification of speckle, and its subsequent masking from the image.

The first system to be detected via direct imaging orbits the star HR 8799 (Figure 4.7). This system contains several massive, young gas giants orbiting at separations of 24, 38 and 68 AU. This detection was made possible by the development of both adaptive optics and the ADI technique at the Keck and Gemini telescope at infrared wavelengths (1.1 to 4.2 μm).

Direct imaging has yielded a handful of Jovian-type planets at distances of 10 AU and above from their host star. Planets that are potentially habitable are unlikely to reside at these distances (see next section), and future instruments will have to be able to resolve much smaller angular separations, and much smaller contrasts (Males et al., 2014).

4.1.6 Detecting Life on Exoplanets

Biosignatures

Humanity's growing ability to characterise the nature of exoplanets, particularly their atmospheric inventory, may eventually yield a detection of extraterrestrial life. *Biosignatures* (sometimes biomarkers) are features in exoplanet data that indicate said exoplanet hosts a biosphere.

The Earth's atmosphere exists in a state of thermodynamic disequilibrium, where this disequilibrium is typically measured by considering the Gibbs free energy of

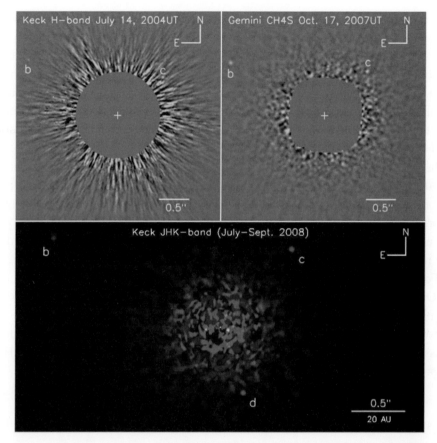

Figure 4.7 The detection of several exoplanets via direct imaging around HR 8799. The artefacts on the image surrounding the star are due to the Angular Differentiation Imaging (ADI) technique, used to remove speckle from the images. The above two images are single-wavelength band images, and the bottom image is a three-colour image from three infrared wavelengths. Taken from Marois et al. (2008).

the current state, and the Gibbs free energy of a final equilibrium state. The combined presence of O_2 and CH_4 indicates this disequilbrium, as both should react to remove at least one from the atmosphere:

$$CH_4 + 2O_2 \rightarrow CO_2 + 2H_2O \quad (4.49)$$

As both are still present in the atmosphere, they are being continually replenished – in Earth's case, from biological organisms. Therefore, it is argued that the joint presence of CH_4 and O_2 (along with a small amount of CO_2) is the most convincing evidence – accessible from interstellar distances – that Earth possesses a biosphere (Lederberg, 1965; Lovelock, 1965; Lovelock and Kaplan, 1975).

Of course, this is predicated on observations of modern Earth. The Earth's atmosphere during the Precambrian (over 500 million years ago) was much less oxygenated than the present day, and it has been demonstrated that the Earth's thermodynamic disequilibrium has decreased with decreasing O_2 (Krissansen-Totton et al., 2018). More concerningly for using atmospheric disequilibrium as an indicator, the level of atmospheric disequilibrium on Earth is almost entirely driven by O_2 levels – in effect, a measure of disequilibrium is a measure of O_2, which as we will see below is not necessarily a guarantee of biology.

In any case, the concept that biological organisms have altered the Earth's atmospheric composition has led many to consider how such changes in exoplanet atmospheres might be considered a biosignature. A principal issue with this approach has been the growing list of *false positives*, where a biosignature is produced by abiotic processes. What might be considered a biosignature on Earth does not necessarily guarantee it is a biosignature on other worlds.

For example, a great deal of early effort was expended considering the detection of O_2 and O_3, key bioactive molecules in Earth's atmosphere (O_3 being produced by photochemical reactions with O_2). While these are certainly biosignatures for Earth, they are not biosignatures in general. For terrestrial planets orbiting low-mass M dwarf stars, the increased levels of UV radiation permit photolysis of H_2O to produce O:

$$H_2O + \gamma \rightarrow 2H^+ + O + 2e^- \qquad (4.50)$$

This can form O_2 by three-body recombination:

$$O + O + O \rightarrow O_2 + O \qquad (4.51)$$

and O_3 by a similar reaction. Photodissociation of CO_2, SO_2 and any other oxygen-bearing gases also provide a source of atomic oxygen (Domagal-Goldman et al., 2014; Tian et al., 2014). We should therefore be extremely wary of O_2 as a biosignature, not just as a false positive, but as a *false negative*. On Earth, early life did not produce oxygen in large quantities until the arrival of cyanobacteria. Even then, the production of oxygen via photosynthesis likely preceded the large-scale oxygenation of the atmosphere by billions of years, as various geophysical processes acted to suppress it (Meadows et al., 2017, and references within). This would constitute an inhabited planet that did not exhibit the desired biosignature.

With oxygen's status as a biosignature highly dependent on context, there has been a substantial effort to compile a list of potential biosignature gases. Systematic efforts from several research groups have attempted to compile lists of atmospheric gases that could ostensibly qualify as biosignatures. The most widely discussed are those of Seager et al. (2013, 2016), which attempted to build a list via combinatoric exploration of possible molecules. This has resulted in a database of some 14,000 molecules that contain up to six atoms (excluding hydrogen), that are stable and volatile at standard temperature and pressure. Of these, some 2500 are composed of the classic biogenic elements (NCHOPS), with the majority being halogenated compounds (containing, e.g., bromine or chlorine), and a small sample of inorganics. Around a quarter of the biogenic compounds are produced by terrestrial life, as well as a small fraction of the halogenated compounds.

This approach has yielded 622 compounds[7] that might legitimately be searched for in exoplanet atmospheres for evidence of life. While many of these compounds are only present in trace form in Earth's atmosphere, it seems clear that they may be present in larger quantities in other terrestrial planet atmospheres, as they will have a different initial atmospheric mix and thermodynamic state.

Despite this, our issues with false positives and negatives remain. Adopting a Bayesian formalism, we can describe the events

- B – 'Molecule B is detected in an exoplanet's atmosphere'
- L – 'The exoplanet has a biosphere'

and ask what is the probability the planet has a biosphere, given that we detect molecule B, $P(L|B)$? We must then construct a likelihood $P(B|L)$, and hence

$$P(L|B) = \frac{P(B|L)P(L)}{P(B|L)P(L) + P(B|\neg L)P(\neg L)} \qquad (4.52)$$

where $\neg$ indicates the negation, i.e., $\neg L =$ 'The exoplanet does not have a biosphere'. We should therefore be careful about the probability of a false positive $P(B|\neg L)$ when attempting to compute our posterior. The prior $P(L)$ will also play a significant role. The probability of a false negative is then:

$$P(L|\neg B) = \frac{P(\neg B|L)P(L)}{P(\neg B|L)P(L) + P(\neg B|\neg L)P(\neg L)} \qquad (4.53)$$

As priors play such an important role in this calculation, it is clear that any attempt to determine the presence of a biosphere will rely on attempts to find multiple coexisting biosignatures, with the findings from one signature being used to inform the prior of the next, and most likely the formation of conditional probabilities:

$$P(L|B_1, B_2) = \frac{P(B_1|L, B_2)P(L)}{P(B_1|L, B_2)P(L) + P(B|\neg L, B_2)P(\neg L)} \qquad (4.54)$$

This will require careful modelling of how biosignatures interact with the atmosphere, surface and interior of terrestrial planets if appropriate likelihood functions are to be derived.

---

[7] http://seagerexoplanets.mit.edu/ASM/index.html

Biosignatures are not limited to spectral lines in the atmosphere – the Earth also produces surface biosignatures that are detectable in reflected light. For example, the vegetative red edge is a sharp increase in reflectance between 0.67 and 0.76 μm, produced by photosynthetic organisms containing the chlorophyll a pigment. There are a variety of other biological pigments that also produce differing reflectance patterns, but more research is needed to fully characterise them for the purposes of exoplanet observations (Catling et al., 2017; Schwieterman et al., 2017).

Technosignatures

In a similar vein, we can define a technosignature as a feature in astronomical data that indicates the presence of technology. In practice, this definition covers essentially any SETI signal of interest, from radio emission to Dyson spheres to interstellar probes, and SETI might rightly be renamed the Search for Extraterrestrial Technosignatures.

It is instructive, however, to make a narrower definition and demand that a technosignature be unintentionally emitted from an exoplanet. Examples can include pollution from chlorofluorocarbons (CFCs), which have no abiotic source (Lin et al., 2014), artificial lighting (Loeb and Turner, 2012), or the waste heat produced by technological civilisations, as was attempted by the GHAT survey (Wright et al., 2014).

Technosignatures tend to have one significant advantage over biosignatures, in that they have few or no false positives. Of course, it is quite clear that almost every technosignature can be prone to false negatives – not every civilisation pollutes, or produces waste heat.

4.1.7 The Exoplanet Population

Figure 4.8 shows the currently known exoplanet population, plotted according to their radius, and the equivalent radiation they receive from their star. Currently, there are some 3,800 confirmed exoplanets, with around 1,300 of those being detected via the Kepler Space Telescope, making it the current champion of exoplanet detection.

Given our understanding of the various detection methods at play, it should be immediately clear that selection bias needs to be considered. Radial velocity measurements are more sensitive to more massive planets at closer distances to the parent star, as the magnitude of the gravitational force depends on both. Transit measurements are more sensitive to larger planets, and have an improved detection probability as the distance to the star decreases. Gravitational microlensing is sensitive to intermediate semi-major axes, and direct imaging is sensitive to the hot planets at large separations – in practice, direct imaging is currently only

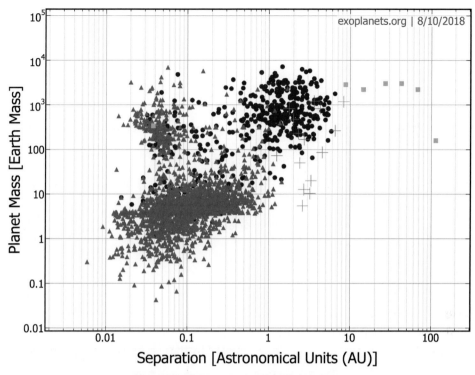

Figure 4.8 The exoplanet population, as a function of planet mass and orbital semi-major axis, produced by the Exoplanet Data Explorer (exoplanets.org, accessed August 10, 2018). Triangles indicate exoplanets detected via transit, circles detections via radial velocity, crosses via microlensing and squares via direct imaging.

effective at detecting very young planets that are still self-luminous due to thermal contraction.

At the time of writing, several future space missions are expected to greatly enhance both the quantity and quality of the exoplanet dataset. The Transiting Exoplanet Survey Satellite (TESS) will survey the entire sky, albeit of stars somewhat brighter than those targeted by Kepler. This decision allows ground-based radial velocity missions to achieve follow-up observations of TESS detections. The all-sky approach will also give us the first true sense of whether planet formation varies from place to place in the Milky Way. As I write, TESS has just launched and is en route to its destination.

The PLATO mission (PLanetary Transits and Oscillations of stars) is perhaps the one mission best described as Kepler's successor. It will study stars significantly fainter than TESS, and will still be able to survey around half of the sky by the end of its mission. Like TESS, it will significantly increase the population of known exoplanets.

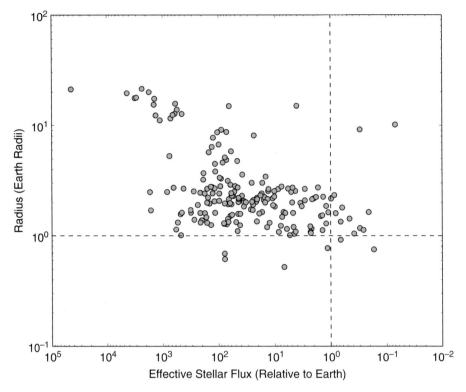

Figure 4.9 Exoplanet radius (in Earth radii) versus effective stellar flux received, where 1 = the stellar flux received by the Earth. Dashed lines indicate values for the Earth. Data courtesy of the NASA Exoplanet Archive.

CHEOPS (Characterising Exoplanet Satellite) is a characterisation mission. To be launched in 2017, it will use transit spectroscopy (amongst other techniques) to target known exoplanet systems to study transit depths with high precision. This will reveal atmospheric data about planets from Neptune to Earth radii around relatively bright stars, and larger planets around fainter targets. This high-quality information will inform future missions such as the James Webb Space Telescope (JWST).

To date, there are only a handful of confirmed terrestrial planets that receive stellar radiation at levels similar to the Earth (Solution R.1, Figure 4.9). They typically orbit low-mass stars, as terrestrial planets produce larger transit and radial velocity produce larger transit and radial velocity signals around these hosts. Some notable current candidates include Proxima Centauri b, an Earth-sized planet orbiting our nearest star with a period of around 11 days (Anglada-Escudé et al., 2016). The TRAPPIST-1 system, a very low-mass star ($0.08 M_\odot$) with seven planets, has three Earth-sized bodies that satisfy some of the conditions for habitability (Gillon et al., 2017); Kepler-186f, a 1.1 $R_\oplus$ planet orbiting a $0.5 M_\odot$ star, and LHS-1140b, a 6.6 $M_\oplus$ super-Earth orbiting a nearby low-mass star, observed both via transit and radial velocity, a somewhat rare occurence for what are now commonly described as 'potentially habitable' objects. Due to current selection bias, all of these objects orbit low-mass stars, which could present a threat to their habitability (Solution R.9), as will become clear in the following sections.

4.2 Habitable Zones

The previous section has shown us that planetary systems are extremely common in the Universe, so much so that in the Drake Equation, $f_p \sim 1$. However, we remain in a state of high ignorance regarding many of these systems. Transit searches give us a planet's silhouette, and if we are lucky, some atmospheric composition information (if our modelling is good, and the atmosphere is not too cloudy). Radial velocity searches give us a mass estimate, which can be used in combination with the transit data to give an estimate of the bulk composition – but as we'll see, the trace elements are extremely important for *habitability*, a word poorly defined and fraught with bias.

While we still languish in this era of the unknown, and intimate knowledge of extrasolar terrestrial planets remains outside our grasp for the next decade or two, what can we do to estimate the habitability of worlds outside the Solar System?

In the meantime, we can fall back on the classic *habitable zone* paradigm, which has deep roots.

4.2.1 Circumstellar Habitable Zones

The modern version of the zone was first espoused by Huang in 1959, who is thought to have given the concept its name, but hints of the concept can be found as far back as the early 1950s (Strughold, 1953), where several authors were interested in the 'ecosphere' or 'liquid water belt', which in retrospect is perhaps a more apposite name.

Essentially, the circumstellar habitable zone (CHZ) is defined as follows. If we assume that a given planet is *Earth-like*, then it orbits inside its parent star's CHZ if the radiation it receives allows liquid water to exist on its surface. The radiative flux F received by an Earthlike planet at an orbital separation r from a star (which emits energy at a rate given by its luminosity L) is:

$$F = \frac{L}{4\pi r^2} \tag{4.55}$$

Once this flux hits the upper atmosphere of the Earth, it is either absorbed or reflected back into space. The atmospheric *albedo*, α, is a measure of the probability that a photon will be reflected. The absorbed flux $F_{absorb} = (1 - \alpha)F$, with α typically being a function of wavelength. The surface also has an associated albedo, and this must be taken into account when calculating the expected surface temperature of a world.

The atmospheric *opacity*, κ, is a measure of the probability that a photon will be absorbed within a parcel of gas. The mean path length a photon will travel in a medium of density ρ is:

$$\ell = \frac{1}{\rho \kappa} \qquad (4.56)$$

The opacity is also a function of wavelength. Like the albedo, it is sensitive not only to the temperature of the atmosphere, but also its constituents. The molecules, atoms and ions that compose a planet's atmosphere will interact in different ways with photons according to the properties of their constituents (in the case of molecules, atoms and their associated bonds).

Habitable zone calculations hinge upon the inverse square law described in equation (4.55), but we can already see that to calculate habitable zone limits, we must start with a detailed description of the atmosphere. This puts strong constraints on what habitable zone limits can actually represent for a large and highly heterogeneous sample of exoplanets.

Typically, habitable zones are delineated by an inner and outer radial boundary, and as such the zone is an annulus. Equation (4.55) ensures that interior to the annulus, the radiation received is too high, and the Earth will begin to lose its water via *runaway greenhouse*. In this scenario, as the planet begins to warm, the increasing water vapour content in the atmosphere increases the atmospheric opacity, instigating a positive feedback loop of evaporation and opacity increase until the oceans are evaporated. This steam enters the upper atmosphere, and is photolysed by UV radiation into hydrogen and oxygen. The hydrogen then escapes, and life on the planet's surface becomes utterly dessicated on a timescale of a few hundred million years (Barnes et al., 2013).

Exterior to the CHZ, the radiation is too low to sustain liquid water. At this outer limit, even the formation of CO_2 clouds is not sufficient to keep the planet warm. In fact, beyond a certain cloud density, the albedo of the planet increases, limiting the cloud's ability to trap heat, hence the moniker 'maximum greenhouse' applied to the outer CHZ limit. Once a planet exceeds maximum greenhouse, the entire planet freezes. This freezing process is also party to a positive feedback loop, referred to as 'ice-albedo' feedback. The albedo of ice sheets is higher than the albedo of oceans. Hence, as the planet begins to freeze, its albedo increases, reducing the net flux absorbed at the surface, accelerating the freezing process until the entire planet's surface water is frozen.

4.2 Habitable Zones

We therefore have two extreme criteria which define our CHZ boundaries – at what flux does a runaway greenhouse occur, and at what flux does a runaway freezing event occur? It is in the answers to these questions that lie the true devilry of habitable zone calculation. To determine what flux limits trigger these two events requires us to make quite sweeping assumptions about the planet's atmospheric composition (generally $N_2/O_2/CO_2$), the planet's surface gravity and atmospheric scale height, not to mention the atmospheric temperature-pressure profile and ocean coverage. With very few exceptions, habitable zone calculations assume an Earthlike body. With this assumption in place, radiative transfer calculations can take a star with a given spectral energy distribution, compute this spectrum's effect on the planet's atmosphere, and conclude whether such a planet may possess liquid water at its surface.

How the CHZ is calculated for the Solar System has evolved significantly over the years. A watershed moment comes with the work of Kasting et al. (Kasting, 1988; Kasting et al., 1993). They attempt to calculate the habitable zone inner and outer boundary using a one-dimensional altitudinal model of the Earth's atmosphere. At every atmospheric layer, radiation from the Sun is reflected, absorbed and re-emitted by the Earth's atmosphere depending on the stellar spectrum, the atmospheric albedo's dependence on wavelength and the pressure-temperature profile of the planet (again assumed to be Earthlike). By fixing a desired temperature for the planet's surface, this model can then compute the required stellar flux to produce this temperature. As the star's luminosity L is known, and the model delivers F, the distance to the star can be inferred from equation 4.55, and this constitutes one of the habitable zone boundaries. As better modelling of the atomic physics surrounding absorption and re-emission of photons by molecules progresses, these habitable zone boundary calculations improve and refine their answers (Kopparapu et al., 2013, 2014).

However, we also know that the stellar luminosity is not constant. Sunlike stars increase in luminosity as they age – for example, a simple approximation to this behaviour, fitted from stellar structure calculations, is (Schröder and Connon Smith, 2008):

$$L(t) = \left(0.7 + 1.44(\frac{t}{Gyr})\right) \frac{L}{L_\odot} \qquad (4.57)$$

where Gyr = 1 billion years, and $L_\odot$ is the present-day luminosity of the Sun. The continually increasing luminosity results in a habitable zone continually shifting outward. The *continuous circumstellar habitable zone* (CCHZ) is the region that remains in the habitable zone throughout some time period of importance (e.g., the main sequence lifetime of the star, or the estimated timescale for a biosphere to develop; see Solution R.10).

The observant will note that, in all the above, *Earth-like* is a descriptor begging to be rigorously defined. It is worth reminding ourselves that, to our current knowledge, the Earth is singular and remarkable. The temperature of its surface (and its available liquid water and other necessary ingredients for life) is dependent on a vastly complex agglomeration of feedback systems and interfaces (see Solution R.4). Indeed, the biosphere plays an important role in the regulation of temperature through the carbonate-silicate cycle, which we will touch on later. An inhabited world can tune itself to become more habitable.

Also, there are other ways of providing habitable worlds with sufficient energy to produce liquid water either on the surface or below an ice layer. *Tidal heating* can be one such way. If a world orbits a bigger body at relatively close proximity (such as a satellite orbiting a planet), then the difference in gravitational force it experiences across its surface deforms the body. This deformation can extract orbital and rotational energy, and is eventually dissipated as heat in the interior, where the typical heating per unit area is given (roughly) by:

$$U_{tidal} = \frac{21}{38} \frac{\rho_m^2 R_m^5 e_m^2}{\Gamma Q} \left(\frac{GM_p}{a_m^3}\right)^{5/2} \qquad (4.58)$$

where ρ_m and R_m are the satellite density and radius, a_m is the satellite's orbital semi-major axis, e_m is the satellite's eccentricity, M_p is the host planet mass, and Γ and Q are the elastic rigidity and tidal dissipation factors respectively.

We see this heating mechanism in action in the moons of the Solar System. Jupiter's moon Europa has an icy surface, but gravity anomaly studies by the Galileo probe and other missions have shown that a subsurface ocean lies beneath the ice. By contrast, Io experiences such drastic tidal heating that catastrophic volcanism coats its entire surface, making it an extremely challenging environment even for extremophilic life (see later sections).

At the time of writing, extrasolar moons (exomoons) are yet to be observed – the current favourites for detection methods are via transit timing/-duration variations due to the planet-moon barycentre (Kipping, 2009a,b; Kipping et al., 2012, 2013), and microlensing (Bennett et al., 2014). Various researchers are now attempting to determine the exomoon habitable zone (EHZ), which considers stellar radiation, planetary radiation (which can be significant in the infrared if the host planet is a gas giant), tidal heating and frequent eclipses of the star by the planet as energy sources and sinks (Heller and Barnes, 2013; Heller et al., 2014; Forgan and Kipping, 2013; Forgan and Yotov, 2014).

While there are likely to be many worlds superficially similar to the Earth in terms of mass and radius and parent star, there are likely to be very few that we would consider to be anything like an 'Earth twin', as many press releases might want you to think. The precise and continually evolving arrangement of

4.2 Habitable Zones

biomass, mineral and volatiles ensures that our world, like all habitable worlds, will remain unique and precious. So how exactly should we classify the 'Earthiness' of other planets? In time-honoured fashion, I will sidestep this seemingly intractable question and leave it as an exercise for the reader.

4.2.2 Galactic Habitable Zones

A considerably less developed concept of habitability pertains to galactic scales. Rather than considering whether individual planetary systems contain habitable worlds, we can instead consider whether regions of space are amenable to forming planetary systems that will yield habitable worlds. These Galactic Habitable Zones are less restrictive, and focus on two aspects: can planetary systems with habitable planets be formed, and are they safe against astrophysical hazard (Solution R.11)? Both of these aspects rely heavily on star formation (and subsequent planet formation), which we will briefly describe below.

Star Formation

The gas inventory of galaxies (sometimes called the interstellar medium or ISM) can be separated into two phases: a hot, low-density phase, and a cool, dense phase. Stars are formed from the cool, dense component. Gas can be converted from one phase to the other via compression. Once the gas achieves a critical density, it can begin to cool effectively, allowing it to grow denser still.

This compression can be achieved either by galaxy spiral arms (e.g., Bonnell et al., 2013), shockwaves generated by supernovae, or by the powerful ionising radiation of nearby stars, which generates strong radiation pressure (e.g., Bisbas et al., 2011). By one or more of the above, the gas forms a *molecular cloud* of turbulent gas, which typically adopts a filamentary structure. Regions of the cloud (typically at the junctions where filaments meet) begin to shrink to scales shorter than the local *Jeans length* λ_{Jeans}, and can then begin collapsing under their own self-gravity:

$$\lambda_{\text{Jeans}} = \sqrt{\frac{\pi c_s^2}{G\rho}} \qquad (4.59)$$

where ρ is the local gas density, and c_s is the sound speed of the gas. These prestellar cores, still embedded in parent filaments (see Hacar et al., 2013), collapse to form *protostars* with *circumstellar discs*. These discs exist thanks to the initial angular momentum of the gas providing centrifugal support against collapse. The typical disc size depends on the angular momentum of the initial core (Lin and Pringle, 1990). For a spherical core of mass, M_c, initial radius $R_c = \lambda_{\text{Jeans}}$, rotating rigidly at angular velocity Ω_c, assuming angular momentum and mass conservation gives:

$$M_c R_c^2 \Omega_c = M_c r_{disc}^2 \Omega_{disc} \tag{4.60}$$

where r_{disc} is the disc radius, and Ω_{disc} is the disc angular velocity. If we assume the disc is Keplerian, then

$$\Omega_{disc} = \sqrt{\frac{GM_c}{r_{disc}^3}}. \tag{4.61}$$

Substituting this expression into equation (4.60) allows us to rearrange for r_{disc}:

$$r_{disc} = \frac{\Omega_c^2 R_c^4}{GM_c}. \tag{4.62}$$

It is common practice to define the core's initial conditions in terms of two dimensionless variables:

$$\alpha_c = \frac{E_{therm}}{E_{grav}}, \quad \beta_c = \frac{E_{rot}}{E_{grav}}, \tag{4.63}$$

where E_{therm}, E_{rot} and E_{grav} are the core's initial thermal, rotational and gravitational potential energy respectively. For our rigidly rotating sphere, we can define the rotational energy by calculating the moment of inertia I:

$$E_{rot} = \frac{1}{2} I \Omega_c^2 = \frac{1}{5} M_c R_c^2 \Omega_c^2 = \beta_c E_{grav} \tag{4.64}$$

The gravitational potential energy of a uniform-density sphere is

$$E_{grav} = \frac{3}{5} \frac{GM_c^2}{R_c}. \tag{4.65}$$

We can now substitute β_c for Ω_c to give

$$r_{disc} = \frac{3 G M_c \beta_c R_c^4}{G M_c R_c^3} = 3\beta_c R_c. \tag{4.66}$$

Typical observed values for β_c vary from 2×10^{-4} to 1.4, with a median value of 0.03 (Goodman et al., 1993), so a typical disc will have a maximum radius of around 10 percent of the original core radius. As the core radius is typically of order 0.1 pc (around 25,000 AU), then the disc is typically less than 2,500 AU. In practice, not all of the mass collapses, and the disc radius might have an upper limit of 1,000 AU or less, depending on the initial core mass.

If the cloud has a relatively strong magnetic field, which is well coupled to the gas, this can remove angular momentum through a process known as *magnetic braking*. In the ideal MHD limit where the fluid conductivity is infinite, magnetic braking is extremely efficient at extracting angular momentum from collapsing prestellar cores, and can completely suppress the formation of discs.

4.2 Habitable Zones

In practice, the conductivity is not infinite. The coupling of field to fluid is less than perfect, and we must consider *non-ideal MHD* (Seifried et al., 2013; Tomida et al., 2015; Wurster et al., 2016) (see box for details). It is clear from observations that discs do form with typical outer radii between 100 and 1,000 AU, with the disc radius for stars less than a few solar masses tending towards 100 AU or less. Much of the material in these discs will lose angular momentum through disc turbulence or spiral arm interactions, and be accreted onto the star. What remains becomes the feedstock for planet formation.

Non-ideal MHD adds three extra terms to the induction equation, relating to the relative drift of positive and negative charges in the fluid, which can enhance or dissipate the magnetic field (see Table 4.1).

The Equations of Magnetohydrodynamics

In so-called *ideal magnetohydrodynamics* (MHD), the magnetic field is 'frozen in' to the fluid and the coupling is essentially perfect. The Maxwell equations are:

$$\nabla \cdot \mathbf{E} = \frac{\rho_c}{\epsilon_0}, \tag{4.67}$$

$$\nabla \cdot \mathbf{B} = 0 \tag{4.68}$$

$$\nabla \times \mathbf{E} = -\frac{\partial \mathbf{B}}{\partial t} \tag{4.69}$$

$$\nabla \times \mathbf{B} = \mu_0 \left(\mathbf{J} + \epsilon_0 \frac{\partial \mathbf{E}}{\partial t} \right) \tag{4.70}$$

And the current J is given by

$$\frac{1}{\sigma_c} \mathbf{J} = (\mathbf{E} + \mathbf{v} \times \mathbf{B}) \tag{4.71}$$

In the ideal MHD limit, the conductivity $\sigma_c \to \infty$, and hence the electric field

$$\mathbf{E} = -(\mathbf{v} \times \mathbf{B}) = 0 \tag{4.72}$$

And therefore the *induction equation* of ideal MHD is

$$\frac{\partial \mathbf{B}}{\partial t} = \nabla \times (\mathbf{v} \times \mathbf{B}) \tag{4.73}$$

And

$$\nabla \times \mathbf{B} = \mu_0 \mathbf{J}. \tag{4.74}$$

In fact, a complete set of equations for ideal MHD, assuming an adiabatic fluid with ratio of specific heats γ is

$$\frac{\partial \rho}{\partial t} + \nabla \cdot (\rho \mathbf{v}) = 0 \tag{4.75}$$

$$\rho \frac{\partial \mathbf{v}}{\partial t} + \nabla P + \frac{1}{\mu_0} \mathbf{B} \times (\nabla \times \mathbf{B}) = 0 \tag{4.76}$$

$$\frac{d}{dt}\left(\frac{P}{\rho^\gamma}\right) = 0 \quad (4.77)$$

$$\frac{\partial \mathbf{B}}{\partial t} - \nabla \times (\mathbf{v} \times \mathbf{B}) = 0 \quad (4.78)$$

$$\nabla \cdot \mathbf{B} = 0 \quad (4.79)$$

where ρ is the fluid density and P is the fluid pressure.

Table 4.1 *Additions to the induction equation due to non-ideal MHD effects. Ohmic resistivity depends on the resistivity, which is the inverse of the conductivity $\eta = 1/\sigma_c$. The Hall effect grows strongest in regions of low electron density n_e, and ambipolar diffusion depends on the local density of ions ρ_i, as well as the drag coefficient Γ describing the coupling of ions and neutrals.*

Effect	Description	Term
Ohmic Resistivity	collisions between electrons	$\nabla \times \frac{4\pi\eta}{c^2}\mathbf{J}$
Hall Effect	non-collisional electron drift	$\frac{1}{en_e}\mathbf{J} \times \mathbf{B}$
Ambipolar Diffusion	ion-neutral collisions	$\frac{1}{c\Gamma\rho\rho_i}(\mathbf{J} \times \mathbf{B}) \times \mathbf{B}$

Planet Formation

We will only briefly summarise planet formation theory here to provide suitable context: the reader is encouraged to consult Armitage (2010) and references within for more details.

Planet formation theories fall into one of two categories, reflecting the direction of growth that protoplanets undertake. *Core Accretion* theory is a bottom-up, hierarchical growth theory. The protostellar disc is initially composed of gaseous material and interstellar dust grains, with typical sizes s of a micron (10^{-6} metres) or less, and a size distribution

$$n(s) \propto s^{-3/2}. \quad (4.80)$$

Core accretion proposes that these grains are the progenitors of planets. The grains initially grow through sticky collisions, where the initial colliders are held together via van der Waals forces. As a result, grains in the early growth phase are extremely fluffy and porous. As collisions continue, the porosity of the grains is destroyed, resulting in harder grains capable of holding themselves together through tensile strength of the material. At this stage, the grains are approximately millimetres in size.

As the grains continue to grow, the gas in the disc begins to exert an increasingly strong drag force:

$$F_D = -\frac{1}{2} C_D \pi s^2 \rho \Delta v^2 \qquad (4.81)$$

where Δv is the particle velocity relative to the gas. The drag coefficient C_D depends on the relationship between the grain size s and the mean free path λ_{mfp}. If the grains are small compared to λ_{mfp}, the grains are in the *Epstein regime*, and the drag can be considered by considering individual collisions between grains and gas molecules. The drag coefficient is therefore

$$C_D = \frac{8}{3} \frac{\Delta v}{\bar{v}} \qquad (4.82)$$

where $\bar{v}$ is the mean thermal velocity of the gas, which is of the order of the gas sound speed, $\bar{v} = (8/\pi)^{1/2} c_s$. As the grain size begins to increase, the drag switches to the *Stokes regime*, where the gas must be modelled as a fluid, and the drag coefficient depends on the Reynolds number Re, a measure of turbulence in the gas:

$$Re = \frac{|v|L}{\nu}, \qquad (4.83)$$

where $L = 2s$ is a suitable lengthscale and ν is the molecular viscosity. The drag coefficient decreases with increasing turbulence:

$$C_D = 24 Re^{-1} \quad Re < 1 \qquad (4.84)$$
$$C_D = 24 Re^{1/2} \quad 1 < Re < 800 \qquad (4.85)$$
$$C_D = 0.44 \quad Re > 800. \qquad (4.86)$$

Grain motion is suppressed by gas drag on the friction timescale

$$t_{\text{fric}} = \frac{m \Delta v}{F_D} \qquad (4.87)$$

This causes the grains to both settle towards the midplane and drift radially inward. This inward drift peaks when the grains reach sizes of a few centimetres to a metre (at distances of a few AU from the star).

The probability of collisions resulting in successful growth also reduces dramatically when grains begin to reach centimetre/metre sizes. If we consider a grain of mass m hitting a target of mass M, we can define the specific energy of the impact

$$Q = \frac{mv^2}{2M}, \qquad (4.88)$$

i.e., the kinetic energy per unit target mass. If this Q becomes larger than a critical value, the collision will no longer result in growth. There are two key thresholds: the critical specific energy for shattering, Q_S^*, and the critical specific energy for dispersal, $Q_D^* > Q_S^*$. A collision that results in shattering, but not dispersal can still result in growth, if a *rubble pile* is formed after the collision which can then coalesce. Below Q_S^*, moderate-velocity collisions may still fail to grow grains due to bouncing as an outcome (Windmark et al., 2012).

Both Q_S^* and Q_D^* vary greatly with impact angle and composition of the material, but a simple two-powerlaw parametrisation defines the key regimes (Leinhardt and Richardson, 2002; Korycansky and Asphaug, 2006):

$$Q_D^* = q_s \left(\frac{s}{1\,\text{cm}}\right)_s^\beta + q_g \rho_d \left(\frac{s}{1\,\text{cm}}\right)_g^\beta \tag{4.89}$$

At low s, the resistance to dispersal is borne by material strength, and the dispersal energy decreases with s ($\beta_s < 0$). As the material's gravitational field becomes significant, the dispersal energy increases with s ($\beta_g > 0$). Q_D^* typically finds its minimum for $s = 0.1 - 1$ km. This, of course, neglects the velocity of impact, which typically affects q_s and q_g without altering the powerlaw indices (see also Stewart and Leinhardt, 2009).

The increasing relative velocity of colliding 'pebbles' results in increasing numbers of collisions that end in fragmentation rather than growth, while the gas drag efficiently extracts angular momentum and sends the pebbles inwards towards the star. Much of modern core accretion theory has been focused on surmounting these significant barriers to planet formation, inward radial drift and growth past the fragmentation regime. Solutions to this commonly invoke local pressure maxima in the gas, which can reverse the drift direction, such as spiral arms (Rice et al., 2004), or even vortices, which can trap grains and reduce their relative velocities in the correct circumstances (Barge and Sommeria, 1995; Lyra et al., 2009). If the radial drift produces a dust fluid with density comparable to the local gas density, *streaming instabilities* can be set up (Youdin and Goodman, 2005; Johansen et al., 2007) where essentially the gas drag is modified by the presence of other dust grains, allowing other grains to 'slipstream' and promote sticking collisions.

Similarly, recent advances in *pebble accretion* theory allow gas drag to assist bodies to accrete via gravitational attraction. As smaller bodies pass close to a larger accreting body, a combined action of aerodynamic drag and gravity can allow the smaller bodies to reduce their relative velocity sufficiently to accrete (see, e.g., Lambrechts and Johansen, 2012 and Ida et al., 2016 for details).

If the pebbles can successfully grow beyond these barriers, the self-gravity of the agglomerate can ensure that fragmenting collisions result in rubble-piles that re-collapse into an object of similar size. This continual fragmentation and recollapse can allow the object to become a *planetesimal*, a body of approximately kilometre size.

Once a significant population of planetesimals has been formed, the true planet formation epoch can commence. In this regime, collisions generally result in growth. While the collision may result in the shattering of one or either bodies, the rubble pile that forms typically has a strong enough gravitational field that it can reassemble into a body possessing the majority of the mass of the two collider bodies. The gas in the disc offers negligible drag on the planetesimals. With billions of bodies participating in collisions and growth, a small number of bodies will experience good fortune, and more successful growth events. This is a positive feedback process, as the gravitational field of larger bodies is able to focus more bodies into its path to accrete them. The sphere of influence of a body M orbiting a star of mass M_* can be described by the Hill radius R_H:

$$R_H = a \left(\frac{M}{3M_*} \right)^{1/3}, \qquad (4.90)$$

where a is the semi-major axis of the orbit of body M. As bodies increase their mass, their gravitational field begins to influence a larger and larger range of the disc. Typically, a precious few bodies grow at the expense of the others. In this *runaway growth* phase, tens to hundreds of protoplanets can grow to masses near to that of the Moon or Mars.

This runaway growth proceeds until either the local supply of bodies inside the Hill radius is exhausted, or the Hill radii of neighbouring bodies begin to overlap, and protoplanets must compete for their supply. These few protoplanets then participate in *oligarchic growth*, where the protoplanets continue to grow, but their growth is checked by their neighbours, as each protoplanet attempts to feed on the remaining planetesimals (Kokubo and Ida, 1998; Thommes et al., 2003). Smaller protoplanets are commonly ejected or destroyed during this oligarchic phase.

As these protoplanets continue to grow, their spheres of gravitational influence begin to overlap, perturbing each others' orbits. As their orbits begin to cross, the *giant impact* phase begins. The protoplanets, now approaching Earth's mass, begin to collide and grow even larger. The most successful cores (which reach around ten Earth masses) begin to accrete significant amounts of gas from the surrounding disc, and hence become gas giant planets.

All this activity occurs against the continuing loss of gas from the disc. The stellar irradiation of the disc upper layers results in photoevaporation of the gas. This, and other forces, destroys the disc's gaseous component on timescales of 1–10 Myr. This tells us that planetesimals must form quickly, if they are to have sufficient time to assemble gas giant planets.

The second category of planet formation theory is usually referred to as *Disc Instability* or gravitational instability theory. During the early stages of the circumstellar disc's life, the mass of the disc is comparable to the mass of the protostar, and as a result disc self-gravity becomes important. When the Toomre Parameter

$$Q = \frac{c_s \Omega}{\pi G \Sigma} \sim 1, \tag{4.91}$$

the disc can become gravitationally unstable, and as a result produce large-scale spiral density waves (in a process very similar to that which generates spiral arms in galaxies). These density waves produce weak shock waves in the disc (known as *transonic* or *low Mach number* shocks), which heat the disc, increasing Q, and hence reduce the amplitude of the instability. The competition between heating and cooling in the disc can result in a marginally stable state where the instability is self-regulated (Paczynski, 1978).

However, this self-regulation can be broken by a variety of factors, and the uncontrolled instability results in *fragmentation* of the disc (Gammie, 2001; Rice et al., 2011; Mejia et al., 2005; Boley et al., 2007; Rice et al., 2005). Large chunks of disc gas and dust, with masses of a few Jupiter masses (Rafikov, 2005; Boley et al., 2010; Forgan and Rice, 2011; Rogers and Wadsley, 2012), collapse into bound objects at quite large distances from the star (typically beyond 30 AU), within a few thousand years.

This produces gas giants (and low-mass stars) quite easily, but these fragments can migrate rapidly inward through gas drag, as the disc is still very young and massive. This can result in their complete disruption, or partial disruption of the outer envelope.

Disc fragments are essentially a sampling of disc dust and gas. Dust will drift towards the pressure maximum at the centre of the fragment, and can continue to grow via collisions on its descent. This can result in the formation of a solid core at the centre of the fragment. If a core can form before tidal disruption, then the outer envelope may be stripped to leave relatively massive terrestrial planets on orbits comparable to planets formed via core accretion (Nayakshin, 2010b, 2011, 2010a). However, modelling attempts to produce large populations of planets by disc fragmentation show that low-mass rocky planets are typically rare (Forgan and Rice, 2013; Forgan et al., 2018a). It seems clear that while both planet formation mechanisms are possible, and probably occur to varying degrees in the Universe, core accretion is the process by which *habitable* planets are formed.

Estimating Galactic Habitability

The chemical composition of these discs is therefore crucial to the final character of the planets. The extremely hot and dense conditions that existed in the minutes after

4.2 Habitable Zones

the Big Bang allow *nucleosynthesis* to take place: the creation of heavier chemical elements by the fusion of lighter elements. However, Big Bang nucleosynthesis makes for a poor periodic table of elements – only hydrogen, helium and a very small amount of lithium and beryllium are produced. As we will see, life on Earth requires more than this.

As we will discuss in detail in section 10.2, stars can fuse lighter elements into heavier elements through nuclear fusion. Massive stars can create all the elements as far up the periodic table as iron, and their death throes as supernovae produce even heavier elements. This means that for the Galaxy to form habitable terrestrial planets, it must have **at least one** prior epoch of star formation. This early generation of stars can produce the necessary chemical elements for life, and then bequeath these elements to nearby regions of space through supernovae explosions. This expelled material can then re-collapse to form a new generation of chemically enriched stars and planets. From measures of metallicity in stellar populations, it seems clear that the Sun belongs to at least the third generation of star formation since the Big Bang.

However, too many supernovae can also be hazardous for life. Planets that exist in a region of space where supernovae explode nearby, and frequently, will suffer catastrophic loss of ozone from their upper atmospheres, and presumably mass extinctions or even sterilisation (see section 11.3). It has been estimated that a supernova explosion within a distance of ten parsecs of the Earth would be sufficient to completely destroy its biosphere. Thankfully, ten parsecs is relatively close by in astronomical terms – remember, the Milky Way is some 30 thousand parsecs across, and there are no supernovae candidate stars detected within ten parsecs.

From some simple calculations of how the Galaxy's supernova rate is expected to change with time, and a description of the elemental abundances in the Galactic disc, a simple annular model of heightened Galactic Habitability arose (Gonzalez et al., 2001; Lineweaver et al., 2004). This model has been increasingly questioned as our understanding of galactic structure and the expected effects of supernova radiation on the ozone layer has improved (Gowanlock et al., 2011). Estimates of the extent of this Galactic Habitable Zone vary widely, with some authors claiming that no such zone even exists (Prantzos, 2007).

However, supernovae are not the only danger for habitable planets. *Gamma ray bursts* can sterilise biospheres up to a thousand parsecs from the detonation site (see section 11.2). They appear to be found preferentially in regions of low metallicity (Piran and Jimenez, 2014), which leads some SETI scientists to suggest they play the role of a *global regulation mechanism* for biospheres and intelligence, which has prevented intelligent life from developing in large quantities before the present (we will return to this concept in section 18.2).

The local stellar environment can reduce planetary habitability without the need for violent explosions. Close encounters between planetary systems can disrupt the orbits of distant asteroids and protoplanetary bodies, bringing them into collisions with habitable worlds and destroying their biospheres. While studies have shown that this effect is not particularly destructive in environments like the Solar Neighbourhood (Jiménez-Torres et al., 2013), it can become devastating in dense stellar environments such as ancient globular star clusters.

It is also worth remembering that galaxies like the Milky Way are formed via hierarchical merger events that result in bursts of star formation and can completely alter the morphology of any Galactic Habitable Zone. It is unlikely that simple annular models are unlikely to capture this, and the final picture of galactic habitability will always be more complex (Forgan et al., 2017). What is clear is that galactic habitability models remain too simplistic to be fully effective, and that a great deal more work is required in this area.

4.3 The Earth as a Habitable Planet

Even if a planetary body forms in conditions that are initially favourable for habitability, there are no guarantees that the planet will remain habitable. As we will see throughout this book, there are many external factors that influence the planet's energy budget, and hence its habitability (Solutions R.2 and R.3).

The Earth's land-ocean-atmosphere system is subject to both negative and positive feedback systems that can regulate or destabilise the surface temperature. The water vapour in the Earth's atmosphere mediates a positive feedback loop on the greenhouse effect. If the atmosphere cools, the saturation pressure for water vapour begins to decrease. If the relative humidity remains constant, then atmospheric concentrations of water vapour also decrease, reducing the greenhouse effect and allowing the atmosphere to cool further. Conversely, if the atmosphere heats, atmospheric concentrations of water vapour increase, boosting the greenhouse effect. We have already seen this effect when considering the circumstellar habitable zone.

We also saw the ice-albedo positive feedback loop, where a slight increase in ice cover boosts the surface albedo, allowing the surface to cool and produce greater ice cover. Positive feedback mechanisms enhance climate instabilities, amplifying changes in surface temperature.

Thankfully, negative feedback loop systems also operate to oppose such unstable oscillations. The simplest is the relationship between the outgoing flux and the surface temperature:

$$F_{out} \propto T_{surf}^4 \qquad (4.92)$$

As T_{surf} increases, F_{out} increases, acting to reduce the temperature and establish an equilibrium. A crucial negative feedback system, which incorporates the biosphere

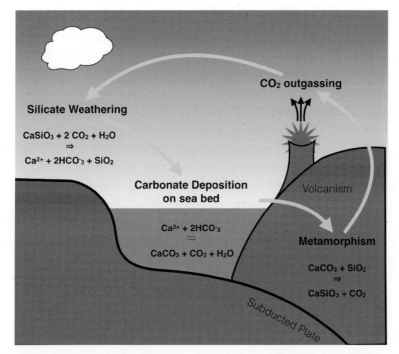

Figure 4.10 The carbonate-silicate cycle. The partial pressure of atmospheric CO_2 is controlled by the production of carbonic acid (H_2CO_3), which weathers silicates and carbonate ions into the ocean, which are then incorporated into marine organisms. Upon death, these organisms return to the sea bed, and the carbonates/silicates are subducted. Under metamorphism, this releases CO_2, which is returned to the atmosphere by volcanic outgassing to complete the cycle.

in its operation, is the carbonate-silicate cycle, which regulates atmospheric CO_2 (see Figure 4.10).

At the beginning of the cycle, atmospheric CO_2 dissolves into rainwater to form carbonic acid (H_2CO_3). This slightly acidic rain lands on silicate rock, weathering it, transporting silicate and carbonate ions to rivers and eventually the oceans. Marine organisms then incorporate the carbonates into their shells. When the organisms expire, their shells return to the seabed (along with the silicates).

This material is then transported into the planet's interior via subduction of tectonic plates. As oceanic plates slide beneath continental plates, the carbonate and silicate minerals can recombine, releasing CO_2 into the mantle. This CO_2 is then expelled back into the atmosphere via volcanism, and the carbonate-silicate cycle is complete. The complete atmospheric budget of CO_2 is cycled through over approximately 200 Myr.

The carbonate-silicate cycle is a negative feedback loop due to the temperature dependence of weathering. If the temperature increases, the weathering rate increases, removing CO_2 from the atmosphere at a higher rate than replacement via volcanism, reducing the greenhouse effect and dropping the temperature. Conversely, a drop in temperature drops the weathering rate, allowing volcanism to boost CO_2 levels and restore the temperature.

The carbonate-silicate cycle is one of the most crucial negative feedback systems in the Earth's climate. As the Sun's luminosity has slowly increased over time, the cycle has adjusted to this slow increase in received flux by modulating CO_2 levels accordingly. It is also a crucial factor in Earth's escape from the fully ice-covered scenario. If no rock is accessible to rainwater, then weathering ceases. Volcanism continues to pump CO_2 into the atmosphere, boosting the greenhouse effect until the ice begins to melt.

Note the role of organisms in sequestering carbonates in this cycle. There is also a rather separate organic cycling of carbon through photosynthesis, where atmospheric carbon becomes sequestered in photosynthetic organisms. The fraction of the carbon budget in these organisms is low compared to the total, and is also largely controlled by oxygen levels, which for much of Earth's history have been too low to regulate climate effectively.

This negative feedback system can be forced by releases of carbon from regions where it is currently sequestered. For example, as deforestation proceeds, the ability for photosynthesis to fix carbon out of the atmosphere is weakened locally. Droughts and desertification can also release long-sequestered carbon resources from peatlands, which rely on anaerobic conditions to trap carbon, and prevent its release on contact with phenolic compounds (especially phenol oxidase). It is thought that peatlands possess a carbon store around 50% of the current atmospheric store (Freeman et al., 2001), and hence a widescale release of this carbon would result in a significant increase in atmospheric CO_2.

A similar, and even more effective burst of greenhouse gas release can come from the melting of arctic permafrosts, which contain CH_4 clathrates (methane trapped within ice crystals). As the ice melts, CH_4 can be released in large quantities, significantly reducing the outgoing longwave radiation of the Earth and resulting in rapid warming.

All of what we have discussed here depends crucially on the relationship between the Earth's surface and its interior (governed by plate tectonics), and the relationship between the Earth's atmosphere and outer space (governed for the most part by the Earth's magnetic field). We will delve into the prospects for plate tectonics and magnetic fields for other terrestrial planets in sections 9.1 and 9.2.

We therefore see that even a living planet can undergo significant climate shocks, with or without a technological civilisation present, simply from the interaction of both positive and negative feedback mechanisms.

4.4 Good and Bad Planetary Neighbours

Another aspect of what makes the Earth 'Earthlike' is its astrophysical context. This extends beyond the parent star towards its planetary neighbours. As discussed already, planetary orbits are free to be extremely eccentric. When a planetary system has more than one planet, the evolution of their orbits cannot be calculated analytically, and *chaotic* behaviour can appear in the system. A slight perturbation of initial conditions can produce quite different results. The evolution of a world's orbit under the gravitational tugging of its neighbours can have profound consequences for its habitability (Solution R.5).

Compared to the extrasolar planet population, the Earth's orbit is relatively close to circular, and its rotation axis is reasonably stable – the angle it makes with the orbital plane (the *obliquity*) remains close to 23 degrees. Most of the Solar System planets and dwarf planets orbit with relatively small eccentricities, and therefore their orbits do not cross (with the exception of Pluto and Neptune).

However, the orbital elements of all the Solar System bodies change with time, including the Earth. As the Solar System is stable, the changes in orbital elements are reasonably small, and they are also periodic. These changes are now obvious to us, as we can run numerical simulations of the Solar System and witness millions of years of orbital evolution on a laptop, but they were not quite so clear to those who proposed their existence, and that these cyclical variations were responsible for cycles of climate evolution upon the Earth.

4.4.1 Milankovitch Cycles

The Earth's orbit and rotation has four fundamental cycles (Goudie, 1992): the first is oscillation in eccentricity (with an oscillatory cycle lasting approximately 96,000 years), precession of periastron (with a periodicity of 21,000 years), oscillation and precession of obliquity over 40,000- and 23,000-year periods respectively.

The effect of eccentricity on a planet's climate is relatively clear. Planets on elliptical orbits move closer to, and further away, from the host star over the course of an orbit. Thanks to the inverse square law, the level of insolation also changes. The periastron and apastron of an orbit with semi-major axis a and eccentricity e are:

$$r_{peri} = a(1 - e) \tag{4.93}$$

$$r_{ap} = a(1 + e) \tag{4.94}$$

We can determine the time-averaged flux received from a star with luminosity L to be

$$F = \frac{L}{4\pi a^2 \sqrt{1 - e^2}} \tag{4.95}$$

In other words, eccentric orbits receive a higher level of flux on average (although the instantaneous flux varies more over an orbit than it does for circular orbits).

The precession of perihelion does not affect the instantaneous flux received by the entire planet, but it does affect the timing of the Earth's closest approach to the Sun. More specifically, it determines the season in which closest and most distant approaches occur. We can specify the orbital longitude at which the Northern Hemisphere experiences the winter solstice $\lambda_{solstice}$, and the longitude of periapsis λ_{per}.[8] If $\lambda_{solstice} \approx \lambda_{per}$, then the Northern Hemisphere approaches winter when the Earth is closest to the Sun, and is hence milder. If $\lambda_{solstice}$ is out of phase with λ_{per}, then the Northern Hemisphere can approach winter with the Earth significantly further away from the Sun (if the eccentricity is high enough).

Finally, spin obliquity δ_0 dictates the strength of seasonal variations on the Earth. The insolation received at a given latitude Λ at an orbital distance r is

$$S = \frac{L \cos Z}{r^2}, \qquad (4.96)$$

where the zenith angle Z (which measures the Sun's height in the sky) is:

$$\cos Z = \sin \Lambda \sin \delta + \cos \Lambda \cos \delta \cos h. \qquad (4.97)$$

h and δ are the hour angle and declination of the Sun respectively, defining its position on the celestial sphere.[9] The solar declination as a function of orbital longitude is given by

$$\sin \delta = \delta_0 \cos (\lambda - \lambda_{solstice}) \qquad (4.98)$$

If $\delta_0 = 0$, then the solar declination is fixed, and the zenith angle does not change throughout the year. As δ_0 increases, the change in Z is more pronounced over the year, resulting in the Sun spending more and less time above the horizon, producing the seasons with which we are familiar.

Simply tilting or 'rolling' the Earth will increase and decrease δ_0, but the rotation vector is free to move in two dimensions, resulting in obliquity precession or 'wobble' (right panel of Figure 4.11).

While the conceptual link between orbit and glaciation owes its genesis to earlier authors, most notably Adhemar and Croll (Crucifix, 2011), it was best articulated by Milankovitch (1941), and as such, after their verification in deep ocean core data (Hays et al., 1976), these oscillations are commonly referred to as Milankovitch cycles.

---

[8] Note this is not the argument of periapsis, although the two are related through the longitude of the ascending node: $\lambda_{per} = \omega + \Omega$.

[9] The Sun's hour angle is measured by $h = LST - \alpha_\odot$, where LST is the local sidereal time and α is the right ascension.

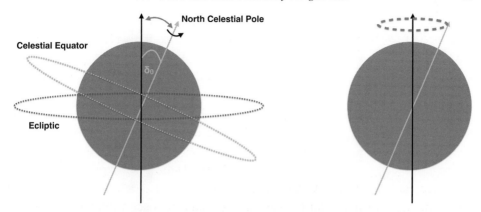

Figure 4.11 The obliquity δ_0 of the Earth. There are two Milankovitch cycles associated with the obliquity. In the first, the magnitude of the obliquity can change (left). In the second, the obliquity precesses (right).

Two things are hopefully clear at this point:

1. Milankovitch cycles interact with each other to boost climatic changes. If obliquity shifts coincide with perihelion shifts, then the resulting dichotomy between 'mild' and 'extreme' hemispheres can be significant, a phenomenon confirmed by studies of glacial termination timings.
2. The Earth system also interacts with Milankovitch cycles to boost climatic changes.

For example, if ice caps are induced to melt through increased insolation, large quantities of fresh water are released into the North Atlantic, resulting in a reduced salinity, affecting the ocean's density and temperature, and hence ocean circulation patterns and the process of heat-exchange with the atmosphere. Changing atmospheric composition such as fluctuations in CO_2 levels will also intensify or damp perturbations to the climate.

4.4.2 Planetary Neighbours as Asteroid Shields

The planets of the Solar System perturb more than the Earth's orbit. A very small percentage of the Solar System's mass not in the Sun, planets or dwarf planets resides in the form of icy/rocky debris, which clusters into several families. Some of these families are explicitly associated with one of the planets, for example the Trojan families that reside in the Lagrange points of the giant planets.

Of those not associated with any planets, we can identify three principal groupings (see Figure 4.12):

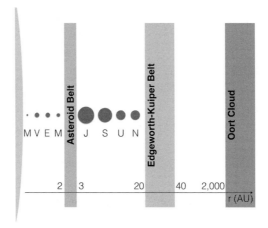

Figure 4.12 The positions of the principal asteroid and comet families in the Solar System, relative to the eight known planets.

1. The main asteroid belt that resides between Mars and Jupiter (2.2–3.3 AU)
2. The Edgeworth-Kuiper Belt of short-period comets (SPCs) that resides in the plane of the ecliptic beyond Neptune, between 20 AU and 40 AU approximately, and an accompanying scattered disc of objects on more eccentric orbits in that region,
3. The Oort Cloud, a spherically symmetric halo of long-period comets (LPCs) that extends from 20,000 AU to perhaps as far as 100,000 AU or more (i.e., halfway to the nearest star system, α Centauri).

Dynamical interactions with the planets can change the orbital parameters of these bodies, resulting in their being placed on crossing orbits with the Earth and impacting its surface (see Chapter 8). Given that these bodies can be tens of kilometres across, this is clearly dangerous for any organisms present during impact (Solution R.6).

It was believed for some time (quite universally) that the presence of Jupiter was a boon to the Earth's biosphere. Jupiter's gravitational field dominates the Solar System, in principle allowing it to capture, destroy or remove comets being thrown towards Earth from the Oort Cloud, essentially acting as a shield. If planets as massive as Jupiter failed to form, cometary material would not be expelled from the inner Solar System, also boosting the impact rate (Wetherill, 1994).

Further numerical integration of Jupiter's efficacy at removing eccentric bodies with orbits crossing into the habitable zone (Laakso et al., 2006) hinted at Jupiter's relative inability to act as a shield. A more definite conclusion was reached in a series of papers (Horner and Jones, 2008a,b; Horner et al., 2010), which considered the three main populations in turn. For each family, a series of N-Body simulations were run, with the mass of Jupiter varied to test its effect on the bodies.

In the case of the asteroid belt, it was found that fewer bodies were thrown into the Earth's environs if Jupiter was absent. In the case of objects perturbed from the Kuiper Belt, the effect of Jupiter's presence is comparable to if Jupiter was absent. For the Oort Cloud, it is indeed the case that more massive planets act as more effective shields.

In all cases, Horner et al. find that the systems where Jupiter is reduced to 20 percent of its current mass show significantly increased impact flux into the inner Solar System. While their results are contingent on all the other planet's properties being held constant, it does seem to be the case that if exoplanet systems have significant amounts of debris material, then the precise configuration of the planetary system will have a strong influence on the impact rate on potentially habitable planets.

4.5 Does the Earth Need the Moon To Be Habitable?

Our knowledge of moons is (at the time of writing) exclusively limited to the solar system. The gas giants possess a dazzling variety of moons, from the swept-up debris of the irregular satellites to the much larger regular satellites. Even although the regular satellites can be quite large (Saturn's moon Titan dwarfs Mercury in size), the mass ratio of satellite to planet remains quite small for the giant planets (i.e., the typical satellite mass does not exceed 0.01 percent of the parent body's mass; see Figure 4.13).

The terrestrial planets possess far fewer satellites than their giant cousins. Mercury and Venus possess no moons at all. Mars's two moons, Phobos and Deimos, are again quite small. The Earth's Moon is unusually massive compared to its parent, being nearly 1 percent of the Earth's mass.

This oddness encourages a knee-jerk reaction – is this a unique feature of the Earth? Is our unusual moon a necessary condition for our world to be habitable (Solutions R.7 and R.8)?

A first response is to consider several related questions, which we will address below.

4.5.1 What Influence Does the Moon Have on Earth's Habitability?

The Moon's influence on Earth and its inhabitants can be split into two categories: dynamical and radiative. The Moon's gravitational field continually exerts itself on the Earth. This force produces tides in the Earth's interior and its surface. In the interior, this force acts as a braking mechanism on the Earth's spin, slowing down its rotation and pulling on its spin axis, which we define by the obliquity and precession angles. This reduction in the Earth's spin angular momentum is

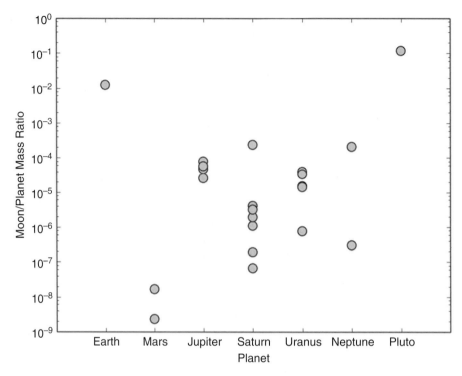

Figure 4.13 The moon-to-planet mass ratio for the major satellites in the Solar System. Note the dichotomy between the Earth's Moon, and Pluto's largest moon Charon, which possess mass ratios of order 10^{-2}, and the regular satellites of the gas and ice giants, with mass ratios of order 10^{-4}. Data from Jet Propulsion Laboratory Planetary Satellite Physical Parameters database, http://ssd.jpl.nasa.gov

accompanied by an increase in the Moon's orbital angular momentum, resulting in the lunar orbital semi-major axis increasing at a rate of 3.8 cm per year.[10]

Slowing down the Earth's rotation appears to have had a beneficial impact on its climate. A faster rotation rate inhibits jet streams in the troposphere, as well as other large-scale zonal flows of air. This in turn prevents the mixing of vertical layers of ocean by these winds, and this has a subsequent impact on the ocean's ability to absorb fluctuations in atmospheric temperature, which can then be transported towards the poles. Generally speaking, a faster rotating Earth would have been less able to distribute heat from the equator to the poles, which can encourage ice-albedo feedback and may well have initiated glacial periods (Hunt, 1979). We see this suppression of latitudinal flows in Jupiter – the distinctive bands of its atmosphere are owed to its rapid rotation period of approximately 10 hours.

[10] http://nssdc.gsfc.nasa.gov/planetary/factsheet/moonfact.html

The direct influence of tides on the Earth's oceans by the Moon has further benefits. The continual shift of water to and from the shore produces a constantly changing habitat. This smooth, continual change in environment at the coast induces a unique oscillating selection pressure on life residing there. We can see this in the great diversity of species that occupy terrestrial coastlines. It is theorised that this ever-changing habitat may have played a key role in the transition of life from the oceans to the land, supplying a 'proving ground' for ambitious organisms that could explore beyond the waters for a few hours and retreat to safety without exerting a great deal of energy.

It is even argued that lunar tides are responsible for the origins of life itself. The lunar tide would have been significantly stronger around 4 Gyr ago, moving large bodies of water away from the coastland to reveal moist salt flats. Biochemists argue that these environments provided fertile ground for nucleic acid assembly during low tide, and disassembly at high tide. The regularity of these tides may even have provided a rudimentary circadian rhythm for self-replicating reactions to work to.

4.5.2 What If the Moon Never Existed?

How would the Earth's rotation evolve in the absence of the Moon? The length of Earth's day would likely be significantly shorter, due to the lack of tidal dissipation. Solar tides would still operate, but with a strength roughly 40 percent of the Moon's. The solar and lunar tides act in opposition twice a month, when the Sun and Moon reside at positions perpendicular to each other relative to the Earth (the first and third quarters of the Moon's lunar cycle). The resulting *neap tide* is approximately 60 percent of the typical tide strength (see Figure 4.14). The corresponding spring tide (at new or full moon) can be 140% of typical tide. One should therefore expect that a Moonless Earth would experience tides around 2/3 of a neap tide, without variation.

This weak tide would have failed to slow the Earth, giving it a rotation period as short as 8 hours.[11] This rapid rotation would have inhibited the climate's ability to absorb temperature fluctuations. The weak tides might have assisted in the formation of prebiotic molecules, but with greatly reduced ability. The spread of self-replicating molecules across the planet may also have been inhibited by the poor redistribution of heat.

What about the direction of its spin? Laskar and Robutel (1993) computed the evolution of the Earth's obliquity and precession angle both in the presence and

---

[11] This, of course, assumes that no other catastrophic events occurred to affect the Earth's rotation. Venus, for example, possesses a spectacularly slow rotation rate, retrograde to its orbit, seemingly due to a turbulent impact history during the early stages of its formation.

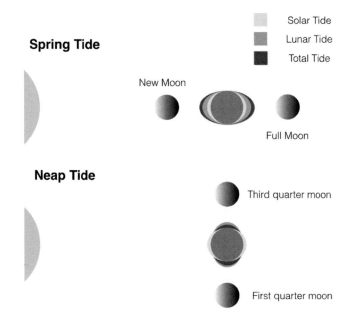

Figure 4.14 The spring and neap tides. Depending on the relative orientation of the Sun and Moon, the lunar and solar tides can co-add to produce the spring tide, or add orthogonally to produce the neap tide.

absence of the Moon, with various different starting values of the obliquity (from 0° to 170°). By Fourier analysis of the resulting oscillations, they were able to identify starting values of ψ that resulted in gentle, regular oscillations, as we see regular variations in Earth's obliquity today. Other starting values led to chaotic behaviour, where the obliquity varies rapidly with large amplitudes. Laskar and Robutel (1993) found that when the Moon was present, obliquity angles between 60° and 90° resulted in chaos, where the obliquity could lurch by 30° in just a few million years. If the Moon is removed, this chaotic zone extends from 0° to 80°!

This means that a moonless Earth would have been subject to stronger changes in its seasons. These changes would most likely have been enhanced by the Earth's faster rotation rate due to the lack of tidal dissipation from hosting the Moon. Laskar and Robutel's (1993) results suggest the amplitude of these obliquity perturbations could have been large, swinging from one end of the chaotic zone to the other. Later direct integrations of moonless Earths showed that the obliquity variations failed to explore the entirety of this chaotic zone. The moonless Earths still achieved large swings of $\sim 20°$ to $25°$ over hundreds of millions of years, but this is much smaller than might have been expected. Interestingly, a retrograde spinning moonless Earth experienced even smaller swings.

4.5.3 Are Large Moons Common in Terrestrial Planet Formation?

We currently have poor observational constraints on the typical properties of satellites of terrestrial planets. For the time being, we must suffice ourselves with theoretical expectations.

Satellite formation has two principal routes, depending on the mass of the host planet. Satellites of the gas giant planets are consistent with formation in a disc, in a microcosm of planet formation (Mosqueira and Estrada, 2003a,b; Canup and Ward, 2006; Ward and Canup, 2010). The mass of the gas giant is sufficient to maintain a circumplanetary disc of material, which is drawn from the surrounding circumstellar disc. These models produce satellite-to-planet mass ratios of around 10^{-4}, which is in accordance with the principal moons of Jupiter and Saturn (see Figure 4.13).

Satellites of terrestrial planets are thought to form much later in the planet formation process. In the final assembly phase, rocky protoplanets experience a series of giant impacts which grow their mass. These impacts have a variety of outcomes, depending on the mass of the two colliding bodies, their incoming momentum and the angle of impact. In the case of the Earth, it is believed that the Moon was formed during a giant impact with a fellow protoplanet, often dubbed 'Theia' (Cameron and Ward, 1976). This neatly explains the fact that the Earth's spin angular momentum and the Moon's orbital angular momentum are well correlated.

Our understanding of the specific details of this impact event has evolved significantly with time. In the original thesis, Theia was approximately of Mars mass, and a similar compositional mix. In other words, Theia must have been a very nearby protoplanet, formed from essentially the same protoplanetary disc material as the Earth. Belbruno and Gott III (2005) suggest that Theia may have formed in one of the Earth's *Lagrange points*. Lagrange points are defined (in the reference frame rotating with the Earth) as regions where the gravitational forces of the Earth and Sun, and the centrifugal force due to the Earth's motion cancel (see Figure 4.15). Belbruno and Gott III (2005) show that Mars mass impactors can accrete in either the L4 or L5 points, and then impact the Earth at relatively low velocity.

Alternatively, one can propose a scenario where Theia was a much larger impactor (around an Earth mass), have the bodies impact at high velocity and produce similar compositional mixes in both bodies, as the impactor donates large amounts of its mantle material to both bodies (Canup, 2012). This would produce a very high angular momentum system compared to the Earth-Moon case, and as such would require additional torques to remove this. For example, the evection resonance, between the Moon's precession of periapsis and the Earth's orbital period, can temporarily boost the Moon's orbital eccentricity and drain the Earth's spin angular momentum (Ćuk and Stewart, 2012).

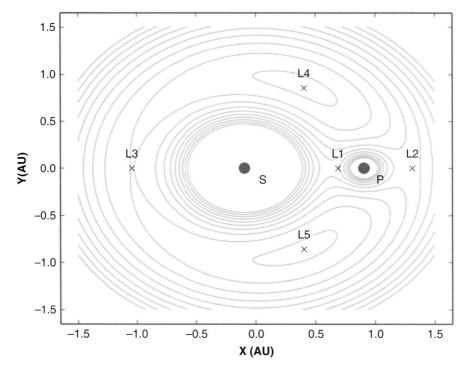

Figure 4.15 The Lagrange points of a star-planet system (in a reference frame rotating with the planet). The planet (P) has a mass that is one-tenth of the star (S) for illustrative purposes. The contours indicate equipotential surfaces, computed from the effective potential of both the star and planet gravity and the centrifugal force. L1, L2 and L3 are all nominally unstable points (potential maxima), although they can be successfully orbited with a minimum of station keeping. L4 and L5 are stable (potential minima). Belbruno and Gott III (2005) propose that the Moon could have formed in either of the Earth's L4 or L5 points.

In any case, it appears terrestrial planets require a secondary body to be accreted nearby and eventually join in a giant impact scenario. There is a possibility that extremely fast rotating planets could extrude a satellite thanks to centrifugal forces (Binder, 1974). While not thought to be the case for our Moon, it may well be so in other planetary systems.

Can we make any inferences about the likelihood of large moons for warm, wet terrestrial planets? At best, we can run a number of simulations of our own Solar System's formation and determine how frequently our Earth-Moon system is formed (O'Brien et al., 2014; Jacobson and Morbidelli, 2014). What is clear from this work is that the motion of the giant planets into the inner Solar System (around 1.5 AU) and back outwards to their current positions (the so-called *Grand Tack*) plays a vital role in determining the available material to grow the terrestrial planets and the nature of subsequent giant impacts. It may well even be the case that this phenomenon is the principal cause for why our inner Solar System looks so different from the tightly packed, short-period multiple planetary systems such as Kepler-11 and TRAPPIST-1 (see Batygin and Laughlin, 2015).

Could this also mean that our Earth-Moon system is relatively unique? Until we can begin detecting exomoons, we will be forced to make educated guesses.

5
Life Is Rare

It may well be the case that the number of worlds broadly suitable for life is very large. Current extrapolations suggest that there are likely to be many terrestrial planets in the Milky Way (e.g., Lineweaver, 2001; Zackrisson et al., 2016, and references within) and we might expect that many of them are rich in water and other biologically active compounds.

But it is still possible that life itself is rare, and that simply having a habitat is insufficient to ensure that such a place will become inhabited. To answer this question, we must consider carefully: How did life appear on the Earth? And why does it continue to persist?

5.1 What Is Life?

Perhaps most pressingly, we must consider what exactly *is* life? Most attempts at simple definitions fail, simply because they either include or exclude entities or phenomena in the 'alive' category in a manner that is intuitively unsatisfying.

For example, the simple criterion that life must be able to move excludes plants, and includes cars and rain. More sophisticated definitions typically include the following requirements (Koshland Jr., 2002):

1. a program (DNA)
2. improvisation (i.e., response to the environment and external stimuli)
3. compartmentalisation (e.g., cell membranes)
4. energy (metabolism)
5. regeneration (and growth)
6. adaptability
7. seclusion (control over local/internal chemistry)

The list must be at least this long to allow the satisfactory exclusion of non-living entities like fire (which satisfies many of these criteria, but not all). A classic rebuttal to this definition is that of the virus, which possesses many of the above requirements but does not metabolise.

Thermodynamic definitions of life are particularly popular amongst physicists (most likely because its description is in a language that they easily understand). The ability of organisms to reduce their internal entropy (at the cost of increasing external entropy) has been taken by some, most famously Erwin Schrödinger, as a clear sign that the reduction of entropy (or alternatively, the increase of negative entropy or *negentropy*) is an indicator of living systems.

Simple heat engines can be used to shunt entropy from hot to cold reservoirs. The efficiency of this exchange is limited by Carnot's theorem. If the temperature of the hot reservoir is T_H, and the cold reservoir has temperature T_C, then the maximum efficiency is

$$\eta_{\max} = 1 - \frac{T_C}{T_H} \tag{5.1}$$

This is as true for complex engines as it is for organisms, but the fact that artificial constructions can reduce the internal entropy of a system makes this definition problematic. Of course, this is quite suitable for SETI scientists, who are interested in both natural and artificial consumers of negentropy!

A more straightforward heuristic requires the application of evolutionary theory, i.e.,

Life is a system which replicates, mutates and changes according to Darwinian evolution.

This tends to be favoured by astrobiologists due to its simplicity. It is important to note that the definitions of life described here give no indications as to how to observe or detect extraterrestrial life. The thermodynamic definition comes closest, by noting that biological systems will deposit entropy into their environment, and that systems out of equilibrium, with significant entropy, may indicate the presence of a vigorous biosphere.

More specifically to terrestrial life, we can define some fundamental components/attributes of all living organisms:

- *Nucleic acids*, which are used to store genetic information.
- *Amino acids*, monomers which combine to form short polymer chains known as peptides, and eventually longer polymeric chains called *proteins*, which possess unique and specific functions.
- Energy storage, in the form of adenosine triphosphate (ATP). ATP is also a key monomer in RNA and DNA synthesis.

Of course, demanding that extraterrestrial organisms possess these precise properties is perhaps too exacting, but it remains important to note what is universal amongst all currently known forms of life.

5.2 The Origins of Life

Origin myths for life and intelligent life are as old as humanity, but a full scientific description of how life originated on the Earth (*abiogenesis*) is currently a source of debate.

The fossil record provides definitive proof of the existence of microorganisms 3.5 Gyr ago, i.e., around 1 Gyr after the formation of the Earth (Brasier et al., 2006). Geochemical evidence (principally high ratios of $^{12}C/^{13}C$) suggests that biological processes had already begun by 3.8 Gyr ago (Mojzsis et al., 1996) or perhaps even 4.1 Gyr ago (Bell et al., 2015).

We can also look to the Last Universal Common Ancestor (LUCA) for clues. We do not have a sample of LUCA – we can only infer its properties by analysis of the phylogenetic tree of life. Broadly, a phylogenetic tree is constructed by comparing the physical and genetic characteristics of species – in essence attempting to determine ancestry by genomic comparisons. Computations of this sort yield what are known as *clades* (groups that contain a common ancestor and all its descendants). Most modern phylogenetic trees are *cladistic*, in that they attempt to identify clades, to bring genetic rigour to the taxonomy of species. In a sense, we can consider LUCA and all life on Earth to form a single clade (Figure 5.1).

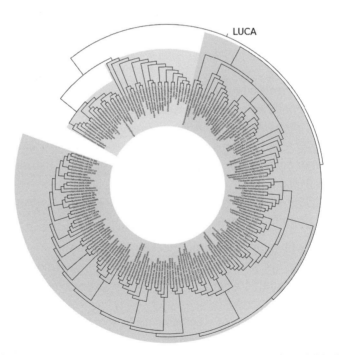

Figure 5.1 The phylogenetic tree of life, with the three domains of life from top left, clockwise: *Archaea*, *Eukarya* and *Bacteria*. The Last Universal Common Ancestor (LUCA) is represented by the notch at the upper right of the outermost boundary. This tree was generated using the IToL tool (Letunic and Bork, 2016).

5.2 The Origins of Life

LUCA resides at the root of the phylogenetic tree (with the root's location determined by a process known as phylogenetic bracketing). LUCA is unlikely to be the first organism that ever existed, and is indeed probably quite far from it. However, it does give us important information about the conditions in which the first organisms (the ancestors of LUCA) likely thrived. The properties of LUCA remain a source of heated debate (Glansdorff et al., 2008). Most attempts to reconstruct the protein families that trace back to LUCA suggest a thermophilic (heat-loving) physiology, with recent studies suggesting LUCA required an environment rich in H_2, CO_2 and Fe (Weiss et al., 2016).

The precise means by which 'non-living' chemical compounds began to organise into macromolecular structures, which separated themselves from their surroundings via some form of membrane, and then began to self-replicate is not known. More specifically, we do not know the order in which the various subsystems composing a living organism were 'invented' by Nature.

We can categorise these subsystems into three:

1. **Informational** – Data storage systems, such that replication procedures have appropriate instructions for defining a child organism's attributes and properties.
2. **Compartment-forming** – Systems which allow the organism to be separated from the environment, and to store the organism's physical attributes.
3. **Metabolic** – Systems which extract and store energy generated from the local environment, and produce essential compounds for survival and replication.

The many different theories espoused for the origin of life on Earth order the formation times of each subsystem differently. We will not undertake an explanation of all the various theories, but it is worth noting that many fall broadly into one of two classes, which we can glibly label 'vents' or 'pool' theories. However, all of modern research into abiogenesis owes a great debt to Oparin-Haldane theory, and the concept that an appropriate collection of chemical reactants, suspended in an appropriate aqueous solution, and judicious free energy could give rise to self-assembling, self-replicating molecules.

5.2.1 Oparin-Haldane Theory and the Miller-Urey Experiments

The Oparin-Haldane theory, formulated in the 1920s, is perhaps the most well-known 'canonical' theory of life's origins on the early Earth. At the time, it was believed that the Earth's atmosphere was strongly reducing after its formation (i.e., it gave electrons freely to molecules). This system would be composed largely of organic molecules with large amounts of hydrogen (e.g., methane) and very little

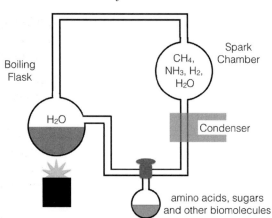

Figure 5.2 A schematic of the Miller-Urey experiment. Water is evaporated and added to a spark chamber containing what were thought to be gases present in the early Earth's atmosphere. Electric discharge was passed through this chamber, and the products were condensed and collected for study. This experiment produced both amino acids and the precursors of nucleotides for RNA and DNA.

amounts of free oxygen. With judicious amounts of ultraviolet radiation, it was believed that these organic molecules would readily produce biological precursor molecules such as amino acids.

Miller and Urey developed these ideas into a groundbreaking experiment in the 1950s (Miller, 1953, Figure 5.2). Their apparatus was composed of two connected flasks. In the first flask, a quantity of liquid water was added and placed under heat, to represent Earth's ocean. This flask was connected to the second, allowing water vapour to mix with the 'atmosphere', simulated by a mix of hydrogen, methane and ammonia.

By passing an electric discharge through the water-ammonia-methane mix, this resulted in the production of amino acids and several of the precursors of nucleotides which compose RNA and DNA. These results excited contemporary observers, but we now look at these results with less optimism.

Firstly, it should be noted that the relative concentrations of some key prebiotic molecules were disappointingly low, especially long-chain fatty acids, which form the basis for biological membranes today and were absent from the Miller-Urey yield. The sugar ribose, which is a principal component of both RNA and DNA, was present only in small quantities. The Earth's atmosphere during this early period is no longer believed to be quite as reducing as was thought in the 1950s, but there is strong geological evidence, particularly from sulphur isotope fractionation in ancient sediments (i.e., over 2.4 Gyr old), that the Earth's atmosphere and oceans were essentially anoxic (oxygen-free) (Farquhar et al., 2014, and references within).

Along with other lines of evidence, it seems clear that the Earth's atmosphere was most likely mildly reducing, with intermittent methane haze cover (Zerkle et al., 2012), which is encouraging for the production of prebiotics such as amino acids. Oparin-Haldane theory is therefore a sensible first iteration towards a final theory of abiogenesis, of which there are now many candidates.

5.2.2 'Vents' Theories

The 'vents' theories state that life began in hydrothermal vents (Wachtershauser, 1990). Volcanic hydrothermal vents are chimney-shaped fissures in the crust, typically found at the underwater boundaries of tectonic plates. This provides a source of superheated water in close proximity to magma, where local water temperatures can reach $\sim 400°C$ due to the extremely high pressure.

Vents theories are attractive for several reasons. LUCA appears to be thermophilic, suggesting an acclimatisation to hot environments. Such deep vents are extremely well sheltered from asteroid impacts or noxious atmospheric conditions. Also, the vents naturally produce steep gradients in temperature, pH and chemical concentrations, ideal for initiating quasi-metabolic processes. In particular, the redox reaction of iron and hydrogen sulfide into pyrite appears to be a means of fixing carbon, while pyrite is proposed to be a template for polymerisation of amino acids.

In short, volcanic hydrothermal vent theory suggests metabolic systems are formed first. These are *autocatalytic* chemical reaction networks, where the reactants that enter the network are produced in larger quantities at the end, providing a mechanism by which organisms can grow and maintain themselves. Metabolic systems should also find ways in which energy can be stored and released, which we will discuss later. As the peptide-rich fluid diffuses into cooler waters, this metabolism is then followed by informational systems, and finally the formation of compartments such as protocell structures (the proto-organism relies on indentations in the seabed to provide a 'shell' until it can form its own membranes).

More recently, alkaline hydrothermal vents have become a promising variant to this theory. Alkaline vents are typically found in close proximity to spreading plate boundaries (e.g., the deep-sea alkaline vent field known as the Lost City, which lies 20 km from the Mid-Atlantic Ridge). These alkaline vents are warm rather than hot (of order 70°C), and are highly porous, rather than chimney-like. This is of particular interest, as this sponge-like system of 10–100 μm-sized pores bears a greater resemblance to cellular structure. As the ocean water mixes with the highly alkaline fluid inside this pore network, steep redox, heat and pH gradients are set up. Together with plentiful sources of H_2 and Fe, the scene seems set for very promising prebiotic chemistry, with pre-made cell-like containers (Sojo et al., 2016).

5.2.3 'Pools' Theories

The idea that life formed in a surface pool of enriched water, the so-called 'primordial soup', is long-lived. The most exciting modern incarnation of this theory comes from John Sutherland's group at the University of Cambridge. Their models consider a repeating drying-wetting pool or stream to achieve their synthesis of many of the precursors for biochemistry. The amino acids cytosine and uracil (key components of RNA) were shown to be synthesisable on the early Earth, using only acetylene (C_2H_2) and formaldehyde (CH_2O) as raw materials (Powner et al., 2009). This suggests that, if the other two nucleobases of RNA (adenine and guanine) could also be formed, then the assembly of RNA could take place in the primordial soup. Acetylene has been detected in interstellar space, although this is only weak evidence that it would have been present on the early Earth.

In more recent advances, using an even simpler setup, precursors of ribonucleotides, amino acids and lipids were formed, using a combination of hydrogen cyanide (HCN), hydrogen sulfide (H_2S) and ultraviolet radiation at approximately 250 nm (Patel et al., 2015). This synthesis requires only raw materials that were likely to be abundant on the early Earth, although it does require the convergence of different synthesis paths (i.e., the mixing of different products after multiple sets of reactions). This could be achieved by the mixing of different streams, or by the washing of reactants into pools by rainwater.

5.2.4 Multiple Origins of Life?

The fossil and phylogenetic evidence does not rule out the possibility that there were multiple abiogenesis events on the Earth. Given our lack of knowledge regarding events before LUCA, it is quite possible that there were as many as ten individual abiogenesis events, and as such ten phylogenetic trees on the Earth, with only one surviving the Precambrian to the present day (Raup and Valentine, 1983). Not all of these trees would have co-existed, and our tree's survival does not guarantee it is the best of all possible trees – perhaps merely the most fortunate (Solution R.13).

5.3 The Major Evolutionary Transitions in Terrestrial Organisms

Once organisms appear on a planet and begin establishing a biosphere, how does this system develop? Following Smith and Szathmary (1997), we can identify several major transitions in the evolution of life on Earth. A common trend among these transitions is a growing need for interdependence between what were originally separate systems. Note that these transitions are not necessarily in the

5.3 The Major Evolutionary Transitions in Terrestrial Organisms

'correct' chronological order, and as we have already discussed, their nature and precise ordering are the source of a great deal of debate.

Figure 5.3 shows the division of geologic time into eons, eras, periods and epochs. Following the Hadean eon, during which the Earth completed its formation and cooled from its initially molten state, the earliest signs of life emerged in the Archaean eon.

Around 500 million years ago, the fossil record suddenly fills with many forms of multicellular life. This event is referred to as the *Cambrian explosion*, and marks the beginning of larger, more complex life forms dominating the Earth's surface and oceans. The right panel of Figure 5.3 records the geological periods where mass extinctions occurred, and lists the geological epochs (or sub-periods) since the last major mass extinction, which resulted in the death of the dinosaurs and the age of mammalian dominance. As we will see later, anatomically modern humans arrive towards the end of the Pleistocene epoch, with most of modern civilisation crammed into the end of the Holocene, which encapsulates the modern day.

5.3.1 Free-Floating Self-Replicating Molecules to Molecules in Compartments

Before the formation of compartments (protocells), self-replicators are required to be able to sustain themselves. Collecting populations of self-replicator molecules in compartments allows co-operation between different molecules. This allows natural selection to operate, with both molecular co-operation and competition inside the compartment as self-replicators vie for dominance.

Some good candidates for the first compartments are *vesicles*, which are simple fluid-filled membranes found both inside modern cellular structures, and as a means of transporting biological material away from cells. Vesicles can form spontaneously when lipids or other organic molecules are added to water. Hydrophobic molecules are repelled by the water and form a bilayer membrane which can absorb and emit organic material, especially if the material is entrained on grain surfaces (such as the volcanic clay montmorillonite, which would have been ubiquitous in the highly volcanic early Earth).

At this stage, it is interesting to note that cellular compartmentalisation is not by any means a universal transition. Viruses are essentially free strands of RNA or DNA contained in a protein sheath, and hijack cells to use their internal machinery.

5.3.2 Independent Replicators to Chromosomes

Inside any compartment, the genes represented by the self-replicators can be reproduced independently of each other, ensuring that intracompartmental competition

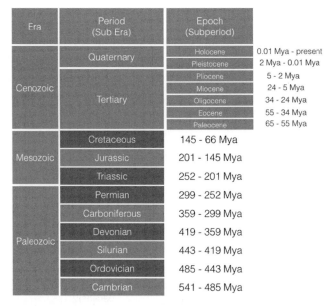

Figure 5.3 Top: The geological eons and eras, covering the time period from the formation of the Earth to the present day. Bottom: Geological periods and epochs (sub-periods) and periods, covering from the Cambrian period to the present day. Mass extinctions occur during the five highlighted periods (Ordovician, Devonian, Permian, Triassic, Cretaceous).

is rife. The transition to chromosomes means that genes are linked together. Therefore, if one gene is replicated, all are. This enforces co-operation between genes in the compartment, and instead shifts the competitive arena to compartment vs. compartment.

RNA, DNA and Chromosomes Nucleic acids are the information carriers of biology. Ribonucleic acids (RNA) can both store genetic data and act as an enzyme, while deoxyribonucleic acid (DNA) is principally a storage facility, with proteins playing the role of enzyme. This dual function of RNA leads many researchers to think that RNA preceded DNA. This concept is referred to as *The RNA World*, and we will explore this further here.

Firstly, let us review the structure of DNA and RNA. DNA (deoxyribonucleic acid) is composed of two strands enmeshed in a double helix (see Figure 5.4). Each strand is composed of a deoxyribose-phosphate backbone, where the sugar groups and phosphate groups are linked alternately. Attached to the sugars are nitrogenous bases or nucleobases. The four bases of DNA are adenine (A), thymine (T), guanine (G) and cytosine (C). The deoxyribose-phosphate-base combination is referred to as a nucleotide.

In the double helix configuration, the two strands are connected together at the bases, forming *base pairs*. The bases interact via hydrogen bonds, and hence only purine bases (adenine, guanine) can connect to pyrimidine bases (thymine, cystosine). Strictly, only adenine connects to thymine (AT, TA) and guanine only connects to cytosine (GC, CG). The ordering of the four bases GCTA along the strands is the means by which DNA stores information. These letters are read in triplets known as *words*. These words are capable of coding $4^3 = 64$ possible

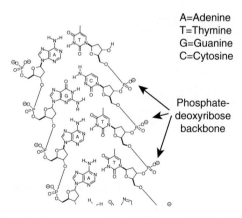

Figure 5.4 The chemical structure of a DNA strand. The four bases of DNA, adenine (A), thymine (T), guanine (G) and cytosine (C) are indicated.

amino acids (the true number coded for is only 21). It is remarkable that AT base pairs and GC base pairs are geometrically equivalent (Orgel, 1973). This exactitude acts as a form of error control – attempts to form the double helix structure with say an AG base pair immediately fail due to this property.

In cells, the entire genome is typically stored in a set of *chromosomes*. The *ploidy* is the number of sets of chromosomes in a cell – haploid organisms only contain one set of chromosomes. Like most organisms, humans are *diploid* – their chromosomes are paired (a total of 23 pairs = 46 chromosomes). We define a *gene* as a stretch of the DNA sequence that codes for a protein or RNA molecule. Humans carry two (usually differing) copies of each gene, one from each parent (known as *alleles*).

RNA is quite similar to DNA in some respects, but has important differences. It is a single-strand molecule, which uses ribose-phosphate groups instead of deoxyribose-phosphate groups to compose its backbone. It also has four nucleobases, but replaces thymine with uracil (U).

Genetic information stored in DNA cannot be directly transmuted into proteins and enzymes. RNA is the intermediary – through *transcription*, the words in a DNA template strand are converted into a complementary set of words on stand of *messenger RNA*. The triplet words on a mRNA strand are referred to as *codons*. These codons then undergo *translation* into the desired amino acids, which can then be assembled into a polypeptide chain.

Experiments with very short DNA chains (6 base pairs) show that it can be replicated without enzymes (von Kiedrowski, 1986), but these are significantly shorter even than the shortest chains found in modern organisms. DNA can produce a variety of complex enzymes by using RNA as an intermediary, but it cannot replicate itself without fairly complex enzymes.

This is the basis for the RNA World hypothesis: DNA suffers from a chicken and egg problem, which is resolved by allowing RNA to be the egg. This leaves a separate problem, which is extremely apposite to our discussion of the likelihood of life.

An isomer of a polymer (such as RNA) is a molecule which possesses the same chemical formula, but has a different arrangement of components. For example, we can create an isomer of any RNA molecule by switching the position of two amino acids. It should be immediately clear that for peptides with even a modest number of amino acids, the number of isomers is enormous. Given the 20 naturally occurring amino acids, if our peptide chain is N amino acids long, then the number of isomers is 20^N. For $N = 200$, the number of isomers is a stunning 10^{260} (!) (Jortner, 2006).

This can be viewed either as a blessing or a curse – the sheer variety of 'codable' structures is at the heart of the rich biodiversity we see on the Earth. However, the parameter space occupied by all the peptides on Earth is but a vanishing fraction of the total possible. For RNA to be a successful catalysis/replication entity, it is estimated that $N = 40$ monomers. If RNA strands are generated at random to achieve this end, then approximately 10^{48} strands would be generated, with a total mass close to one Earth mass!

One can now clearly see the quandary that astrobiologists and SETI scientists face. On one hand, it appears that the raw material for life (especially $NCHOPS$ and amino acids) appears to be common, both on the Earth and in the Solar System, and that metabolic processes could be engaged in some environments we expect on the early Earth.

On the other hand, the sheer combinatorial gulf that separates a random string of monomers from a functional strand of RNA seems almost impossible to bridge. Fortunately, a planetary body provides the opportunity for many billions of individual 'experiments' to be carried out, with a variety of reactants, catalysts and products. The combinatorics worked for the Earth, but we must be careful in assuming it will work (or has worked) at least more than once in the Universe.

5.3.3 Cyanobacteria to Photosynthesis

Photosynthesis is perhaps the most important energy storage process in Earth's biosphere, supporting the vast majority of its biomass, either directly or indirectly as part of the food chain. Photosynthetic organisms are *photoautotrophs* – that is, they can synthesise their own food supply, absorbing photons to convert CO_2 into carbohydrates. The general form of the reaction (given a chemical species A) is

$$CO_2 + 2H_2A + \gamma \rightarrow [CH_2O] + 2A + H_2O \qquad (5.2)$$

where the H_2A molecule acts as an electron donor. In oxygenic photosynthesis, A=O and hence

$$CO_2 + 2H_2O + \gamma \rightarrow [CH_2O] + O_2 + H_2O, \qquad (5.3)$$

i.e., photosynthesis is a primary source of O_2. Water is required as a reactant and is also a product (although note that the reaction involves a net loss of one water molecule).

The reaction has two principal stages: a *light-dependent* stage, where the pigment *chlorophyll* absorbs a photon and donates an electron. This electron is passed through a series of other pigments/molecules in a transport chain to produce the energy storage molecules ATP and nicotinamide adenine dinucleotide phosphate (NADPH). In the second, *light-independent* stage, these molecules are used to draw CO_2 from the air and secure the carbon in sugars (carbon fixation).

Several different metabolic pathways for carbon fixation are possible, all of which rely on the enzyme RuBiSCo (Ribulose-1,5-bisphosphate carboxylase/oxygenase) to catalyse the fixation reaction. One such pathway is C_3 photosynthesis, which directly fixes carbon from CO_2 via the Calvin-Benson cycle. It is the oldest pathway, and the most proliferant.

In hot, dry areas, the dual carboxylase/oxygenase nature of RuBiSCo means that some of RuBiSCo will oxidise, resulting in *photorespiration*, and a reduced carbon fixation rate. Plants must either prevent water loss in these dry climes, or find other ways to fix carbon. The C_4 photosynthesis begins by synthesising 4 carbon organic acids (e.g., malic acid and aspartic acid), which allows CO_2 to be regenerated inside specialised bundle sheath cells, and this CO_2 can then be fixed using RuBiSCo as in the C_3 pathway. Around 5% of plants use the C_4 pathway.

Evidence of carbon fixation is found in Archaean stromatolite fossils around 3.8 Gyr old. This is not concrete evidence for photosynthesis being the cause of carbon fixation, although it is suggestive. We know that there was insufficient O_2 in the atmosphere for oxygenic photosynthesis to be present. Other reductants can play the role of H_2A in our general photosynthesis equation. Indeed, H_2 could provide the required donor electrons, or H_2S, both of which were likely to be more abundant in the Archaean.

Anoxygenic photosynthetic organisms include the *Chlorobium* (green sulphur bacteria), *Rhodobacter* (purple bacteria) and *Chloroflexus* (green gliding bacteria).

Cyanobacteria are believed to be the originators of oxygenic photosynthesis. Microfossils that appear to be cyanobacteria-like date back to 3.5 Gyr ago, but this is at best weak evidence of photosynthesis in this period. Oxygenic photosynthesis must have been present before around 2.6-2.7 Gyr, at the point of the first Great Oxygenation Event.

The evolution of anoxygenic photosynthesis occurred relatively early, and yet is a non-trivial event, given the complexity of the light-independent reactions, and the need for RuBiSCo. Remarkably, the photosynthetic mechanisms used today are close to identical to the mechanisms produced by the most ancient algae. The evolution of photosynthesis most likely required the chance combination of a large number of unicellular organisms, with metabolisms that bore little resemblance to any photosynthetic reaction!

Oxygenic photosynthesis had one sinister advantage over its anoxygenic competitors – it produced a toxic waste product, which most likely destroyed its competition. As the cyanobacteria population exploded through using this new innovation, they soon starved themselves of CO_2. The sequestration of carbon by tectonic subduction allowed oxygen to remain in the atmosphere, bringing O_2 levels to around 7% (30–40% of their present value). This quickly halted cyanobacteria growth. Oxygen would stay at this low value until the arrival of eukaryotic algae around a Gyr later.

5.3.4 Prokaryotes to Eukaryotes

The fossil evidence shows that eukaryotic cells have existed on the Earth for around 1.8–2 Gyr, compared to the 3.5 Gyr (and possibly more) that prokaryotic cells have existed.

The key difference between prokaryotes and eukaryotes is the complexity of their internal cellular machinery. Eukaryotic cells possess a nucleus and mitochondria, which are enclosed in their own membranes, whereas the contents of a prokaryotic cell are thought to be mixed within a single compartment (see Figure 5.5).[1]

The most popular theory for eukaryotic cell development is referred to as *endosymbiont theory*, where the modern eukaryotic cell is assembled from a collection of primitive prokaryotes. The mitochondria, chloroplasts and other related

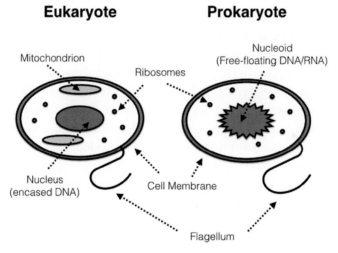

Figure 5.5 A simplified schematic of the eukaryotic and prokaryotic cell. Prokaryotes possess no other organelles except for the ribosomes, the sites of protein synthesis. The nucleoid is an aggregation of RNA, DNA and proteins that is not encased in a membrane. By contrast, eukaryotic cells have a membrane-encased nucleus, and a variety of organelles depending on whether the cell belongs to plant or animal. Almost all eukaryotic cells contain mitochondria, which produce the majority of the cell's supply of adenosine triphosphate (ATP). Plant cells also contain chloroplasts, which undertake photosynthesis.

---

[1] There is evidence of protein structures inside cyanobacteria, which would indicate that prokaryotes possessed a similar form of organelle containment (Kerfeld, 2005).

organelles were most likely engulfed by a larger prokaryote, either as prey for digestion or as internal parasites. As chloroplasts and photosynthesis began to oxygenate the Earth, prokaryotes would have derived a great advantage from possessing an onboard oxygen processing system to produce ATP (mitochondria), and in some cases (i.e., plants) a secondary energy generation system (the photosynthesising chloroplast). Over time, the new hybrid cell adjusts its energy production behaviour to account for its new 'tools', and eventually becomes completely dependent on them for survival.

The fact that all eukaryotic cells have mitochondria, but not all eukaryotes have chloroplasts (or plastids in general) hints at which engulfment happened first. Indeed, the very fact that mitochondria and plastids still retain their own DNA and protein construction machinery is strong evidence that they were once independent organisms.

5.3.5 Single-Cell to Multicellular Organisms

The transition from unicellular to multicellular organisms has occurred at least 25 times in the history of eukaryotic life on Earth (not to mention in the prokaryotes, Parfrey and Lahr, 2013). This is established by considering the multicellular behaviour of organisms across multiple clades.

There are broadly two forms of multicellular behaviour. Aggregative Multicellularity (AM) occurs when multiple single-celled organisms co-operate to produce a single 'fruiting' body. Divisive Multicellularity (DM) is the division of single cells, which is common in animals, plants and fungi. There are several behaviours cells must be able to emulate to be capable of multicellular actions – cell adhesion, communication and differentiation. These behaviours appear to be relatively easy to produce, for example in green algae, if the appropriate selection pressure is applied, suggesting that the genetic variation/mutation required for AM to occur is relatively low.

Unicellular organisms boast an impressive diversity of biochemical processes, far greater than that of multicellular organisms. The key advantage of multicellularity is the entirely new organisational opportunities afforded by effective division of labour. Specialisation of cell types (all sharing the same genetic information) is a powerful innovation that allows for much larger organisms to operate, and seems crucial for intelligence and innovation.

Of course, there is an intermediate stage between unicellularity and multicellularity, as evidenced in the colonies of unicellular organisms. Bacterial mats show differentiation of gene expression and metabolism depending on a bacterium's location inside the mat. Slime moulds (such as *Dictyostelium discoideum*) are unicellular during times of ample food supply – during lean times, the moulds are

5.3 The Major Evolutionary Transitions in Terrestrial Organisms

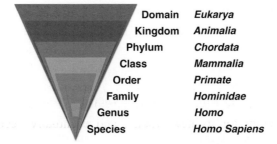

Figure 5.6 The principal taxonomic ranks. For each of the three domains of life (Eukarya, Bacteria and Archaea), organisms are assigned into kingdoms, phyla, classes, orders, families, genera and species.

encouraged to join together to make a multicellular colony, with cell differentation between the interior and the surface of the mould slug.

This level of cell specialisation requires an overall direction for an effective organisation. As a result, evolution has constructed a variety of different *body plans*, characteristic morphologies that give an organism an underlying structure upon which to develop and modify. It is instructive to consider the most widely accepted taxonomic ranking system for organisms, which is summarised in Figure 5.6. Species with a similar body plan typically belong to the same *phylum*, although this is not a strictly observed rule and depends on whether one defines phyla in a phylogenetic fashion or not.

The animal kingdom has approximately 33 recognised phyla – humans belong to the *chordates*, meaning that, amongst other properties, they share a notochord, a stiff rod of cartilage that extends from the organism's head to tail. In the subphylum of vertebrates, this notochord forms the central nucleus of the intervertebral discs (i.e., the spine).

The Energy Requirements of Large Life Forms

In terms of mass and number, the vast majority of organisms on Earth are microbial. It is quite the injustice to refer to them as 'primitive' or 'simple', but these words often find their way into the mouths of SETI practitioners, as they typically chase other forms of life.

As we will see, what we refer to as 'intelligent' life requires a minimum brain size to operate higher cognitive functions. Even the technological and sociological innovations of 'small' organisms such as ants and termites require animals to be *metazoan*, i.e., many orders of magnitude larger than a bacterium.[2]

---

[2] Of course, the danger for chauvinism is great. Microbiologists commonly speak of 'microbial communities'. Biofilms show remarkable coordination without requiring specific signalling, and their macroscopic

There is a relatively strong correlation between the body mass of a mammal and its energy consumption demands, as enshrined in the following power law relationship:

$$BMR = AM_b^x, \tag{5.4}$$

where BMR is the basal metabolic rate, the energy consumption rate required for basic physiological functions (except digestion), expressed in kilocalories per day, and M_b is the body mass in kilograms. The constant of proportionality is approximately 70 kilocalories per day per kilogram, and the exponent x approximately 0.75 (Kleiber, 1932), although the values of both are typically expressed without uncertainties, and the true values for different animals are debated (see, e.g., White and Seymour, 2003). This relationship holds well for modern humans and chimpanzees, with deviations of a few percent at most.

If we think in terms of energy budgets, then we can see that possessing a lower BMR in environments where the local energy supply is roughly constant gives a competitor a higher energy margin. Energy consumed goes to two core categories of function: maintenance, which includes tissue repair, BMR and other physical activities required to procure food, and production, which includes tissue growth and reproduction.

Clearly, for genetic provenance the more energy allocatable to growth and reproduction, the better. There does therefore seem to be an advantage in slightly lower BMR and hence lower body mass.

We can view much of humanity's trajectory through such a prism. Bipedalism offers similar energetic advantages compared to walking on all fours (Sockol et al., 2007), and continuing refinement of this skill through biomechanics is apparent as we consider the changing skeletons of humans throughout the last few hundred thousand years.

Indeed, the production of labour-saving devices can reduce energetic demands immensely. Tools and technology can confer important caloric advantages, and these benefits eventually outweighed the costs of the energy-rapacious brains that conceived them.

But what chemical reactions could produce and store such energy? We can rule out a large number of reactions on this basis. The large quantities of oxygen in the Earth's atmosphere facilitate a plentiful energy source. The origin of Earth's relatively large oxygen supply is photosynthetic organisms. When photosynthetic mechanisms arrived on the Earth approximately 2.7 Gyr ago, the atmospheric O_2 levels increased by a factor of 10,000 (to around 10% of modern levels). Many forms of prokaryotic life found these hugely boosted oxygen levels inimical,

---

properties hinge on the balance between competition and co-operation, like any human society one could name (see, e.g., Nadell et al., 2009).

as oxidation can extract electrons from chemical bonds, inhibiting and destroying cells and enzymes. Anaerobic environments remained, preserving prokaryotic forms.

Eukaryotic cells took full advantage of this new oxygenated environment, developing O_2 respiration as an energy generation source. Indeed, the eukaryotes that had absorbed chloroplasts were active participants in this second Great Oxygenation Event, which boosted O_2 to around 20% of current levels from evidence of increased oceanic sulphate concentrations (Catling and Claire, 2005), but the cause of this is not clear. From this level, atmospheric oxygen increased steadily until the present day.

5.3.6 Asexual Reproduction to Sexual Reproduction

Before the advent of sexual reproduction, the transmission of genetic information was largely chaotic. *Horizontal gene transfer* allows organisms to share and augment genetic data at extremely high speeds. Once sexual reproduction begins in a population, *vertical gene transfer* can occur, allowing the process of speciation to begin. Breeding between individuals begins to sunder them from each other, to the extent that two individuals with a common ancestor can no longer successfully produce fertile offspring, which is the strict definition of said individuals belonging to different species.

It is important at this stage to give a brief summary of our understanding of genetics and evolution. Practitioners of astrobiology and SETI commonly make critical errors in their use of evolutionary theory. In particular, they make fundamental errors in their application to human psychology. The field of *evolutionary psychology* is much maligned, and is plagued by misunderstandings of what makes a trait 'adaptive'.

Modern evolutionary theory relies on several key concepts.

Mutation

When parent organisms produce offspring, the reproductive process is not perfect. Errors creep into the propagation, either during the replication process, or as a result of DNA damage. Mutations take a variety of forms in DNA. Point mutations are simply changes from one base-pair to another at some location in the DNA sequence. Recall that base pairs are composed of a purine (G, A) and a pyrimidine (T, C). Point mutations that switch purine for purine or pyrimidine for pyrimidine are referred to as *transitions*. Point mutations that switch purine for pyrimidine (or vice versa) are *transversions*. Other mutations can increase or decrease the number of base pairs at a given location in the DNA sequence. If the increase or decrease is not a multiple of 3, this results in a *frameshift* mutation. The triplet *words* read off the sequence downstream of the mutation are shifted, analogous to two English words being incorrectly separated (i.e., 'two words' becoming 't wow ords').

Whole sections of DNA can be duplicated and inserted into the offspring's sequence (*tandem duplication*). In extreme cases, entire genomes or chromosomes can be duplicated.

Extreme duplication events can change the ploidy of an organism. Typically, the addition of an extra chromosome to a given pair is fatal to humans (with the exception of chromosome 21, which is responsible for Down's syndrome).

Recombination

Recombination refers to the production of new combinations of genes. Most commonly, this is achieved via some form of sex or reproduction. In humans, this occurs through the production of the male and female gametes (sperm and ovum, both of which are haploid), which then combine to form a fertilised, diploid zygote. This is one of many different recombinative cycles seen in terrestrial eukaryotes, but it serves as a sufficient example of how heritable traits are passed from generation to generation, while still providing a source of genetic variation beyond mutations.

Gene Flow and Random Genetic Drift

Let us imagine a population of organisms of a given species, where we can determine the total number of given alleles of a gene g in that population (the allele frequency P_g). Mutations can clearly affect allele frequencies, as well as recombination.

The carriers of genes (organisms) rarely stay fixed in a single location. Animals move in search of food and mates; seeds, pollen and bacteria are dispersed on the wind and by rainwater (or indeed other organisms). *Gene flow* refers to the changing location of genes between generations.

Genetic drift refers to the stochastic variation of allele frequencies that is guaranteed to occur, even if said variations do not confer an advantage on the individual. This phenomenon is guaranteed to occur by the fact that individuals will leave behind varying numbers of offspring, and hence there will always be a random element to evolution. This is witnessed by the observation that most eukaryotic DNA does not code for proteins, and in fact does not appear to possess any specific function. The combination of mutation and random reproduction (i.e., recombination without preference) is often referred to as *neutral evolution*.

Natural (and Artificial) Selection

The keystone of modern evolutionary theory, *selection* is the concept that genes that confer an advantage to their host are more likely to be propagated into future generations. We can think of this more rigorously by defining the *fitness W* as the number of offspring an individual leaves behind (after one generation). The fitness of any allele is therefore the average fitness of all individuals possessing that allele. Commonly, the fitness is not measured in absolute terms, and is instead normalised by some mean value in the population.

Natural selection prefers inherited variations that increase fitness. It is therefore the only aspect of evolution that results in *adaptation* – the production of traits whose function is to increase fitness, and was therefore (in a loose sense) 'evolved for' that function. Mutation, gene flow and genetic drift are statistically more likely to decrease fitness than increase it, as their actions are random and do not prefer increases in fitness.

Precisely how natural selection increases fitness can be viewed through either one of two concepts: Fisher's Fundamental Theorem, or Wright's Adaptive Landscape. Fisher's Fundamental Theorem states that the rate of increase in mean fitness ($\Delta \bar{W}$) is equal to the additive genetic variance in fitness:[3]

$$\Delta \bar{W} = \frac{V(W)}{\bar{W}} \tag{5.5}$$

To understand Wright's adaptive landscape, imagine a species with two alleles A and B. If we plot the fitness $W(A, B)$ as a function of A and B, we can produce a contour map, with probably multiple peaks and troughs of fitness. Generally speaking, selection results in allele frequency shifts that move towards peaks in W. This is an appealing image, especially to physicists used to understanding motion of particles in a potential in a similar fashion, but there are important complications to this metaphor.

Of course, populations have many thousands of alleles, and many peaks of fitness as a result. But even in a two-allele system, there are assumptions underlying this picture that must be examined. A key assumption is that alleles A and B are randomly associated in organisms (this is also known as *linkage equilibrium*). In practice, the presence of one allele is often conditional on the presence of another (an example of linkage disequilibrium). When linkage disequilibrium occurs, an allele that has little effect on fitness can find itself boosted in the population.

A most important assumption is that $W(A, B)$ is constant in time. Both the physical and biotic environment will evolve, and also the surrounding genes. A given allele can be beneficial if partnered correctly, say with a distinct counterpart (heterozygous), and yet be damaging to the organism if it is paired with itself (homozygotic). All these factors ensure that W will change with time.

---

[3] The additive genetic variance is not the total variance of the population, which is biased by random, noninherited differences in reproductive success of individuals, but rather the variance that can be attributed to the *average* effect of each allele in the population.

Natural selection has no specific purpose – artificial selection has an agent guiding the selection (humans), and therefore assumes the purpose of said agent. Unsurprisingly, rather than excecuting the 'random walks' associated with natural selection, genomes undergoing artificial selection undertake more direct paths towards peaks in Wright's adaptive landscape. Essentially, this is achieved by humans (knowingly or otherwise) artificially inflating the fitness of specific alleles relative to the mean.

Horizontal Gene Transfer

The appearance of vertical gene transfer does not end the action of horizontal gene transfer. This remains an important process in evolution, especially for, e.g., the development of antibiotic resistance in bacteria (Burmeister, 2015).

There are three principal modes of horizontal gene transfer:

- Transformation – Bacteria absorb DNA from the local environment.
- Conjugation – Bacteria directly transfer DNA between each other.
- Transduction – An intermediate organism conducts a transfer of DNA from one organism to another (e.g., a bacteriophage between two bacteria).

Horizontal transfer can be rapid, and can augment later vertical gene transfer, giving rise to extraordinary variability in bacterial genomes (Ochman et al., 2000).

In fact, it now seems clear that horizontal transfer (sometimes lateral gene transfer) is much more common than previously thought, and is not restricted to bacteria – genes can be transferred across domains, even between bacteria and animals (Dunning Hotopp, 2011), and it is likely that eukaryotic evolution would have been quite different without horizontal transfer (Keeling and Palmer, 2008).

5.3.6.1 Evolution is the Interplay of Mutation, Recombination, Gene Flow, Drift and Selection

Each component of evolutionary theory is necessary for it to have explanatory power. For example, we can only understand the phenomenon that deleterious alleles can find themselves fixed in small populations, if we recognise that random genetic drift is more effective in such low numbers (i.e., only a few individuals with the allele in question can cause fixation by having more offspring). Conversely, beneficial mutations that increase fitness can be lost in small populations via the same effect. Gene flows are crucial to the establishment of diverging populations, which pick up differing favourable mutations as a result of geographical differences, resulting in speciation.

This underlines the undirected nature of evolution. Evolution has no 'purpose' or 'goal' – it is a stochastic phenomenon, where stochastic changes to allele frequencies are amplified to varying levels depending on the relative strength of selection compared to the other forces acting on the genome. Therefore, species can possess traits that are not adaptive, but are rather a consequence of a period of strong drift and weak selection, or a correlate of selecting for another trait which is adaptive. We will consider this in more detail in future chapters.

5.3.7 The Brain

Before the brain, information processing was a spatially distributed activity encompassing much of the body, as can be seen in modern jellyfish and anemones. The evolution of central nervous systems allowed organisms' information to be more efficiently organised and processed.

The formation of specialised nerve and neuron clusters near the sensory organs became ever more advantageous, with the first brains growing from the clusters located near the mouth and eyes. Animal brains are principally networks of *neurons*, which consist of a single axon and multiple dendrite connections. The principal differences between animal brains are in the total neuron number, the local density of neurons, the typical number of connections (*synapses*) per neuron, and the overall network structure.

The human brain is principally composed of:

- the *cerebral cortex*, which is primarily responsible for the most complex aspects of human intelligence, including our aptitude for reason, imagination and language;
- the *cerebellum*, which is principally concerned with sensory-motor co-ordination;
- the *thalamus* and *hypothalamus*, which act as routing stations between the various brain regions. The hypothalamus also regulates the production of hormones and neurotransmitters, and appears to be the master control centre for emotions;
- the *limbic system*, the earliest form of the forebrain to develop. Its components include the *hippocampus*, which plays an important role in learning and memory, and the *amygdala*, which controls fear responses, as well as sexual and aggressive behaviour.

Together with the spinal cord (and the brain stem which connects the two), this constitutes the human central nervous system. But how can a system like this, the ultimate quest for SETI searches, be developed through evolution?

Determining the evolution of brains from the fossil record is highly challenging. Brains are soft tissue, and do not fossilise easily. We are forced to rely on *endocasts*. Fossil skeletons can be retrieved when the animal in question is buried in such a fashion that the bone can become mineralised. In an endocast, the soft tissue interior to the skeleton decays and is replaced by an inblown matrix of sands, mud, clays and pebbles. If this matrix remains undisturbed for a sufficient period of time, it too can harden into sedimentary rock.

Endocasts of brains require the cranium to remain uncrushed when the organism is buried, and for complete fossilisation. Depending on the skeleton's journey through the upper crust, the entire skeleton can be unearthed, and the cranium sectioned to extract the endocast, or in exceptionally fortunate cases, the cranium weathers under exposure to the elements, yielding the endocast brain like a nut from its shell.

Inferring brain properties from endocasts of 'lower vertebrates' can be challenging, as the endocast often includes extra cartilage and bone material, and can fill the entire cranium, which in general is somewhat larger than the brain.

The oldest endocast brain known to science is that of a *chimaeroid* fish, dating to some 300 Myr ago, showing that identifiable brains are already in place by the Carboniferous period. It is likely that the brain's origin actually goes all the way back to the Cambrian period, to the last common ancestor of all bilaterally symmetric animals, the urbilaterian (Hejnol and Martindale, 2008).

Determining precise brain volumes is crucial to studies of its evolution as an information-processing organ across species. In modern animals, we see (on average) a relation between the body weight P and the brain weight P:

$$E = kP^{2/3} \tag{5.6}$$

where k is a proportionality constant fitted to the clade of interest (for mammals, $k = 0.12$). The 2/3 exponent suggests a relationship between surface and volume. Therefore, for a given species i with a body weight/mass described by P, we can use the above equation to compute an expected brain weight/mass E_e. We can then compare this to actual measurements for the brain E_i, and compute what is known as the *encephalisation quotient*:

$$EQ_i = \frac{E_i}{E_e} \tag{5.7}$$

The EQ for humans is around 6.5, with bottlenose dolphins scoring around 5.5, macaques around 3.15, chimpanzees 2.63 and gorillas 1.75 (Cairò, 2011).

Taking these measures at face value is dangerous, and implies some odd results (thin humans are smarter than fat humans), but as an average for inter-species comparison, this points to a continuum of intelligence, connected to the relative brain size.

This makes some intuitive sense, as the volume of the cerebral cortex grows linearly with brain volume, and the total number of cortical neurons grows as $E^{2/3}$ (as larger brains have lower cortical density). Brains clearly confer an advantage relative to the demands of the organism's environment. In general, the higher the information-processing demands of the organism, generally the larger the brain (relative to its size).

This rule is imperfect, especially when we consider the cranial capacity of the *homo* genus. Over most of the 2 million years of the genus, the absolute brain size of hominids has increased, from a volume of around 600 cm^3 for *homo habilis* to around 1,496 cm^3 for *homo sapiens sapiens*. Whether this points to a **relative** increase in brain size (or EQ) is less clear, due to confusing factors such as sexual dimorphism (different body sizes for males and females), and the quality of endocasts or cranial samples. That aside, it is clear that through a combination of increased brain size and reduced body mass, there is a significant change in EQ between the *Australopithecus* and *homo* genii.

Neanderthal brain sizes are within the sampling uncertainties for *sapiens* brain sizes. Due to their larger bodies, Neanderthal EQ typically lags *sapiens*'s by about 1 or so (Mithen, 2011). We have two organisms with EQ and global brain size/body size properties that are difficult to distinguish, and yet *sapiens* dominates, despite being physically weaker.

We must therefore point to the internal brain structure as well as its total size. A fundamental rewiring of the brain, particularly the development of Broca's area in the frontal lobe (Cooper, 2006), is proposed as a cause for the sudden explosion in linguistic faculty around 70,000 years ago. This demonstrates that when it comes to intraspecies competition, it is not only size that matters.

5.3.8 Solitary Individuals to Social Units

Creatures with brains, operating in an environment surrounded by creatures of the same and different species, must make decisions on how to act that best suit their own interests. From a Darwinian point of view, natural selection will ensure that organisms that process information efficiently, and make decisions that result in optimal numbers of surviving offspring will thrive.

In this picture, every organism is in competition with every other organism (including members of its own species) to ensure its genetic inheritance survives. How can co-operation evolve in the midst of this struggle?

We can formalise these questions by considering the mathematics of *game theory*, which studies the gameplay of multiple *players*. Each player i adopts a *strategy* S_i, which results in a numerical *payoff* $p(S_1, S_2 \ldots S_n)$ for each player. For two-player games, this can be represented in a payoff matrix p_{ij}, where i represents the strategy of player 1, and j the strategy of player 2. The most famous example of a formal two-player game is known as the Prisoner's Dilemma (Table 5.1).

Table 5.1 *The payoff matrix for the Prisoner's Dilemma. Prisoners can either co-operate with their comrade (C) or defect (D). The Nash equilibrium of this game is (D, D).*

Prisoner's Dilemma	C	D
C	5,5	1,7
D	7,1	3,3

Imagine that two prisoners have committed an infraction, and are called individually to the governor's office. The governor asks them to tell the truth about their actions. Each prisoner has two options: they can either co-operate with their fellow prisoner by lying to the governor, ($S_i = C$), or they can defect to the governor and blame the other for the infraction ($S_i = D$). In this formulation the maximum payoff to either player is $p(C, D)$ or $p(D, C)$, i.e., to defect while one's comrade is co-operating. Indeed, if either player defects, the other player's best strategy is to also defect.

The *Nash equilibrium* of any game can be defined as the combination of S_1, S_2 such that no player would wish to change their strategy. This is quite apparent in the Prisoner's Dilemma. The Nash equilibrium is (D, D) – if either player decides to co-operate instead, they will see their payoff reduce from 3 to 1. Note this does not maximise the total payoff, which is given by (C, C).

The Prisoner's Dilemma may seem like a contrived example, but the structure of the payoff matrix is common to many circumstances throughout nature. Two organisms can either co-operate or not, and in many circumstances it is safest indeed not to co-operate. Maynard Smith introduced the concept of the *evolutionarily stable strategy* (ESS), which is an extension of the Nash equilibrium concept to populations of players. An ESS is a strategy that, if adopted by a population, cannot be invaded or superseded by another strategy which is rare. In other words, an ESS will remain a commonly used strategy in a population, and the number of players using non-ESS strategies will be suppressed.

As Axelrod (1984) notes, the Prisoner's Dilemma in its simplest formulation ignores an important fact about realistic game-play. The players can play the game multiple times, and retain a memory of the player's previous strategies. In a repeated Prisoner's Dilemma played N times, there are 2^N possible strategies, i.e., combinations of N choices between (C,D).

In fact, rather than considering all these strategies, we can look at three heuristic choices: CC (always co-operate), DD (always defect) and TFT ('tit for tat'). In TFT, the player plays whatever the opponent played in the previous game.

Table 5.2 *The payoff matrix for three possible strategies in the Repeated Prisoner's Dilemma. Prisoners can either always co-operate (CC), always defect (DD) or enact a 'tit for tat' strategy (TFT) where the player plays whatever the opponent played in the previous game. In a population of players, two evolutionarily stable strategies can be played: (DD, DD) and (TFT, TFT).*

Repeated Prisoner's Dilemma	CC	DD	TFT
CC	20, 20	40, 10	22, 19
DD	40, 10	30, 30	30, 30
TFT	19, 22	30, 30	30, 30

If the opponent played C before, the player reciprocates with co-operation. If the opponent defected, the player plays D as a punishment.

We can see from the payoff matrix for this game (Table 5.2) that (DD, DD) remains an evolutionarily stable strategy, but so is (TFT, TFT), and it is slightly more attractive. In this way we can see how simple strategies can build co-operation between interacting players, given enough interactions. Interestingly, it is worth noting that $(CC, TFT) \equiv (CC, CC)$, in that if one player always plays CC, then TFT rules mean the other player must also always play C.

There are three typical phenomena by which co-operation can be built in a population – kin-directed altruism, enforcement and mutual benefit. From a genetic point of view, it is always preferable for organisms to assist their kin against unrelated organisms, as this increases the odds of the individual's genetic heritage transmitting to the next generation. Co-operation can always be enforced by a dominant individual over a supplicant, but persists only due to persisting dominance.

Finally, mutual benefit can cross kin and even species boundaries, setting up symbiotic relationships between organisms. Smith and Szathmary (1997) considers mutual benefit in the form of the 'rowing' game. If two players are in a boat, both players will always benefit from rowing together than not rowing together. The strategy CC then becomes an evolutionarily stable strategy (see Table 5.3).

The rowing game is a good analogy for animal societies where every individual is capable of breeding (e.g., lions, chimpanzees, baboons, humans). Some social groupings consist of a single breeding pair, with a collection of other adults (using siblings of the breeders) that assist in raising offspring. The helpers remain fertile, and can become breeders either by assuming dominant control of the current group, or leaving to establish a new group.

Table 5.3 *The payoff matrix for the 'rowing game'. Two players are rowing a boat. Players can either co-operate (row) or defect (coast).*

Rowing Game	C	D
C	7,7	0,5
D	5,0	7,7

Eusocial societies contain individuals which are sterile non-breeders. As a consequence, these societies naturally produce a caste system of breeders and non-breeders. Offspring are cared for co-operatively. Insects are the most common example of eusocial society – for example, all ants are eusocial. Eusocial societies are also common amongst bees, wasps and termites.

Parasocial societies represent an intermediate step, where organisms gather together for mutual advantage (say a collection of breeding females). A eusocial society can then form if the number of fertile females begins to drop with time.

Eusociality may also evolve directly from the 'helper' social groups as described in the previous paragraph. Depending on social dynamics, helpers may eventually evolve to become permanently sterile.

Precisely why this happens is a source of debate. A commonly held theory is kin-selected altruism. In honey-bee (*Aphis mellifera*) colonies, worker bees are more closely related to eggs laid by the queen bee than those occasionally laid by other workers. Selfish gene theory would predict that a worker will act in the interests of the queen's egg, not the fellow worker's egg.

Kin-selected altruism (or inclusive fitness) theory usually relies on haplodiploidy, a specific method by which the sex of an egg is determined. Males are formed from unfertilised eggs and are haploid, females from fertilised eggs and are diploid. In haplodiploidy, sisters share more genes with their sisters (75%) than they do with the mother (50%). Therefore, it is to an organism's genetic advantage to assist a mother in rearing sisters rather than leaving the nest to raise daughters. Many eusocial organisms are haplodiploid, but many are in fact diplodiploid (e.g., termites). This has encouraged other explanations of the evolution of eusocial societies that move away from inclusive fitness (Nowak et al., 2010).

Social arrangements are a common trend in all metazoan organisms, ranging from the solitary-but-social behaviour of jaguars, to the subsocial behaviour of animals that rear their young, to the parasocial and, at the highest level of social organisation, the eusocial colony. It is a somewhat controversial view that modern humans constitute a eusocial society. Foster and Ratnieks (2005) argue that the 'grandmother effect', where post-menopausal women help their children to rear their own grandchildren, satisfies the non-reproductive helper definition of eusocial societies. Combined with the exceptionally variegated division of labour in human society, and the commonality of caste systems in human culture, there is some evidence to back their argument, although they freely admit that the definition is imperfect.

5.3.9 Social Units to Civilisation

In keeping with our inability to define life and intelligence (see section 6.2), a single accepted definition of civilisation evades us.

Some common criteria for defining a complex society as a civilisation include social stratification, urbanisation and separation from the natural environment, and a means of symbolic communication (e.g., writing).

Let us consider a termite mound. The mound's structure contains a sophisticated air conditioning system, where interior heat circulates air upwards, to capillary-like vents in the upper surface of the mound. The air inside the capillaries cools, and exchanges CO_2 for O_2. As a eusocial colony, termites exhibit social stratification. With several million organisms inside each colony, they live with a population density that humans would easily recognise as urbanised. Termites lack a symbolic communication system, but do possess sophisticated chemical messaging systems, supplemented by vibro-acoustic communication, and visual systems among the higher castes (see Bagnères and Hanus, 2015, and references within).

Just like intelligence, there appears to be a continuum between animal social structures and civilisations. Termites almost completely meet the criteria for a civilisation, despite not possessing even rudimentary self-awareness. We could in fact argue that the colony acts more like a super-organism than a society, with energy consumption, respiration, waste disposal and internal physico-chemical signalling, attributes that all large organisms rely on to function.

If we test a range of animal societies by the above common criteria for civilisation, the presence of a symbolic communication system appears to be the distinguishing factor – language is what makes human social groupings civilisations, and may be the rate-limiting step in ETI development (Solution R.20, R.21). As we will see in upcoming sections, the appearance of language coincides with a significant change in the prospects for human civilisation.

5.3.10 Biology to Post-Biology?

It is commonly argued that advanced civilisations will be *post-biological* – this can either mean that ETIs have discarded biological constraints and have moved their intelligence into mechanical containers, or that ETIs have managed to create true artificial intelligence separate from their own, or perhaps a convergence of both biological and machine intelligence (see Chapter 15).

The motives for such a transition are wide-ranging. For some humans, shedding the weakness and vulnerability of the body is extremely attractive, especially if such a transition secures a form of immortality. Most importantly for SETI scientists, the transition to a post-biological state completely alters the expected signals received from ETI. Perhaps more fundamentally, the definition of 'habitability' will be extremely different, and most likely not conform to any biological standards. We might therefore expect that biological and post-biological civilisations occupy disparate locations in the Milky Way (see Chapter 22).

Indeed, it is now argued by many that post-biological intelligence most likely outnumbers biological intelligence, given that post-biological intelligence will be able to successfully nullify many of the risks that biological systems fall prone to. If that is so, then targeting biologically habitable planets is likely to be the wrong approach for SETI (Shostak, 2010).

Is post-biology a guaranteed end-state for intelligence? Dick (2003) argues for the Intelligence Principle:

The maintenance, improvement and perpetuation of knowledge and intelligence is the central driving force of cultural evolution, and to the extent intelligence can be improved, it will be improved.

In other words, once intelligence evolves, culture immediately accompanies it. As culture evolves, it will find post-biological solutions to biological problems, and hence societies will become increasingly post-biological with time. Therefore, a society with a larger L is more likely to be post-biological. If we expect SETI efforts to be most successful with the most long-lived civilisations, it can then be argued that it is most likely that our first evidence of ETI will be post-biological.

As even Dick notes, assuming the Intelligence Principle requires some large accompanying assumptions, mainly that ETIs are much older than humanity, and that cultural evolution has a goal of increasing intelligence/knowledge (Dick, 2008, 2009). One might make a weak argument for a civilisation increasing its negentropy, and hence its ability to store information, but it is far from iron-clad.

It is also commonly assumed that at some level, sophisticated general artificial intelligences can be built, which is a subject of great dispute (see Chapter 15). Even an intermediate post-biological stage (where human brains reside in machine bodies) makes great assumptive leaps in considering the sustainability of flesh/silicon interfaces.

6
Intelligence Is Rare

6.1 The Evolutionist Argument against Intelligence

The most famous rejoinder from evolutionary biologists against the prevalence of intelligent life (and the SETI endeavour) comes from George Gaylord Simpson. In *The Nonprevalence of Humanoids* (1964), Simpson argues from a probabilistic foundation that the appearance of humans on Earth was a highly contingent event, depending on 'long chains of non-repetitive circumstances'.

As we have seen in the previous chapter, the evolution of life from its beginnings as a collection of self-replicating molecules has acquired multicellularism, sexual reproduction and large metazoan forms with brains, largely through a series of environmental accidents acted on by selection. Simpson argues along the same lines as Stephen Jay Gould, who uses the analogy that one cannot 'rewind the tape', i.e., if the history of Earth was replayed, any slight change to the environment makes a cascading series of changes, resulting in a fundamentally different ecosystem (see also Beatty, 2006).

By this token, we should certainly not expect to see humanoids on other planets, and non-humanoid intelligence is far from guaranteed. It is certainly not an endgoal of evolution – evolution has no goals.

Dobzhansky (1972) makes a careful distinction between the origin of life, and the evolution of intelligence. Primordial life may have appeared many times, on many different planets quite divergent in properties from the Earth, but this is no guarantee that they will produce intelligence. At the time, some 2 million species had been categorised on Earth. Dobzhansky argued that on this limited dataset intelligence is estimably a one in 2 million phenomenon.

As an illustrative (and overconservative) example, he proposed an organism with 1,000 genes, possessing ten possible alleles of each gene. The genomic landscape is composed of some 10^{1000} possible genomes! Consequently, if these organisms were placed on millions of planets, we can infer that only a very small fraction of this genomic landscape would ever be explored. If intelligent organisms occupy a relatively small area of this landscape (which appears to be the case, given Earth only has one technological civilisation), then one can conclude that intelligence is an exceptionally rare, contingent, marvellous phenomenon.

Mayr revisited and expanded these arguments in a series of short articles (1993; 1995), further noting that many civilisations existed on the Earth without radio technology or other attributes visible at interstellar distances. These anti-SETI arguments took root in public thinking, and during this period NASA SETI projects were defunded by the United States Congress.

There are several counter-arguments to these critiques. Simpson's description of physicists/astronomers as evolutionary determinists was a fairly gross generalisation, as Carl Sagan was quick to protest. The search for intelligence does not require that intelligence be humanoid – one thinks of Fred Hoyle's classic story *The Black Cloud*, which considers an intelligent organism that evolved in interstellar space, and not on a planetary surface. Sagan also notes that attempts to define a probability of intelligence $P(I)$ from the number of intelligent species $N(I)$ (and non-intelligent species $N(\neg I)$):

$$P(I) = \frac{N(I)}{N(I) + N(\neg I)}, \tag{6.1}$$

make important assumptions which should be challenged. Either we can extrapolate from our single sample of evolution, in which case we expect from the Principle of Mediocrity that intelligent beings are a relatively common outcome given 'enough' time and evolutionary freedom; or we cannot use our single sample, and hence the above estimation of $P(I)$ is invalid.

An evolutionary counter-argument is championed by Conway Morris (2003a) – that of *convergent evolution*. Across many different phyla, one can observe the same molecular or anatomic tools being developed independently to address environmental challenges. The most famous is the compound eye, which has evolved independently several times in vertebrates, cephalopods and cnidaria (e.g., jellyfish). This convergent evolution is thought to be due to the physics of photodetection, which places heavy constraints on the necessary machinery (Fernald, 2006; Kozmik et al., 2008).

Under convergent evolution, the genomic landscape remains very large, but species tend to follow paths that produce similar results. In other words, multiple genotypes (sections of a genome that produce a macroscopic trait or characteristic) reproduce a specific phenotype (said trait/characteristic). This reduces the vast genomic landscape to a more tightly constrained roadmap (Conway Morris, 2003b).

Conway Morris goes on to argue that if life evolves in environments similar to ours, then evolution will 'invent' similar solutions for organisms to function capably within those environments (for the most ardent expression of this concept, see Flores Martinez's [2014] description of 'Cosmic Convergent Evolution'). Convergence can even apply to social structures, which as we will see later are perhaps a crucial source of selection pressure to produce intelligence.

This reasoning suggests that Earth-like planets produce Earth-like organisms. It is **not** an argument that Earth-like planets produce humans *per se*. At a fundamental level, one cannot avoid the contingent history of our species, as we will see.

6.2 What Is Intelligence?

As with the concept of 'life', the concept of 'intelligence' also frustrates the scientific community. A naive definition, which admittedly persisted over centuries of academic discourse, elevates humanity as the only intelligent species upon the Earth. As we will see in subsequent sections, this definition of intelligence has been largely discarded, and the concept of intelligence being a binary yes/no phenomenon no longer holds credence in the scientific community.

SETI practitioners tend to sidestep this issue by focusing on *technology* rather than intelligence. It seems more likely that intelligent life will be detected by its manipulation of matter and energy into tools and information, rather than its biological footprint, which for the most part is difficult to distinguish from 'primitive life'. In any case, if we wish an unequivocal detection of intelligent life rather than a detection of biological systems, the best discriminant appears to be technology.

SETI's working definition of intelligence, if somewhat unspoken, is therefore 'an entity which constructs technological artefacts potentially visible from interstellar distances'. Given that our first highly visible artefacts (i.e., radio signals) were constructed in the twentieth century, SETI searches would have failed to find human intelligence had it looked at Earth only a few hundred years ago (even the slightly raised CO_2 level and small amounts of pollution from the early Industrial Revolution would have been at best inconclusive).

This definition is clearly unsatisfactory, as it only describes humans as they have been in the very recent past. *Homo sapiens* has a provenance extending back at least 200,000 years, and anatomically these ancestors are indistinguishable from us.

SETI's working definition also has unwelcome implications for non-human intelligence. We see many examples of tool use and language in the animal kingdom (as we will discuss later), and SETI scientists have worked with animals such as dolphins to attempt to bridge communication barriers between species, to hopefully learn how this can be done if a species from another world wishes to speak with us. It is clear that SETI's working definition suits astronomical surveys, but not much else.

With these caveats in mind, we should now consider the only species that satisfies the current criteria SETI applies for intelligence – *homo sapiens*. We will consider humanity's journey to civilisation from this viewpoint, to see what can be gleaned regarding the development of ETIs.

6.3 How Did Humans Become Intelligent?

Modern humans (*Homo sapiens sapiens*) have the distinction of being the last existing species in their genus, *Homo*. During the early days of our species's existence, there were at least six other human species/subspecies roaming the Earth. It is instructive (especially in the context of SETI) to consider why only one 'intelligent' species survives to the present day.

The earliest undisputed evidence for the *Homo* genus dates to around 2 million years ago, from fossil evidence of *Homo erectus* obtained in Africa and Eurasia, suggesting an early crossing from the former to the latter around 1.8–1.9 Myr ago (Klein, 2009). The ancestors of *Homo erectus* – *Homo habilis* and *Homo ergaster* – were present on the African continent some million years prior, along with the still more ancient *Australopithecus* genus, from which the entire *Homo* genus derives (Spoor et al., 2007). It is important to note that bipedalism is already present in *Australopithecus*.

Tracing the exact branching points of speciation is by definition challenging – evolution is a continuous process, and intermediate forms are not always easy to find in fossil records. What can be generally agreed is that the migration of *H. erectus* across the Eurasian continent eventually resulted in speciation into *H. antecessor* (found in Spain), *H. heidelbergensis*, and *H. rhodesiensis*. Other species such as *H. floresiensis*, found in Indonesia, have quite baffling origins (see Figure 6.1).

The exact chain of speciation remains unclear, but it seems the case that *H. heidelbergensis* (and potentially *H. antecessor*) eventually speciated into the subspecies of *Homo sapiens*: *Homo sapiens neanderthalensis*, *Homo sapiens Denisova*, and the only surviving species – *Homo sapiens sapiens*.[1] Extremely low genetic variation in our species compared to other primates sets a lower bound on the gene pool of approximately ten thousand adults. Note that this may not be the initial population size, which can be a good deal larger and still limit the available genetic material. In any case, such a limited stock suggests that our direct ancestors experienced an evolutionary bottleneck. This is most likely due to the migration of *H. sapiens sapiens* from Africa into Eurasia some 100,000 years ago.

This is a particularly convoluted chain of inheritance, sensitive to the geographical terrain that our ancestors attempted to conquer, not to mention the

---

[1] Note that there is disagreement in the anthropological community regarding whether Neanderthals, Denisovans, etc. are a subspecies of *Homo sapiens*, or a distinct species within the *Homo* genus. We will stick with the subspecies convention in this book.

6.3 How Did Humans Become Intelligent?

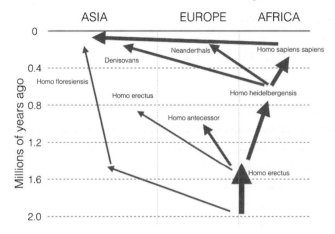

Figure 6.1 A schematic of the evolution and migration of hominid species. *Homo erectus* is known to have spread from Africa through Asia, and speciated into *antecessor* in Spain. *Homo floresiensis*'s origins are unclear, but is found in remote regions of Indonesia. *Homo heidelbergensis*'s location of speciation is also unclear, but spread throughout Africa, southern Europe and southern Asia. *Homo sapiens sapiens* left Africa approximately 60,000 years ago, encountering and interbreeding with already present subspecies of *Homo sapiens*, *Homo sapiens ssp neanderthalensis* and *Homo sapiens ssp denisova*.

hominids already present when a new species arrived. The evidence for interbreeding between the subspecies of *sapiens* is evident in our DNA today, showing what could either be positive or negative interaction between various intelligent species.

We say intelligent in this context, because essentially all the members of the *Homo* genus could be said to be technologically advanced beyond the ability of their cousins – chimpanzees and other members of the *Pan* genus, with a last common ancestor some 4–5 million years ago, and the *gorilla* genus some 2–3 million years before. The patrons of the genus, *Homo erectus*, were using stone tools, and may well have mastered fire (Berna et al., 2012). It was certainly a tool in hominid repertoire by 300–400 kyr ago, and hearths with heated stone and sediment are commonly found in Neanderthal sites (James, 1989; Roebroeks and Villa, 2011).

We can identify some crucial technological and cultural milestones along the road from the beginning of the *Homo* genus to the present day (Figure 6.2). The pace of cultural evolution was initially as slow as that of biological evolution – stone tools made by early *H. erectus* are indistinguishable from those made by Neanderthals tens of thousands of years later. This glacial pace was likely set by the inability of hominids to communicate effectively between generations.

Approximately 70,000 years ago, a fundamental change occurred in the behaviour and remains of humans. The first evidence of language dates from this period, and the cultural inheritance of humanity truly begins. Note that the physical

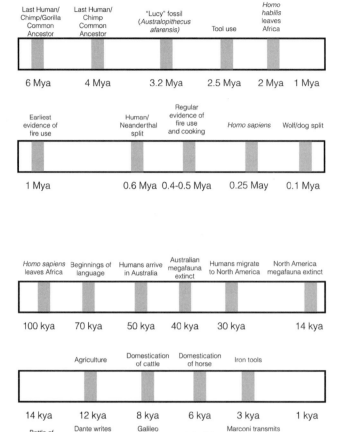

Figure 6.2 A timeline of selected key human milestones, from the last common ancestors with gorillas and chimps to the Renaissance, and the technological achievements that have made humanity visible at interstellar distances.

properties of *Homo sapiens sapiens* had not altered significantly from their first emergence around 250 kyr ago. The brain size or dexterity of these humans would not be particularly different from that of you or me, but yet we can trace an important cultural demarcation, and arguably the beginnings of what we now deem civilisation.

The arrival of humans in each new continent is strongly correlated with the death of so-called 'megafauna', which is strong evidence for humans hunting the latter to extinction. Art, agriculture and domestication are quickly followed by the forging of iron, and the march of civilisation rapidly accelerates through the last 3,000 years to the present day. Note that much of human activity was invisible at interstellar distances. It is generally agreed that the Industrial Revolution is a watershed between an effectively invisible human civilisation and a technological footprint that intelligent species could detect using methods known to humans today.

Agriculture is typically taken as the first indication of what we would now refer to as human civilisation. The earliest signs of a transition from the hunter gatherer lifestyle towards agrarian techniques comes from fossils found in the Levant dating to approximately 9500 BC. Carbonised fig fruits are among the first finds suggesting cultivation. While these fruits are likely from parthenocarpic (seedless) varieties, it is reasonable to believe that they were grown as intentional cuttings along with other crops (Kislev et al., 2006). The transition from foraging to agriculture begins with the appearance of the founder crops (in their wild form) shortly after 9500 BC – emmer and einkorn wheat, barley, lentils, peas, chickpeas, bitter vetch and flax. Plant domestication appears to be complete by at least 7000 BC, judging by evidence of cereal cultivation in Syria (Willcox et al., 2008), although there is some discussion as to how rapid this transition was (see Kennett and Winterhalder, 2006, and references within).

The cause for the initial transition is debated, and likely has different causes and drivers for different regions. Accidental seeding of crops by hunter gatherers during foraging may well have encouraged repeated yields. An interplay of inherited knowledge/behaviour and unwitting artificial selection for better growing strains resulted in crop domestication. Such behaviour may have been exacerbated by climate change. Long dry spells during this critical period favoured annual crops that die, leaving dormant (edible) seeds and tubers.

At the same time, animal domestication continued apace. The earliest success in domestication was the dog. Wolves naturally followed hunters, scavenging kills and finishing off wounded prey. Feeding wolves and rearing pups began a process of artificial selection that culminated in the somewhat bizarre panoply of canine breeds we see today. Cats learned that farming attracted large amounts of mice for them to predate, and soon became firm friends of humans (although with a good deal less modification). Domesticated goats, sheep, cattle and pigs appear in the Fertile Crescent (eastern Turkey, Iraq and southwestern Iran) around 10,000 years ago. Sheep and goats appear to have been first, possibly due to their ability to inhabit more rugged and less fertile terrain. Again, different regions show great variations in timescale for cultivation, which can be explained by biogeography and the availability of specific genetic stock.

Was the change in lifestyle from forager to farmer adaptive? There is some evidence to suggest that it wasn't. Particularly, evidence for reduction in adult height after agriculture suggests that farming reduced the nourishment of human societies rather than increased it. The hunter gatherer diet was never overflowing with calories, but the need to forage encouraged humans to eat a variety of foods, with subsequent nutritional benefits. It could be argued that farmers, who focused their cultivation efforts on a limited set of foodstuffs, invited malnutrition. Farmers also required extra caloric input per calorie derived compared to hunting and gathering, at least in these early days. Finally, the adoption of agriculture usually coincides with an increased incidence of plant and animal-borne diseases, and a reduced life expectancy (see Latham, 2013, and references within).

This argument might lead one to suggest that agriculture is 'the worst mistake in the history of the human race', as Jared Diamond would title his 1987 essay.[2] Of course, the above narrative is highly oversimplified. Not all populations experienced height reduction post agriculture. Diseases were a byproduct of agriculture, but repeated exposure resulted in immunity from the worst. Agriculture may have required more work to achieve the same caloric output as hunting and gathering, but it also resulted in a crucial social innovation – craft specialisation, or division of labour.

The flow of food from agriculture was much more predictable than from foraging. Crops can and do fail in bad years, but in a general sense food production now had a schedule that humans could plan against. This planning freed some individuals from much of the work required to produce the food – in some cases, freeing them completely from such tasks. A well-fed human could continue to forage and bring a diverse range of extra calories back to the group for added benefit, or do something completely different. Craftsmen of objects and ideas are at the centre of all civilisations, and they were now free to emerge thanks to the predictability of agricultural output. Of course, the political structures that grew around agriculture were typically strongly hierarchical, with a small cadre of elites sharing the most valuable crops. For most of humanity's existence, the majority of its 'citizenry' were farming peasants. We will return to the issue of political structures in section 6.7.

Over time, improvements in crop strains resulting in greater seed sizes (see e.g., Kluyver et al., 2017), and labour-saving devices and techniques yielded more and more calories per input calorie. Larger and larger populations of humans could be supported, allowing dense communities to form in fertile regions. In the hunter gatherer era, the world population of humans reached approximately five million. With the advent of agriculture, this population surged, and population densities soared. Hunter gatherer groups, which required large domains to procure sufficient amounts of food, were probably forced out of these grounds by sheer numbers of farmers, who occupied increasingly smaller domains. Immunity to livestock-borne

---

[2] http://discovermagazine.com/1987/may/02-the-worst-mistake-in-the-history-of-the-human-race, retrieved 30/11/2017

disease also conferred an advantage on the farmers – as the numbers of livestock grew, and farmlands encroached on foraging territory, the hunters most likely died of infections from which they had no protection.

Modern civilisations now rely on exceptional variegation of labour, which is only possible if humans are not required to produce their own food. The vast majority of citizens of developed countries are utterly disconnected from food production, and as a result free to produce other goods, services and products for the economy.

Can we argue that craft specialisation is essential to other intelligent species? Thermodynamically speaking, the maintenance of highly ordered structures and the computations associated with thought and action require free energy from the environment. Agriculture has allowed humans to outsource the problem of finding free energy to other agents. We can probably say with some certainty that alien civilisations have also carried out a similar form of outsourcing, the nature of which may resemble agriculture, or may not. It is certainly true that humanity would have struggled to produce the technological advances it has without the 'freedom from survival' that most of us take for granted.

6.3.1 Why Be Intelligent at All?

It is by no means guaranteed that possessing a larger brain (or cerebral cortex) is of advantage to an organism. Fitness is a highly relative concept, depending on the circumstances that an organism finds itself in. In the charmingly titled article *The Evolutionary Advantages of Being Stupid*, Robin (1973) relates how maximal diving time of cetaceans, a key measure of their ability to fish in deep water, is inversely correlated with brain size. Bottle nose porpoises, with a high cerebralisation factor, can only dive for around 5 minutes. The harbour seal, on the other hand, can dive for around 20 minutes, due to its brain/body weight ratio being some ten times smaller.

Anthropologists can identify many aspects of increasing hominid intelligence that were at least maladaptive, if not downright dangerous.

The growing hominid brain was increasingly energy-hungry. The modern human brain is approximately 5% of the body's mass, but consumes around 20% of the total energy produced. In a world where food remained relatively scarce over millenia, it seems rather odd that hominid brains continued to demand more and more of the total energy supply. The regular use of fire around 400 kyr ago, and the invention of cooking, allowed early humans to process consumed food and store energy with less exertion, but this did not wholly cancel out the rapacious demands of the brain.

The continuing growth of the brain necessitated a similar growth in the human skull. Without a corresponding growth in the birth canal, the risk of maternal death in childbirth increased. The attendant need for infant humans to undergo considerable training in the early years of their existence also placed increasing burdens on human parents.

All these factors might suggest that the growing intelligence of hominids went (at least in part) against the grain of natural selection (Solution R.18). A pessimistic view might suggest that of the half-dozen or so intelligent hominids that have existed on the Earth, only one has survived, and hence intelligence only results in limited improvement to a species's fitness.

Of course, other hominids are no longer extant primarily due to the domination of *Homo sapiens sapiens* over its cousins, but it speaks to the fact that intelligence is not a necessary goal or endpoint of evolution. According to natural selection, we should expect that maladaptive traits are eventually removed from a population, unless they are correlated with adaptive traits (i.e., the corresponding alleles are in linkage disequilibrium), or (for example) they are a consequence of low population counts and gene drift. We must therefore either conclude that:

1. The particular flavour of *Homo sapiens sapiens* intelligence was highly adaptive, or
2. Intelligence is a *spandrel* – a trait that is the byproduct of the evolution of another characteristic.

The success of humans in dominating the Earth and its biosphere suggests that the first option must eventually have been the case. However, there may well have been protracted periods where intelligence, or aspects of intelligence were spandrels (see, e.g., Pinker, 1999, for examples).

6.4 What Is the Role of Language?

Whatever our definition of intelligence or civilisation, communication between individuals appears to be fundamental to their development. An encoded message from an ETI will presumably contain some form of syntax or grammar, and as such we should think carefully about language, both in its role as a mediator between individuals in a species, as well as in its possible role as a mediator between species.

Note that throughout we use a specific definition of language as

a purely human and non-instinctive method of communicating ideas, emotions and desires by means of voluntarily produced symbols.

(Sapir, 1921)

6.4.1 The Origin of Human Language

A satisfactory theory that describes the origin of human language must answer two key questions:

1. How and why did humans, and not other species, evolve language?
2. How and why did language evolve with its current properties, and not others?

The reader will immediately notice that the word 'language' as it is used above implicitly refers to human communication, and not the other forms of communication we observe in non-human species. Before addressing these key questions, we should therefore elucidate the properties of human language that make it unique in the biosphere.

A set of properties of human language that can be universally agreed upon is difficult to achieve. According to Bouchard (2013), there are two distinctive properties that can draw good consensus amongst linguists. The first is that human language uses *Saussurean signs*. A 'sign' can take the form of a word, image, sound, odour, gesture, act, flavour or object. These things become 'signs' when they are invested with meaning. In other words, a sign signifies something other than itself.

The definition and properties of these signs are the province of *semiotics*. The Saussurean model of a sign is one of several possible models for the construction of a sign. In this paradigm, the sign is composed of the *signifier* (one of the forms above that the sign takes), and the *signified* (the concept to which the sign refers). There is a philosophical distinction between the *signifier* as a material entity (for example the sound waveform registered by the human ear as 'door') and a psychological impression (the brain's translation of the waveform to the listener). It is common in the literature to consider the signifier as both material and psychological, although Saussure's definition was emphatic that the psychological impression was the only entity of the signifier.

The second property of human language generally agreed upon is that human language is *recursive*. The definition of recursive in this context is specific to linguistics, and doesn't neatly map into concepts of recursion in mathematics. It is best described with an example. The following sentences show how a truly recursive sentence can be constructed:

- The cat meowed
- The cat [that the dog chased] meowed
- The cat [that the dog [that the man hit] chased] meowed.

In the above, we use brackets to outline how the structure is embedded recursively. In principle, this embedding could carry on *ad infinitum* – in practice, working memory prevents us from parsing sentences much longer than this with clarity, and multiple sentences become the norm.

These two basic properties – a symbolic system that establishes some core semantics, and a means of generating infinite combinations of symbols – are sufficient to express a functionally infinite number of concepts, scenarios, emotions, or whatever is required by a human. We should now ask three questions, rather than two:

1. How and why did humans, and not other species, evolve language?
2. How and why did language evolve Saussurean signs?
3. How and why did language evolve recursion?

There are two broad camps in the theory of language's origin. In one camp, language is a product of cultural evolution, and in the other language has a biological, perhaps genetic origin.

Possible Cultural Origins of Language

It is quite obvious to modern humans, as highly practised and effective communicators of human language, that it is a tool of extremely high evolutionary utility. Effective communication allows humans to perform tasks more efficiently. Early humans with improved communication skills would experience more success in hunting and foraging. Sharing solutions to complex problems gives the fortunes of any social group a tangible boost.

Individuals within a social group will also experience a boost in their fortunes. Navigating the social landscape to one's own benefit (what Bouchard, 2013, calls 'the Machiavellian intelligence in social life') clearly has benefits for one's genetic inheritance.

Dunbar (1998) gives a more precise definition of language as a form of vocal grooming. Manual grooming is common amongst primate groups, and facilitates the build-up of trust and maintains alliances. As groups grow in size, manual grooming becomes less and less feasible – vocal grooming provides a useful substitute, albeit one less trusted by potential breeding partners (Power, 1998). The interplay between increasingly complex vocal grooming, modified and mediated by the strategies of females in their selection of partners based on the information carried by vocal grooming, drove the evolution of grooming into symbolic speech.

Language may well have been a key parameter for females in mate selection (Bickerton, 1998), and likely played a role in child-rearing as a tool for soothing children when physical contact may not have been possible (Falk, 2004).

A neo-Darwinian model of language proposes that cultural replicators, like biological replicators, are the principal agents of evolutionary change, and that the presence of these replicators (*memes*) drove their representation in symbolic form at an early stage. The interaction between replicators and their symbols is ultimately responsible for the appearance of advanced syntax and language at a later date.

A common criticism of cultural origins of language is that for language to have utility to early humans, it must first *exist*. Most of the advantages described above require some form of language to already be present. If that condition is satisfied, then these are plausible explanations for why a proto-language continued to grow in complexity from simple starting conditions, to satisfy the growing needs of larger and larger social groups of humans. Also, arguments that suggest language is a product of sexual selection must address the fact that very young, prepubescent humans are highly effective communicators, long before they enter the reproductive domain of puberty.

We should also acknowledge the large communicative potential of expressions absent of language. It is said (cf. Mehrabian, 2017) that when it comes to the expressions of feelings and attitudes, approximately 7% of the transmitted meaning uses language, with the rest being conveyed by paralinguistic features such as tone, and other means such as gestures, facial contortions or other bodily functions such as crying or laughter.

Finally, most of the above solutions make no predictions as to the nature of human language, merely that it be sufficiently complex to represent our minds, the minds of our companions, and the world around us.

From a SETI point of view, this may not be a particular disadvantage – perhaps alien languages are not predicated on Saussurean signs and recursion. Language in the Milky Way may be drawn from a distribution of various properties. If we ever manage to collect a sample of alien languages to compare with, we may even be able to use an *anthropic principle* to argue that we observe human language to be this way because it is we who are present to observe it.

6.4.2 Possible Biological Origins of Language

In this category, the early use of signs is believed to have contributed in some sense to the natural selection of our ancestors. These theories are often referred to under a general banner of *Universal Grammar* (UG) – the concept that language is an innate function of humans, and distinct from all other aspects of human cognition. In this model, language is another complex organ, much like the compound eye, but lacking obvious physical representation. Both the seeing organ and the language organ evolved through a series of intermediate stages, with each stage conferring an advantage to its host and being transmitted through genetic inheritance.

Depending on the linguist/neurolinguist, there is some debate as to the actual nature of the language organ. In the strongest form of UG, the language organ is 'real' in the sense that it is a specific region of the brain that runs the 'language software'. This is commonly identified as Broca's area, and is built on studies of patients with brain damage localised to this area. Other theories of UG posit that the language organ is actually a distribution of brain areas. In its weakest form, language is not particularly localised in the brain, and the organ is effectively 'virtual'.

Regardless of the form of the organ – real, virtual or some hybridisation of hardware and software – the biological origin of language implies that symbolic communication had selective effects on our ancestors. In a form precisely analogous to the compound eye, humans began with primitive communicative tools (sometimes called 'protolanguage'), and then built upon this in discrete, minute increments. Broadly speaking, we can identify a few key elements of the language organ, which in rough chronological order would begin with discrete symbolic signals, which are essential to all human languages. As the lexicon of symbols grows, with an eventual innovation that symbols themselves can be concatenated to produce new, more refined symbols, the protolanguage develops symbols which describe relationships ('on', 'in front', 'better than'). The beginnings of grammar include the separation of symbols into semantic categories (noun, verb), then a deeper syntactic complex.

Each of these innovations arrives as the result of incremental changes to both software and hardware. In the same way that a more effective visual organ has obvious benefits to the organism, a more effective language organ allowed our ancestors to navigate both their physical and social landscapes with increasing success. As human social structures grew more complex, language becomes a crucial organ for acceptance and survival, just as efficient locomotive organs and sensory acuity are crucial for being able to quickly escape predators.

The sequence by which language evolved is still debated – Pinker and Jackendoff (2005) espouse a gradualist view of acquisition, beginning from a protolanguage composed of symbolic signs and adding functionality slowly. Bickerton is most famous for adopting a two-stage view, where the protolanguage, consisting of nouns and verbs placed in a structureless fashion, persists for a long epoch as the brain undergoes a series of pre-adaptations. Once the brain is primed to receive it, syntax is suddenly injected into the language. Bickerton's principal argument for the two-stage theory is that selection pressure for symbols differs from the selection pressure for syntax.

Selection pressure clearly is important in this theory, as we can look to the bonobos, who exhibit highly complex social structures without the need for symbols. Hominids were clearly more put-upon to develop signs due mainly to the contingency of food supplies. Hunter-gatherers, working in rapidly growing and dissolving groups to find food that was:

- spread over a large geographical area,
- from a wide range of sources, each of which were subject to failure,

would have placed increased selection pressure on hominids to exchange information quickly about the local food supply (not to mention possible danger).

The frailties of the niche could also have encouraged other traits, such as *anticipatory cognition* (i.e., planning for the future).

The selective forces that drove the creation of syntax are less clear. After all, a hominid that communicates complex syntactic statements can only communicate with other syntactically advantaged hominids, which in the initial phases of its evolution are likely to be few and far between. Chomsky (2005, 2007) argued for syntax as a means of organising thought, which will deliver advantages regardless if one cannot be understood by anyone else.

6.4.3 What Can We Say About the Language of ETIs?

The emergence of language in hominids seems to have been driven either by natural selection, cultural selection, or a bootstrapping of both, with cultural selection growing in importance as language itself became more developed. Equally, we might even argue that language itself had a backreaction on human intelligence. Being able to express a complex concept allows its refinement, especially with an interlocutor. The call and response of language would refine our ancestors' *theory of mind*, i.e., their construction of the interlocutor's thought processes.

If language in hominids emerged as the result of a set of selection pressures, it seems sensible to begin with this assumption for the communicative structures of all ETIs. The structure of any alien language will depend very strongly on the environment it was forged in. Human language is highly contingent on the social structures that humans prefer to make, as well as the relationships between humans and their environment.

The structure of human languages permits a functionally infinite number of expressible statements with a relatively limited vocabulary. It seems safe to expect that any alien language is likely to share this property. One might even suggest that alien languages will consist of a finite number of signs/symbols, and relationships between those symbols. Without an example of a *bona fide* alien language to study, we cannot be certain. At best, we can extrapolate from our understanding of non-human communication.

6.4.4 Non-Human Communication

At a basic level, communication is an essential component of mating for many species. Courtship rituals communicate the male's willingness to mate with a female, and the quality of the ritual (amongst other factors) determines the female's odds of mating with said male.

Such a ritual cannot really be said to be language. We can explain the ritual from its beginning to end as a series of stimuli, and responses to stimuli. One's actions modify the behaviour of the other.

True communication should not only modify the behaviour of your circumlocutor, but modify their understanding. When two humans communicate, one human is activating a representation in the other human's mind. For example, by describing an object, the listener forms an internal representation of the object, guided by the speaker's description.

If we now discuss, for example, the calls of vervet monkeys, are we seeing animals generating representations in each other's minds? Vervets possess three acoustically differing alarm calls, each for a different predator. If a vervet hears an alarm call corresponding to 'leopard', it will run into the trees. If the alarm call corresponds to 'eagle', it will run for cover in the bushes.

Is this an example of a simple response to stimuli (hear a specific noise, go to the trees), or of genuine representation (hear a specific noise, realise a leopard is coming, and act accordingly)? Experiments seem to indicate the former more than the latter (Shettleworth, 2010, Chapter 12).

Bottlenose dolphins (*Tursiops* genus) use a series of burst sound pulses, clicks and whistles to communicate with each other. Each dolphin possesses a signature whistle to identify itself. These whistles are honed over time with learning, and tuned to match those of their parent community (Fripp et al., 2005). Dolphins can remember the signatures of fellow dolphins for decades. While there is no defined dolphin language, approximately 100 sounds are used in context-specific environments, i.e., during hunts or mating. The frequency of signature whistling correlates with mating itself, or stress depending on the environment. Copying another dolphin's signature whistle is correlated with strengthening of social bonds (King et al., 2013).

Wild chimpanzees do not seem to possess such clear auditory communicative skills, but show a rich pattern of gestural communication, with at least 66 different gestures (Hobaiter and Byrne, 2011). Intriguingly, some 24 of these gestures are also common to gorillas and orangutans. These are used in a highly intentional manner, suggesting a possibility of more advanced communication (which has been attempted on several occasions; see the next section).

Other communication modes (pheromone signalling in ants and termites, colour changes in cuttlefish, waggle dances in bees) do not appear to be more than stimulus and response, but these are presumably valid modes for more advanced communication (given appropriate time and selection pressure).

In short, we may expect ETI's communication to be based on a symbolic communication system very similar to our own. However, our symbols are derived from an oral tradition, and the nature of our oral tradition has shaped our symbolism. An ETI with a more cetacean ancestry may encode its bursts, whistles and clicks in a very different system from our writing. Indeed, our first message from an intelligent species may be an electromagnetic encoding of pheromone smells!

Attempts at Communicating with Non-Human Intelligences

Primate Communication   Since the 1960s, various researchers have attempted to teach chimpanzees American Sign Language (ASL). This focus on sign language was influenced by earlier attempts at teaching chimps human vocalisation, and the growing realisation that chimpanzees have vocal apparatus unsuitable for human-like sound generation. In 1965, Washoe, a female chimpanzee, was selected for the first attempts to teach ASL. Washoe was a captured chimpanzee, originally for use by the United States Air Force as part of the space program.

Gardner and Gardner (1969) took possession of Washoe, attempting to raise her in an environment as analogous to that of a human infant as is possible with a semi-wild animal. A range of human companions (fluent in ASL) partook in various games and play to encourage Washoe to sign back. Any vocal communication that Washoe could not imitate was forbidden in Washoe's presence. Through a form of gestural 'babbling', and guided 'moulding' of Washoe's hands to form the correct gestures, she eventually began to consistently sign, with her signs eventually evolving into distinctive ASL signs, which included nouns such as 'hat' and 'toothbrush', as well as more abstract signs like 'sorry', 'clean' and 'in/out'. In the first 22 months, Washoe consistently and spontaneously used 30 signs without explicit prompting.

Follow-up experiments with multiple chimps show that Washoe was not by any means exceptional in her ability to use ASL (Fouts, 1973). What remains hotly debated in the literature is whether Washoe and her compatriots were really using ASL as a language, or that the chimps were learning a set of stimuli to generate a response.

The 'Nim Chimpsky' experiment, designed under the philosophy that chimps raised as humans would act as humans (given their significant shared DNA heritage), held a higher standard for language acquisition. While Nim was able to sign 125 ASL gestures, the team behind the project were more concerned with the sequences of signs that Nim generated. If Nim could generate sentences and display an understanding of grammar, then this would be harder evidence that chimp ASL was indeed language, and not simply a 'clever trick' to yield a reward. Their analysis showed Nim's number of signs per sentence failing to grow with vocabulary, a common trait in human infants, and no appreciable lexical structure indicating grammar (Terrace et al., 1979). Critics have also suggested Nim was relying on subtle non-verbal cues from enthusiastic experimenters to generate the correct stimulus for a treat (the so-called 'Clever Hans' bias).

Koko the gorilla is perhaps the most famous of primate signers (Patterson, 1978). Koko eventually learned over 1,000 signs, and most interestingly has shown an ability to communicate regarding objects not present. Koko's reported ability to sign about signing (meta-language), and invention of new signs by combination (ring = finger-bracelet[3]) are more encouraging indications of a true linguistic capacity.

The academic community is divided on whether primates can truly acquire and use a language to communicate. Both sides cite problems in each other's methodologies (see, e.g., Patterson, 1981), and it is clear that humans *want* to be able to speak to primates. As our closest cousins, a conversation with them would yield fascinating and important insights into ourselves.

Dolphin Communication   Delphinids score highly on many indicators of intelligence, from their physical attributes to their observed behaviour. Human–dolphin communication is of course hampered by their very distinct physiologies, and the environments they inhabit.

In the late 1950s, John Lilly had observed that dolphins in prolonged contact with humans tended to increasingly vocalise above water. Captives then began to emulate human vocalisations, albeit very loudly and at relatively high speed. Lilly (1962) describes an example where a subject, Elvar, was deliberately spraying a human colleague with water. Upon being told loudly to 'stop it', Elvar responded with a high-pitched, two-phrase tone which, when slowed down, appeared to say 'OK'. Later discussions encouraged Elvar to slow down and lower his frequencies when vocalising.

Buoyed by these experiments, he published the very successful popular science book *Man and Dolphin* (1961), and attended the famous Green Bank Meeting in 1961 that saw the genesis of the Drake Equation. The attendants of the meeting were known as the Order of the Dolphin, as a nod to Lilly's work, which at the time was the closest that humans had ever come to linguistic communication with another (reasonably intelligent) species.

Lilly arranged for a house on the island of St Thomas to be partially flooded so that a teacher, his research associate Margaret Lovatt, and three dolphin students, Peter, Pamela and Sissy, could live and play together. Lilly and Lovatt took different teaching approaches, but both saw the dynamic between Lovatt and the students as a mother–child relationship.

Peter, being a virile young male recently come to sexual maturity, did not see the relationship the same way. To attempt to circumvent these distractions to his learning, Lovatt would often 'relieve these urges manually'.[4]

However, Lilly's experimentation with LSD and the chaotic nature of the 'Dolphinarium' ultimately derailed the experiment. Lilly's legacy in this area lives on

---

[3] http://www.koko.org/progress-plans
[4] www.theguardian.com/environment/2014/jun/08/the-dolphin-who-loved-me

through his students and collaborators, and the ability for dolphins to mimic human speech is thoroughly documented (Clarke, 2014).

But does dolphin vocalisation entail linguistic ability? And are we able to speak Dolphin, as Carl Sagan hoped?

Herman et al. (1984) describes how two dolphins, Phoenix and Akeakamai, were trained in artificially constructed acoustic and visual languages respectively. Once trained, both dolphins showed evidence of comprehension of sentences, both in their semantic and syntactic structures. Wild dolphins, while not demonstrated to be using or understanding syntax, use long sequences of whistles and clicks. Recently, Ryabov (2016) claimed to have observed two communicating dolphins emanating a series of these clicks without interrupting each other, and suggested this is a conversation without overlap, a prerequisite of rudimentary language. However, this work's methodology and conclusions are widely criticised by experts in the field (the reader is advised to turn to Herzing and Johnson, 2015, chapter 5 and references within for more details).

These developments give a small amount of hope that humans can communicate with dolphins in a manner they understand, even if dolphins do not themselves possess an intraspecies communication that we would define as language. Given that humans and dolphins have been co-operating for many hundreds of years in fishing, where dolphins drive fish into waiting nets for a share of the profits (Zappes et al., 2011), dolphins remain one of the best prospects for linguistic communication (if somewhat one-sided) with another species.

6.5 What Is the Role of Art?

Some 17,000 years ago, the Lascaux caves in the department of Dordogne in southwestern France would become the canvas for some of the most striking prehistoric art yet discovered. Their discovery in 1945 would thrill and stun modern humans, attracting such crowds that the caves were finally closed in 1963 due to the damaging contaminants brought in by spectators.

Since Lascaux, approximately 350 European Ice Age cave art sites have been discovered, with the earliest (Chauvet) stretching back some 30,000 years. Common themes displayed in each of these sites include handprints (adult and juvenile) and animal figures (either drawn figures via lines, or brought out by a series of dots). The age of these art sites assures us that we are anatomically identical to the artists. Through careful study of various sites, we can also identify the idiosyncracies of individual artists, and even track an evolution in the forms used.

A neat example of this is the 'subjective contouring' phenomenon, where we see a closed boundary in a drawing, despite the lines used not being in any way connected (a simple example is shown in Figure 6.3). The art at La Pasiega shows

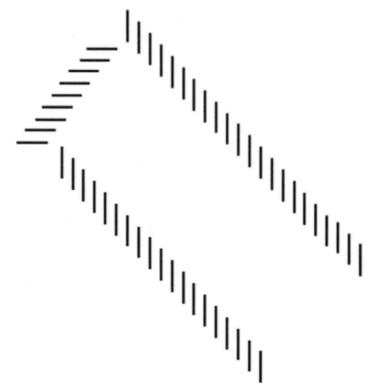

Figure 6.3 Subjective contouring. Despite none of the lines touching each other, the human brain defines an implicit boundary interior to the lines.

a bison with head and neck etched using this subjective contouring technique. This highlights the crucial interactions between hand, eye and brain in the construction of art. The neuroscience of visual perception encourages the use of certain forms, and discourages the use of others.

Cave art is the most striking of the prehistoric cultural ouevre that is handed down to us, and its age indicates that the artist was *Homo sapiens sapiens*, but it is nowhere near the oldest form of art on record. So-called 'portable art' (engraved tools and jewellery) is less well-preserved than a sealed cave, but occasional pieces do survive; for example, the recent discovery (Joordens et al., 2014) of geometrically engraved shells, with $^{40}Ar/^{39}Ar$ dating indicating an age of approximately half a million years. This would suggest that *Homo erectus* take the credit for this craftwork, although at the time of writing, it is the only evidence that *erectus* possessed the cognitive faculty and neuromotor skills to carry out such delicate carving.

Ancient carving and painting gradually takes on a simple syntax, centred around the two concepts of symmetry and repetition. Bilateral symmetry can be found in numerous examples of primitive art. This symmetry is probably inherited from the symmetry of flint hand axes that hominids had been producing (via *knapping*) for at least a million years beforehand. In addition to this, bilateral symmetry is found throughout the body plans of a vast number of biological organisms, including hominids. The desire to reproduce our own symmetries in art appears eminently understandable. The careful incorporation of bivalve fossils in a symmetric fashion in hand axes found in England dating from some 250,000 years ago demonstrates a careful resolve on the part of the hominid knapping the stone (McNamara, 2010).

It is interesting to note that the concept of spatial symmetry is rather straightforward to program into a computer, despite the apparent fact that artificial intelligences are currently incapable of 'judging' art. The perceptual shortcuts offered by symmetric forms have clear advantages to a primate processing visual input from its environment to avoid danger. Perhaps as a result, the modern human continues to appreciate the simplicity of symmetric forms, both in art and in science.

Repetition (and variations of the form) are a key, perhaps ritualistic aspect of early art. In a similar fashion to symmetry, repetition provides an intellectual fast-track to visual perception. A visual input that the brain has received before can be processed quicker than an entirely new input. It is likely that the craving for symmetry stems from this procedure, by allowing the brain to comprehend half the image and then quickly construct the other half from previously processed data. This processing is at the heart of many persistent optical illusions (Sugihara, 2016).

Where Did Human Art Come From? Why do humans engage in art? There is a clear physiological response in humans exposed to art, with marked changes in blood pressure, heart rate, perspiration and even brain patterns (Zeki, 1999). The aesthetically pleasing effect of sunsets seems to be common across most of the human species, with some evidence that even chimpanzees share this aesthetism (as reported by Adriann Kortlandt; see page 101 of Martinelli, 2010). On recovery of sight, blind people commonly describe the sudden perception of colour as a visceral experience. Sacks (1993) relates one account, who was quoted as saying: 'I felt the violence of these sensations like a blow to the head'.

There has been a recent resurgence in attempts to describe the emergence of art in hominids as an evolutionary consequence, perhaps deriving from the play behaviour seen in animals (which itself is evidentially adaptive).

The argument for art itself being adaptive (i.e., having a specifiable evolutionary function) is most forcefully given by Boyd (Boyd and Richerson, 1988; Boyd, 2010). Connecting nonhuman play directly to human art, they draw a parallel – like play, art is behavioural practice. In juvenile animals, play is a training ground for physical techniques and behavioural structures that, if practised, have clear benefits for the adult. Boyd argues that art is itself a training ground for many behaviours that prove useful in human society, for example the ability to communicate concepts in a detailed, relatively sensitive manner, or the ability to conform to expectations (in the cave art example, being able to draw one of the greatly repeated forms, such as the bison or aurochs). The argument, like many arguments in evolutionary psychology, rests on three properties:

1. 'Good Design' or 'Tight Fit' – the properties of the adaptation are well matched to its proposed function;
2. 'Universality' – The adaptation takes hold in all members of the species;
3. 'Developmental Reliability' – the adaptation develops with good fidelity in the organism, either as a juvenile or during maturity.

This argument does not guarantee art is in any way adaptive, and could indeed be a spandrel. Verpooten (2015) gives the example of stone-handling Japanese macaques, which engage in quite specific handling and 'clacking' behaviour with no apparent advantage to their social status or reproductive prospects. Also, identifying an evolutionary function does not guarantee a trait is adaptive. Beneficial traits can arise without having been selected for it – this process is known as *exaptation*. An example of an exaptive trait is the ability to drive a car – it is demonstrable that humans have not been naturally selected for driving, but being able to drive has clear benefits to humans in modern society.

The universality of art seems like a convincing argument for natural selection, but its implication is that there is an 'artistic impulse', with genetic predispositions we could isolate. This does not sit well with the evidence, and it also rules out any contribution from cultural evolution, which is clearly important.

The matter of music may be a different story. Singing is a trait well established throughout the animal kingdom, and animal song is highly communicative. The barrier between language and music is highly blurred – hominids evolved in a soundscape of natural rhythmic noise and animal song. As communication between humans grew in sophistication, the ability to reproduce these noises (what Brown, 2000, calls 'musilanguage') was likely to have benefits (in the same way that faithfully representing animals on cave walls would have). Like art, music is a social glue that transcends biological, intellectual and cultural differences, and may have even participated in sexual selection (see, e.g., Cross and Morley, 2009, for a review).

Finally, the reliability of development of artistic endeavour is perhaps questionable. It is true that pretend play is universal in humans (and many other species), but to what extent do parents encourage the creative impulse in their children? Is art nature or nurture?

The origins of hominid art are bound indelibly with the origins of hominid language. In their primeval forms, both are focused strongly on symbols, and it is not inconceivable that language, art and music share a common origin in the string of adaptations the hominid brain underwent several million years ago. The feedback between creating abstractions of the physical world – through drawings, songs and words – and then relating those abstractions back to the world is likely to have resulted in some form of bootstrapping, where one activity refines the other, building increasingly sophisticated mental representations in the brains of our ancestors.

Our understanding of art, music, language and the connections between suggests that human intelligence owes its existence to the confluence of these activities. We might even say that this is a universal trend for technological civilisations generally, but we cannot be certain. As we continue to uncover the evolution of art, music and language in early hominids, our speculation will become better informed.

6.6 What Is the Role of Science and Mathematics?

Traditionally, SETI practitioners are physical scientists by training. While a bias cannot fail to emerge from a group of individuals thinking along relatively narrow disciplinary lines (Wright and Oman-Reagan, 2017), it still seems reasonably sensible to believe that technology builders must possess a method for acquiring and rationalising knowledge, which we know as science.

The construction of artificial structures relies at a basic level on an understanding of geometry. Even before Newton's laws were fully understood, the required geometric properties of a building have been part of human knowledge for several thousand years. The construction of a successfully standing pyramid may have required several failed attempts, but each failure develops a better understanding in the mind of the architect.

It is safe to say that societies can only advance when their systems of knowledge are well curated and efficiently stored. Indeed, a well-organised complexity is probably the principal driver of problem-solving and technological advances (Tainter, 1988).

Mathematics is an attractive body of knowledge, as it offers the ability to derive theorems that are permanently true. They are also true regardless of location in the Universe. Even if words and concepts are difficult to exchange, it is likely to be the case that when we compare our mathematics to an alien's, we will find common ground. This concept has driven the formation of synthetic languages like Lincos (Freudenthal, 1960).

A scientific civilisation is likely to have measured the fundamental physical constants of the Universe. Of course, it is unlikely that ETIs will use the same physical units as we do, but it is possible to exchange dimensionless transcendental numbers such as $\pi = 3.141\ldots$ or $e = 2.718\ldots$, which will be a key moment of cultural exchange between any two ETIs. A common agreement on a set of physical units will be crucial. DeVito (1991) proposes defining the periodic table of elements (beginning with hydrogen) as a first step. By then defining a set of simple chemical reactions, the Kelvin and the calorie can be defined. This relies on ETIs agreeing with our perspective of chemical elements as an atomic nucleus plus electron cloud, which may not be guaranteed.

The physical laws an ETI might write down will be mathematically equivalent to ours, but potentially structured in a very different format. One could imagine, for example, Newton's Laws, which deal principally with inertial mass, acceleration and force, being recast entirely in terms of momentum, or in the motion of a system's centre of mass. In this example, all three versions are relatively easy to deduce from each other, but this will not generally be the case if we encounter the scientific knowledge of ETIs.

6.7 What Is the Role of Political Structures?

It is a common tenet of SETI to accept that if ETIs are explorers, then they must be scientists, as scientific progress is the driver of human exploration. This ignores the other factors that have driven exploration throughout human history. The voyage of Darwin was facilitated by the British Royal Navy, an organisation forged in combat and the political machinations of the European continent in the three hundred years prior.

The role of exploration in accelerating political progress is relatively modern. One could date its genesis to Columbus's voyage in 1492. The maps before that period did not contain blanks – humans were content with the horizon as it was. Columbus may not have intended to discover a new continent, but in doing so it became clear that beyond the horizon were riches and political power. The Spanish conquests of South America were driven by the desire for gold and other precious resources. Scientists became attached to subsequent exploration missions because they could improve the fleet's navigation, learn and appropriate the culture of indigenous peoples, and identify promising resources. We cannot deny that our understanding of ancient Egypt owes a great deal to the French, who were keen to preserve and understand the country's historical treasures just as Napoleon was keen to subjugate its people.

Perhaps the most important political structure is money. The concept of money is a crucial component in allowing a flexible society to operate. Complex social structures depend on division of labour, and division of labour is successful only if the fruits of one's labour can be exchanged for the fruits of another's. Monetary systems provide the means for that exchange.

6.7 What Is the Role of Political Structures?

Scientific research is a form of labour that requires significant resources. It therefore depends on acquiring funding, either through investment by governments or through monetising its own products. Therefore, scientific research and knowledge acquisition is constrained by political will.

We should expect ETIs to also be bound by similar cost constraints, which can be traced back to thermodynamic principles on energy use. A soft solution to Fermi's Paradox is 'alien politicians restrict resources for scientific exploration and communication'. A message is entirely possible, but it is politically untenable to send one.

Part III
Catastrophist Solutions

We now consider a Milky Way teeming with life, including intelligent life. To reconcile this view with our observations of the Galaxy, we must place the Great Filter into our future, and consider how our civilisation's future may be cut short.

There are two subclasses of solution: either a death by natural causes, or an act of self-destruction. We will deal with accidental death in the beginning of this part. Chapters 12 onwards describe self-destructive solutions to the Paradox.

7
Doomsday Arguments

Before beginning to consider specific catastrophes, it is instructive to consider some arguments that attempt to show humanity has a short shelf-life from probabilistic, Bayesian approaches.

7.1 The Carter-Leslie Argument

Consider the following thought experiment (Carter and McCrea, 1983). The total number of human beings that have ever lived is quite large (around 60 billion, Olum, 2000), but the recent explosion in human population (see Chapter 12) means that, selecting a human at random, you have around a 10% chance of picking a human being who is alive today.

Imagine now that one of two possibilities exist for humanity, which we will call S and L respectively. The human race is either short lived (and the total number of humans ever will be $N_S = 200$ billion), or the human race is very long lived (and the final count will be $N_L = 200$ trillion). In the absence of any data about which version of humanity's future will come to pass, we can assign prior probabilities $P(S)$ and $P(L)$ to both outcomes.

But we are not completely lacking in data – we know that we are the Nth human to have graced the Earth, where $N = 6 \times 10^{10}$. We can use Bayes's Theorem to obtain an updated a posteriori probability. For example, in the case of hypothesis S:

$$P(S|N) = \frac{P(N|S)P(S)}{P(N)} \qquad (7.1)$$

And similarly for hypothesis L. It is more useful to compute a ratio of the two probabilities:

$$\frac{P(L|N)}{P(S|N)} = \frac{P(N|L)P(L)}{P(N|S)P(S)} \qquad (7.2)$$

155

If we assume that we could have been born at any point in human history, then the likelihood functions are simply uniform over all the possible values of N, i.e., $P(N|L) = 1/N_L$ and $P(N|S) = 1/N_S$. Let us assign $R = N_L/N_S = 1000$, and hence

$$\frac{P(L|N)}{P(S|N)} = \frac{1}{R}\frac{P(L)}{P(S)} \qquad (7.3)$$

This means that unless our a priori probability of being long-lived compared to being short-lived is large, then it appears that we are more likely to be a short-lived civilisation than a long-lived one.

Gerig et al. (2013) extend Carter and Nielsen's argument to the set of all possible existing civilisations. In what they refer to as *the universal Doomsday argument*, they consider the fraction of all civilisations that are either long-lived (f_L) or short-lived ($1 - f_L$). Again, given our position N in the human race, we can update our prior probability $P(f_L)$ to obtain the posterior probability $P(f_L|N)$. They show that the likelihood function is

$$P(N|f_L) = \frac{1}{1 + f_L(R-1)} \qquad (7.4)$$

We can immediately see that this likelihood function is inversely proportional to f_L, and therefore we might expect the posterior probability to also be similarly proportional. This analysis suggests that short-lived civilisations are more common than long-lived civilisations.

These types of argument are easily countered. Gerig et al. (2013) note that the details of their results are very sensitive to the choice of prior. If we return to the original Doomsday argument, a natural first approach is to assign uniform priors to both outcomes, which would mean in this case that $P(L) = P(S) = \frac{1}{2}$. This reflects a high level of ignorance – we may be unwilling to say that one outcome is more likely than the other in the light of such little information. This prior selection clearly validates the argument that we shall be short lived.

However, a more appropriate prior would reflect the fact that we are more likely to be born into a more numerous civilisation than a less numerous one. In other words, the probability that I exist, given that I am in scenario s, is

$$P(I|s) = P(s)N_{total}(s) \qquad (7.5)$$

Therefore we should multiply our previous answer for probability ratios by a factor of R:

$$\frac{P(L|N)}{P(S|N)} = \frac{P(L)}{P(S)}, \qquad (7.6)$$

7.2 Gott's Argument

which makes the Doomsday argument less clear-cut (Dieks, 1992; Olum, 2000). This is but a small fraction of the arguments and counter-arguments that compose a sub-branch of philosophy, which are too voluminous to list here. What is unanimously agreed is that the composition of appropriate prior probabilities is essential. To calculate such priors, we must consider the various scenarios for the end of humanity, as we shall in the following chapters.

7.2 Gott's Argument

A similar doomsday concept was espoused by Gott (1993), but its mechanics are quite different to the Carter-Leslie argument, and it is usually referred to as the 'Δt argument'. Consider any event that begins at time t_{begin}, and has a duration given by T. If we assume that we observe this event at a random time t, then we can, for example, compute the probability that we are observing the event within the first or last 2.5% of its life. More rigorously, if we define a variable r such that

$$r = \frac{t - t_{begin}}{T} \qquad (7.7)$$

then

$$P(0.025 < r < 0.975) = 0.95 \qquad (7.8)$$

In other words, the 95% confidence interval is defined between $t = t_{begin} + 0.025T$ and $t = t_{begin} + 0.975T$. If we now define $t_{end} = t_{begin} + T$, and we are now interested in how much future time is left before the event ends $t_{future} = t_{end} - t$, given that we know how much time has already elapsed, $t_{past} = t - t_{begin}$, then we can rearrange the above to show that

$$P\left(\frac{t_{past}}{39} < t_{future} < 39 t_{past}\right) = 0.95 \qquad (7.9)$$

As a demonstration, Gott recounts a visit to the Berlin Wall in 1969 ($t_{past} = 8$). The Wall fell 20 years later, giving $t_{future} = 2.5 t_{past}$, which falls within the 95% confidence interval. If we now consider the lifetime of *Homo sapiens*, $t = 200{,}000$ years, then the 95% confidence interval shall be

$$5.1 \text{ kyr} < t_{future} < 7.8 \text{ Myr} \qquad (7.10)$$

There are strong philosophical objections to this reasoning – for example, if we select a different 5% time interval, rather than the initial and final 2.5%, we can compose a new set of limits. For example, if we exclude the centre, the confidence interval becomes

$$P\left(t_{future} < \frac{19}{21} t_{past} \ \& \ \frac{21}{19} t_{past} < t_{future}\right) = 0.95 \qquad (7.11)$$

This results in an incoherence of belief. Equations (7.9) and (7.11) are derived from effectively the same assumptions, and yet as statements of belief they cannot be held together consistently (Roush and Sherri, 2001). Also, there is a tacit assumption that T is not infinite, and that different divisions of the possibility space (such as the short/long possibility division in the Carter-Leslie argument) will produce different confidence intervals, which again can be inconsistent.

The key difference between the Carter-Leslie and Gott arguments for Doomsday is again the selection of priors. Gott's argument is independent of one's personal prior, and to some level inimical to treatment in a proper Bayesian formalism, whereas the Carter-Leslie argument is Bayesian to the core, and hinges on prior choice. The interested reader is directed to Bostrom's review article on both arguments.[1]

---

[1] www.anthropic-principle.com/preprints/lit/

8

Death by Impact

Planet formation is not perfectly efficient. Planetesimals and other debris commonly survive the accretion epoch, leaving behind the comets and asteroid systems we see in our Solar System and in other star systems. We also know that the typical mass in these belts varies significantly from star to star – in some cases, the belts hold many times more material than those around our Sun.

Gravitational perturbations by the other planets deflect these objects onto trajectories that intersect the Earth's path. The principal impact risk to the Earth originates in the so-called Near Earth Objects (NEOs). These are defined as objects with a periastron less than 1.3 AU. As of 2018, some eighteen thousand Near Earth Asteroids (NEAs) have been catalogued, with various instruments.[1] This figure is likely to continue to increase thanks to surveys such as those carried out by the PAN-STARRs telescopes (Wainscoat et al., 2015). Of these, some 20% have orbits that approach within 0.05 AU of the Earth's orbital path, and are hence referred to as Potentially Hazardous Objects (PHOs).

8.1 Asteroids

Near Earth Asteroids are obtained from the main asteroid belt between Mars and Jupiter, generally the byproduct of asteroid collisions within the belt. These fragments undergo dynamical interactions with the various orbital resonances produced by Jupiter and Saturn, which sends them into close proximity to the Earth. A small fraction (no more than 10%) of NEAs are in fact cometary in origin. The composition of these bodies ranges from highly porous, structurally weak comet fragments to rather solid meteoritic nickel-iron cores, to loose collections or rubble piles, which are most common above sizes of 200 m.

---

[1] https://cneos.jpl.nasa.gov/stats/totals.html

These objects have been detected by a series of ground-based survey telescopes, and the current data is believed to be complete for bodies above 3 km in size. The distribution of size follows a power law, i.e., the number of objects greater than size L is given by

$$N(> L) \propto L^{-2.354} \tag{8.1}$$

The aim of most tracking missions is to characterise all NEAs above 1 km, a population believed to number approximately 1,100. At the time of writing, the detection rate or completeness in this category is approximately 75–80%. Writing in 2000, Rabinowitz et al. predicted that the pace of discovery of NEAs was sufficiently rapid that 90% of all bodies in this dangerous category would be detected by 2020.

Asteroids have densities which range from silicate/chondritic values (2,000 - 3,000 kg m^{-3}), all the way to iron (8,000 kg m^{-3}). The impact velocity is typically 30-70 km s^{-1}. Objects travelling above 72 km s^{-1} can achieve escape velocity and leave the Solar System.

8.2 Comets

While a much lower fraction of the total NEOs, comets can pose a much greater threat to the Earth. Generally, cometary NEOs occupy the larger size regime, and their impact velocities are generally greater (they are scattered into the Earth's path from the Kuiper Belt and beyond, and given more opportunity to accelerate). Comets are commonly found on highly elliptical orbits – this does increase the probability of intersecting the Earth's orbit, but it also increases the heating they receive upon close approach to the Sun, which helps to melt and destroy the body.

The density of comets is clearly much less – for example, the estimated density of recently visited comet 67P/Churyumov-Gerasimenko is 534 ± 7 kg m^{-3} (Jorda et al., 2016).

Cometary bodies also originate from the Oort Cloud, which resides at 50,000–150,000 AU (approximately 0.72 pc). The Oort Cloud is yet to be observed, but its existence can be inferred from the so-called *Oort spike*, which is observed in long-period comets. Oort (1950) was first to note the spike in 19 comets, which can be observed as a function of the inverse of the semi-major axis. This clustering of objects around this value of a, all with high eccentricities (and hence low perihelion distances), suggested that most of the bodies in this spike would soon disappear. The fact that such a large number of long-period comets are observed is sufficient evidence for a large halo of progenitor bodies at high semi-major axis.

Oort Cloud objects have their orbits perturbed either by the Galactic tide, or by the passing of stars close to the Solar System. In simulations, the Galactic tide acting on particles orbiting the Sun is typically given by (Heisler, 1986):

$$\mathbf{F}_{\text{tide}} = (A - B)(3A + B)xx' - (A - B)^2 yy' - \left(4\pi G\rho_\odot - 2(B^2 - A^2)\right)zz' \quad (8.2)$$

This equation relies on a fixed co-ordinate frame (x, y, z), and a rotating frame (x', y', z'). To obtain the rotating frame, we have assumed the Sun's orbit around the Galactic Centre to be circular with angular velocity Ω_0 (see box for description of the Oort constants (A, B)).

The Oort Constants

The Oort constants (A, B) approximate the Galactic rotation curve, i.e., how the angular velocity of stars around the Galactic Centre vary as a function of distance from the Sun (assuming the distance between these stars and the Sun is small). The radial velocity v_r and tangential velocity v_t of a star a distance d from the Sun is

$$v_r = (\Omega - \Omega_0) R_0 \sin \alpha \quad (8.3)$$

$$v_t = (\Omega - \Omega_0) R_0 \cos \alpha - \Omega d \quad (8.4)$$

where Ω_0 is the angular velocity of the Sun, and Ω that of the other star (the angle α between the Sun, the other star and the Galactic Centre is shown in Figure 8.1). A suitable Taylor expansion of $\Omega - \Omega_0$ gives these velocities in terms of the Oort constants

$$v_r = Ad \sin 2\alpha \quad (8.5)$$

$$v_t = Ad \cos 2\alpha + Bd \quad (8.6)$$

where

$$A = \frac{-1}{2} R_0 \frac{d\Omega}{dR}\bigg|_{R=R_0} = \frac{1}{2}\left(\frac{V_0}{R_0} - \frac{dv}{dr}\bigg|_{R=R_0}\right) \quad (8.7)$$

$$B = \frac{-1}{2} R_0 \frac{d\Omega}{dR}\bigg|_{R=R_0} - \Omega = -\frac{1}{2}\left(\frac{V_0}{R_0} + \frac{dv}{dr}\bigg|_{R=R_0}\right) \quad (8.8)$$

A describes the local shear in the velocity field, while B describes the gradient of angular momentum. These tools are sufficient for us to describe the perturbations of the Oort Cloud numerically due to nearby stars.

8.3 The Risk of Impact on Earth

The Earth is continually impacted by around sixty tons of interstellar material daily. As most of this material is composed of relatively small dust grains, which immediately vaporise on impact with the upper atmosphere, this activity goes unnoticed by the majority of humans. For material to survive atmospheric entry and hit the Earth's surface, the approaching body must be tens of centimetres in size (i.e., an object the size of a soccer ball).

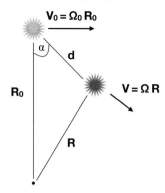

Figure 8.1 Estimating the velocity of nearby stars using the Oort constants. The Sun (top) moves in a circular orbit at radius R_0, with velocity V_0. The velocity V of stars a distance d from the Sun with an angular separation from the Galactic Centre of α can be estimated using the Oort constants A and B (see text).

Generally speaking, the larger the body, the less frequent the impact. Of course, the larger the body, the larger the energy of the impact, and consequently the greater the damage caused. A general rule of thumb is that bodies above 1 km in size begin to pose existential risk for humanity. The body that caused the Cretaceous-Tertiary impact is thought to have been a comet larger than 10 km in size (see Figure 8.2).

The 1 km target for detection by NEO surveys reflects the scale of damage that such a body would inflict upon striking the Earth. The impact energy of an asteroid of size L, travelling at a relative velocity v, is simply given by the kinetic energy

$$E = \frac{1}{2}mv^2 = \frac{\pi}{12}\rho L^3 v^2 \qquad (8.9)$$

where in the second part we assume the impactor is a sphere of radius L and constant density ρ. This is commonly converted into kilotonnes (kt) or megatonnes (Mt) of TNT:

$$1\,\text{kt} = 4.184 \times 10^{12}\,\text{J} \qquad (8.10)$$

This reflects the fact that impact events in recent history have been relatively small ($L \sim 50$ m or less), and as such have been comparable to the detonation of large nuclear bombs at high atmospheric altitude.

For example, the Hiroshima nuclear bomb energy rating was approximately 15 kt. For an impact with energy equal to Hiroshima, produced by an object with $\rho = 2000\,\text{kg m}^{-3}$ and velocity 65 km s^{-1}, the body size

$$L = \left(\frac{12E}{\pi \rho v^2}\right)^{1/3} \approx 3\,\text{m} \qquad (8.11)$$

8.3 The Risk of Impact on Earth

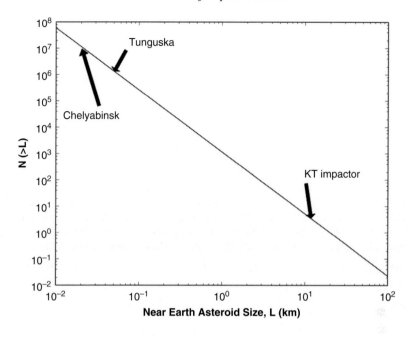

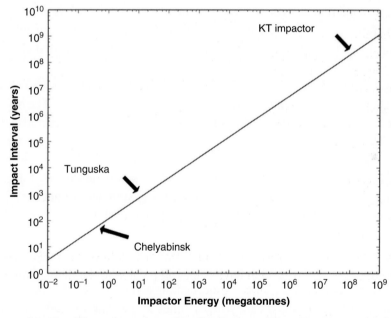

Figure 8.2 Top: The estimated number distribution of Near Earth Asteroid sizes, which decays with a power law slope of approximately 2.354 (equation 8.1). Bottom: The time interval between impacts of a given energy (in megatonnes), which also behaves as a powerlaw, with slope 0.78.

If we recalculate the impact energy for a 1 km body we obtain

$$E = 2.212 \times 10^{21} J = 5.286 \times 10^5 \text{Mt} \tag{8.12}$$

A single impact with these parameters is approximately equal to the total energy release by all nuclear detonations in history (Norris and Arkin, 1996).

8.4 The Consequences of Impact

We can consider two broad categories of impact. In the first category, moderate-sized impactors (less than 100 m) are able to penetrate the Earth's atmosphere, but explode some distance above the planet's surface, with the principal damage to land-based objects occurring thanks to the powerful blast waves driven by the explosion.

We are fortunate to have recorded evidence of such types of impact, referred to as *airbursts*. The Tunguska event (which occurred on June 30, 1908) was an airburst which flattened around 2,000 square kilometres of uninhabited forest. The energy of the explosion required to cause such devastation depends upon its height above the ground, but estimates range from 3 to 30 Mt.

More recent airbursts in populated areas include the Chelyabinsk bolide in 2013. This 20 m meteor exploded in an airburst with an energy of approximately 500 kt. Even at a distance of 18 miles from the burst point, onlookers reported intense skin burns from radiation generated by the burst.

If we are in the 1 km+ regime, then it is likely that these objects will be able to traverse the Earth's atmosphere successfully and impact the surface. The object will be travelling at speeds anywhere between 11 and 72 km s^{-1}, with significant kinetic energy. This kinetic energy is transferred to the surface in a very short timescale. Much of this energy is converted into thermal energy, vapourising the impactor and the local surface/atmosphere, and delivering a significant boost in local seismic energy. The force delivered by the impactor is sufficient to alter the planet's rotation period by about 1 microsecond for a 1 km impact, rising to approximately a millisecond for 10 km impactors (Collins et al., 2005).

The immediate impact zone is compressed to very high pressure, and subsequently heated to very high temperatures. This generates a powerful shockwave which travels away from the impact zone, heating and fracturing the rock in its path. This temporarily produces a cavity many times larger than the km-size impactor (the so-called *transient crater*). The transient crater is unstable, and collapses under gravity to form the final impact crater. In the immediate surrounds of this final crater is a layer of ejecta, thrown up during the impact and rained back down onto the planetary surface as an assortment of fine grains and larger sized 'bombs'.

The shockwave travels in both the surface rock and atmosphere. At very short range, the atmospheric shockwave completely ionises the atmosphere. The plasma

8.4 The Consequences of Impact

in the immediate vicinity of the impact is extremely optically thick, resulting in a brief 'fireball' phase, with an effective temperature of tens of thousands of Kelvin. As the plasma cools and becomes optically thin, the thermal energy trapped within the fireball is finally unleashed on the surrounding area, travelling behind the initial shock.

Beyond the fireball (up to of order 100 km for a 1 km impactor) the atmospheric shockwave remains powerful enough at short range to destroy buildings and infrastructure, as well as crush animal and plant life. At longer range, the winds generated by the shockwave and fireball can still cause significant damage, and sweep large quantities of ionised and radioactive material over continental scales in a matter of days.

If the impactor lands in an ocean, obviously the energy deposited by the impactor will generate substantial water waves. These waves are either generated via rim uplift followed by curtain collapse (*rim waves*), or oscillations caused by the transient crater collapsing, overshooting, collapsing, etc. (*collapse waves*) (Wünnemann et al., 2010). Collapse waves require a sufficient depth of water to instigate (a rule of thumb is approximately twice the impactor diameter).

Both waves are likely to result in the generation of tsunamis. Ezzedine et al. (2015) simulated a 50 m impactor striking several locations off the coast of the mainland United States. For an impact with velocity of 12 $km\,s^{-1}$, and resulting energy of approximately 10 Mt, waves strike land with heights ranging from 3 to 12 m, with subsequent flooding that requires several days to subside. Given that sea-level rises of order a metre place coastal sites at high risk (especially power plants), even a relatively small impact can cause widespread devastation.

Increasing the asteroid size to 200 m (and the impact to 20 $km\,s^{-1}$) is likely to generate waves at least a kilometre high at the impact site, decaying to tens of metres at landfall. A 1 km impactor travelling at this speed will carve out the entire ocean around its impact site, as well as portions of the seabed (~ 5 km deep), with 100 m waves arriving at the nearest coast; 20 m waves eventually arrive at the other side of the planet within the day (Ward and Asphaug, 2003). Continental sites far from the coast are not strongly affected by such impacts – however, given the concentrations of populations around coastal areas, and the atmospheric effects of such a powerful explosion, it seems clear that there is nowhere on Earth that is safe from an impact of this magnitude.

The Torino Scale

The Torino Scale was devised to communicate impact risk effectively to the public. The scale ranges from 0 to 10, with each integer identifying a scenario that combines the probability of impact and the resulting damage caused.

At zero, the probability of a collision of note is zero. This by definition excludes the constant stream of dust and small-scale debris which impacts the Earth and is vaporised in the upper atmosphere.

- A Torino rating of 1 or 2 indicates that a body of notable size will execute a pass near the Earth with 'no unusual level of danger', which still merits attention by astronomers.
- A rating of 3 indicates the probability of localised destruction is greater than 1%, and official action is required if the encounter is to occur within the decade.
- Ratings of 4 expect regional devastation, with a probability greater than 1%. Commonly, closer inspection of objects in categories 3–4 will prompt reclassification back to zero.
- As the rating increases through 5 and 6, the probability of impact becomes more significant, and the possibility of global effects must be considered.
- As the rating increases to 7, 8 and 9, the probability of impact is extremely high, and the resulting damage increases until a global catastrophe is imminent. A 10 rating expects the future of civilisation to be threatened.

Statistically, we expect events with a 10 rating to occur once every 100,000 years. Since the scale was first published in 1997, most objects have received either 0 or 1 on the Torino scale. The notable exception is Apophis, which briefly occupied a 4 rating, before being downgraded back to zero.

8.5 Impact as an Existential Risk to ETIs

We must of course recognise that the particular circumstances of our Solar System do not always pertain to other planetary systems. The precise configuration of asteroid belts and comets is a sensitive function of the gravitational fields present, principally from the star and any giant planets in the system.

The frequency of impacts on Earth suggest that kilometre-sized bodies strike the surface roughly once every 10^5-10^6 years. We could conceive of circumstances where terrestrial planets in other star systems experience extremely large impact rates, either as a matter of course or as a result of recent orbital instability amongst neighbouring planets. We can point to evidence of the latter on Earth. The Late Heavy Bombardment, which left its scars across multiple Solar System bodies around 4.1–3.8 Gyr ago, is thought largely to be due to the orbits of Jupiter and Saturn crossing an unstable resonance (Tsiganis et al., 2005; Raymond et al., 2013).

ETIs that are not capable of space travel are also not capable of defense against the larger asteroids in their home star system. A single impact has the potential to end their civilisation (as it has the potential to end ours). ETIs that occupy more than one planetary body reduce their risk of complete destruction by a single impact, and may even be able to divert incoming impactors.

8.5 Impact as an Existential Risk to ETIs

In a report to the United States Congress in 2007, NASA outlined its thinking on asteroid impact mitigation. The various techniques for mitigation can be subcategorised into 'impulsive' or 'slow push'. Impulsive mitigation techniques will be familiar to viewers of Hollywood blockbusters such as *Armageddon*. In essence, these techniques require either an explosive device to be detonated, or a high-velocity impact from a projectile, either to shatter the asteroid or deflect its path, depending on its mass and tensile strength.

Explosive devices being considered range from the conventional (which are generally ineffective against asteroids that pose any substantive risk) to the nuclear. The location of the explosive is also important – is it placed on the surface of the asteroid, drilled into the subsurface, or exploded at a small standoff distance? One might initially assume that a high-yield nuclear device drilled into the subsurface is the best defence – however, such an explosion is likely to fragment a larger asteroid into many smaller bodies, which would still be large enough to pose substantial existential risk.

Predicting the response of an asteroid to a surface or subsurface explosion requires intimate knowledge of its strength and composition. By contrast, a nuclear explosion detonated at a standoff distance requires much less knowledge while still being able to generate reasonable levels of thrust on the asteroid to deflect its path.

Kinetic impactors may also be suitable in certain cases. For kinetic options to be viable, they must arrive at the target with relative velocities above 10 km s^{-1}, with a roughly linear increase in performance as impact velocity increases.

The 'slow push' techniques require a long time interval in which to modify an asteroid orbit. These push techniques, like the surface explosion techniques, require reasonably detailed knowledge of the asteroid's internal structure and ability to handle stress.

Slow push methods include the so-called 'gravity tractor', where a body of similar mass to the asteroid is placed nearby to provide a competing gravitational force; space tugs that attach to the asteroid and accelerate it away from a collision course; and mass loss methods such as continuous laser fire to 'boil' the outer layers, or asteroid-borne 'mass drivers' that mine the asteroid and throw material from the surface.

We can see just how long a slow push might take with an example. Let us consider a rocket with fuel m attached to an asteroid of mass M moving at velocity $\mathbf{v}$. If we wish to deflect our asteroid (i.e., keeping $|\mathbf{v}|$ constant), then we can quickly write down conservation equations for before the rocket fires, and after it has expelled reaction mass m at velocity v_{ex}:

$$(m + M)\mathbf{v} = M\mathbf{v}' + m\mathbf{v}_{\text{ex}} \tag{8.13}$$

The final velocity is

$$\mathbf{v'} = \frac{m+M}{M}\mathbf{v} - \frac{m}{M}\mathbf{v}_{ex} \qquad (8.14)$$

Assuming the rocket is firing perpendicular to the asteroid trajectory in a 2D plane, we can then define the angular deflection from θ to θ' as follows:

$$\cos\theta' = \cos\theta - \frac{mv_{ex}}{Mv}\sin\theta \qquad (8.15)$$

$$\sin\theta' = \sin\theta - \frac{mv_{ex}}{Mv}\cos\theta \qquad (8.16)$$

This indicates that to deflect the path of a 100 m asteroid by one degree with a Saturn V rocket ($v_{ex} = 2.5\text{km s}^{-1}$) would require a total propellant mass of around 10% of the asteroid. This can be reduced if one is able to use the asteroid itself for reaction mass. It is also clear that a Saturn V cannot do this much work in a short time interval – around 2 million kg of propellant was burnt by the Saturn V at launch, and the above calculation would require several thousand times as much. With each launch requiring months of preparation and refuelling, we can immediately see that this example would require decades of continuously applied force.

At the time of the report, the only technique sufficiently mature for use was considered to be a standoff nuclear explosion (the report authors assumed that little information would be available regarding the asteroid's properties). The implication, however, is clear – as humans refine and enhance spacefaring capability, it is highly possible that we will be able to manipulate NEOs and other asteroids, not only to deflect them from dangerous collisions with our planet, but for other purposes (such as mining). Any spacefaring ETI should also have access to similar mitigation methods, and be able to avoid this particular extinction channel.

9
Death by Terrestrial Disaster

9.1 Supervolcanism

We have seen in the Rare Earth solutions that volcanism is a crucial component of the climate feedback cycles that have kept the Earth's temperature relatively stable, allowing for the flourishing of life. The volcanism that we see on Earth exists in large part due to the properties of the Earth's lithosphere. The brittle nature of the lithosphere has resulted in its fracturing into a set of rigid plates. The plates are free to move on top of the asthenosphere (the lower mantle), under the constraints that the plates themselves impose on each other.

The science of *plate tectonics* is primarily concerned with the boundaries where two plates meet. Several possible configurations exist at these boundaries:

- Divergent (constructive) boundaries. Two plates move away from the boundary, exposing the mantle. This mantle can well upwards, adding new material to the lithosphere. Typical examples are mid-ocean ridges, such as the mid-Atlantic Ridge.
- Convergent (destructive) boundaries. Two plates move towards the boundary, with one plate eventually being *subducted* (pushed under) the other. These subduction zones result in the destruction of lithospheric material as it is melted by the mantle. Typically, oceanic lithosphere is subducted in favour of continental lithosphere. This can result in the uplift of mountains (*orogeny*).
- Conservative boundaries. The two plates move transverse to the boundary, producing *transform faults*.

The principal plate boundaries are the sites of the majority of major seismic events on the Earth, with the Pacific plate boundary being particularly active. Mid-ocean ridges typically form hydrothermal vents as the partially molten mantle is exposed to the seafloor, providing a valuable source of energy for deep-sea organisms.

Subduction zones are typically accompanied by *arc volcanism*. The subducting plate delivers water to the mantle, which reduces its melting temperature and forms magma. The magma's viscosity is high enough to keep it trapped, until pressure build-up forces it onto the crust as an eruption, along with solids (ash, pyroclastic rock, etc.) and a range of gases (principally CO_2 and H_2S). Repeated eruptions eventually form volcanoes with a range of properties depending on the viscosity of the magma, and the accompanying material. Volcanic ridges form on the crust, on top of the overriding plate, either as mountain-like ranges or island chains.

Volcanism is closely connected to the plate tectonics active on the Earth, but it is worth noting that Venus's volcanism is unlikely to result from plate boundaries, as its extremely hot lithosphere is too ductile to break into plates (Mikhail and Heap, 2017). The *shield volcanoes* observed there are formed by weaknesses in the crust, allowing low-viscosity magma to escape.

The Earth's biosphere has depended on volcanic activity to help regulate its atmosphere and act as a general-purpose recycling unit. However, under the correct conditions, volcanoes can avoid regular eruptions and begin to store very large quantities of ash and magma. These so-called *supervolcanoes* can pose a significant existential risk. Supervolcano activity is measured by the Volcanic Explosivity Index (VEI), which is a function of the volume of expelled material E (in km^{-3}):

$$VEI = \log(E) + 5 \tag{9.1}$$

The largest supervolcanoes in the fossil record (VEI 8) expel some 1,000 km^3. The most recent VEI 7 eruption (and the largest since the end of the last Ice Age) is the Mount Tambora eruption, which occurred in Indonesia in 1815. The expulsion of significant quantities of volcanic dust into the atmosphere are thought to be responsible for the 'Year Without a Summer'. The injection of dust into the Earth's upper atmosphere increased its opacity and reduced the sunlight received at the surface. This reduction is linked with a drop in global average temperatures by around half a degree, and the failure of crop harvests around the world, resulting in famine and riots.

For comparison, the more famous Krakatoa eruption in 1883 is classified as VEI 6, and most eruptions of VEI 6 or larger are associated with unusually cool temperatures as volcanic dust increases atmospheric opacity.

The largest recorded supervolcanic eruption during the lifetime of the *Homo* genus is the so-called Toba event, which originated in Lake Toba in Indonesia around 75,000 years ago (according to K-Ar dating). The Toba event's eruptive volume appears to have exceeded 2,500 km^3 (Rose and Chesner, 1990). The literature is divided on the effects of such a substantial eruption on Earth's climate, with some suggesting that this eruption may have triggered the last glacial period. More conservative studies accept that a drop in global average temperature of 15 C

9.1 Supervolcanism

would have occurred for several years, with the climate taking decades to recover to its original average (Robock et al., 2009). Even this conservative case would have had devastating consequences for the biosphere. Hominid groups would also have suffered heavily as a result, leading some to suggest that the apparent genetic 'bottleneck' observed in humans was a result of *homo sapiens* populations dwindling in the decades following the Toba event (although this is hotly debated).

The secondary consequences of volcanic eruptions cannot be underestimated. Cumbre Vieja, one of the most active volcanoes in the Canary Islands, has been a cause for concern for several decades. While its recorded eruptions achieve VEIs of 1 or 2, it is possible that the volcano might undergo a catastrophic failure of its west flank. If this lateral collapse occurs, some several hundred km^3 of rock (a total mass of some 10^{15} kg) would be dumped into the sea. This would trigger a megatsunami event that would sweep the Atlantic. The original calculations of Ward and Day (2001) suggested waves exceeding 10 m breaking onto the eastern coast of the Americas, although other calculations have suggested the wave heights will be significantly lower, if the collapse happens at all (Gisler et al., 2006; Carracedo, 2014).

What is clear is that supervolcanism can have profound effects on the agricultural food chain. Even VEI 7 events can result in mass starvation. It seems likely that another VEI 8 event would result in a wide-scale collapse of the human population. It has been estimated that a VEI 8 event is equivalent in effect to a 1 km asteroid impact, at least in terms of its atmospheric dust and aerosol production (Chapman and Morrison, 1994). Given that the estimated frequency of VEI 8 events are around once every 50,000 years, we can therefore expect VEI 8 events to occur almost twice as frequently as 1 km impacts, and hence such events would pose a greater existential risk (Rampino, 2002, and references within).

Is volcanism of this magnitude common to terrestrial planets? To answer this, we must be able to predict the tectonic regime on exoplanets. The literature is heavily split on the conditions for plate tectonics (or equivalently a mobile-lid tectonic regime).

Briefly, there are two conditions for a sustainable plate tectonics system. Firstly, the lithosphere (crust and upper mantle) must be subject to failure (i.e. cracking) to produce separate plates. Secondly, said plates must then subduct into the mantle, which requires the plates to have negative buoyancy at subduction zones (i.e., the plate mean density should be larger than the mantle density).

Lithospheric failure depends principally on two key factors: the plate's ability to resist deformation, and the convective stresses placed on the plate by the mantle. The plate's resistance to deformation is governed mainly by the plate thickness and its material strength, and the convective stress depends on the viscosity of the mantle. Broadly speaking, a more viscous mantle is a more slow-moving mantle, and hence lithospheric failure is more likely to occur for low viscosities. The mantle viscosity ν_{mantle} is temperature-dependent. Most studies of plate tectonics on exoplanets assume an Arrhenius-like relation (Korenaga, 2010):

$$\nu_{\text{mantle}} = \nu_0 \exp\left(\frac{E}{RT_p} - \frac{E}{RT_r}\right) \qquad (9.2)$$

T_p is the mantle potential temperature (this is used to correct for the effect of adiabatic compression as a function of depth). ν_0 is the reference viscosity at $T_p = T_r$, and E is an activation energy. When $T_p \lesssim 750K$, the viscosity is so excessive that a stagnant-lid regime is almost unavoidable, and plate tectonics cannot occur.

The convective behaviour of the mantle is described by the Rayleigh number Ra, a dimensionless property that describes whether heat transfer is primarily conductive or convective. Mantles with large Ra can produce larger mantle velocities and hence more stress. Once the stress reaches a critical yield stress, the plate fractures. The yield stress is typically of the form

$$\sigma_{\text{yield}} \propto \mu \rho g z \qquad (9.3)$$

where μ is the friction coefficient, ρ is the plate density, g is the gravitational acceleration and z is the depth. Water can significantly increase the hydrostatic pore pressure in the rock, suggesting a crucial role for it in the emergence of plate tectonics (Korenaga, 2010):

$$\sigma_{\text{yield}} \propto \mu \left(\rho - \rho_{\text{H}_2\text{O}}\right) g z \qquad (9.4)$$

The internal Rayleigh number of a super-Earth with mass M can be fitted using the following formula (Valencia et al., 2006; Korenaga, 2010):

$$Ra = Ra_\oplus \left(\frac{\Delta T_p}{\Delta T_{p,\oplus}}\right) \left(\frac{\nu_{,\text{mantle}}}{\nu_\oplus}\right) \left(\frac{M}{M_\oplus}\right)^{1.54} \qquad (9.5)$$

ΔT_p represents the temperature difference between the lithosphere and mantle. Worlds with a cool surface can therefore be more likely to experience lithospheric failure and plate formation (Foley et al., 2012).

Valencia et al. (2007) initially suggested that plate tectonics is easier on super-Earths, due primarily to their increased Ra (and decreased plate thickness). O'Neill and Lenardic (2007) quickly responded by noting that the fraction of convective stress actually driving lithospheric deformation decreases as planet mass increases, resulting in a stagnant-lid scenario.

Further studies have since elaborated the difficulty in estimating mantle viscosity and convective behaviour. Determining the critical yield stress for lithospheric failure is also non-trivial, and requires a deep understanding of crust composition and the nature of fault production (Foley et al., 2012). It is also clear that mantle viscosity is pressure dependent as well as temperature dependent (Stamenković et al., 2012), adding a further dependence on depth z:

$$\nu_{\text{mantle}} = \nu_0 \exp\left(\frac{E + zV}{T_p + T_0} - \frac{E + z_r V}{T_r + T_0}\right) \tag{9.6}$$

where V is the activation volume, which decreases with increasing pressure, and ν_0 is now measured at the reference temperature and depth (T_r, z_r). T_0 is the surface temperature.

Most importantly, the heating sources responsible for mantle convection vary with time, and it has been shown that the starting temperature of the planet affects its final state (Noack and Breuer, 2014).

Essentially, the likelihood of plate tectonics on other planets will depend sensitively on the surface temperature, the plate composition and thickness, the plate's history of weakening events, the presence of water, the mantle's temperature and pressure, the temperature at the core-mantle boundary, the radiogenic heating rate, and likely many other factors. This illustrates the massive challenge in predicting the tectonic behaviour of terrestrial exoplanets, about which we typically know only the bulk mass and radius.

Simulations also indicate that plate tectonics is only a phase in the life of a terrestrial planet, and that depending on the radiogenic decay and the cooling of the plate and mantle, most if not all terrestrial bodies initially begin life as a hot, stagnant-lid body, transitioning to an episodic regime of plate tectonics, before entering a fully plate tectonic regime, and ending in a cool stagnant-lid state, with the duration of each phase highly sensitive to the above list of constraints (O'Neill et al., 2016).

Finally, we should also consider the possibility that exoplanets experiencing tidal heating may generate significant volcanism, even in a stagnant-lid regime (Valencia et al., 2018).

It has been argued that plate tectonics (or at the very least, significant volcanism) is essential for technological civilisations (Stern, 2016). If that is so, then volcanoes may be crucial to life and intelligence but also cause their ultimate doom.

9.2 Magnetic Field Collapse

The Earth is afforded a great deal of protection from stellar activity by a combination of its atmosphere and its magnetic field. The upper atmosphere absorbs ultraviolet, X-ray and gamma ray photons, preventing their arrival at the surface. Very high-energy photons can generate 'air showers' of photons and charged particles, which can cause secondary hazards if the initial input energy is sufficiently high.

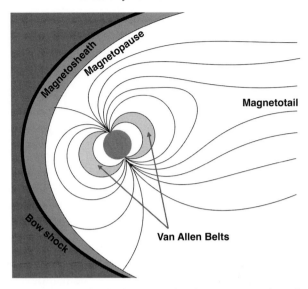

Figure 9.1 The magnetic field of the Earth. The magnetosphere is constrained largely by the solar wind, which produces a bow shock where the two meet. Most of the charged particles cannot penetrate the magnetosheath directly behind the bow shock. Those that do are typically in the Van Allen radiation belts, and occasionally rain onto the magnetic poles producing aurorae.

Charged particles emitted by the Sun (see section 10.1) and other bodies are deflected by the Earth's magnetic field. The *magnetosphere* defines the region of interplanetary space dominated by the Earth's field (Figure 9.1). The solar wind meets the magnetosphere and produces a bow shock. The field lines behind the shock are distorted into a tail-like shape (the *magnetotail*). As a result, most of the Sun's hazardous charged particle flux is funnelled around the Earth, with most of the particles never penetrating the magnetosheath, a turbulent layer between the bow shock and the beginning of the magnetosphere (the *magnetopause*). Some of these inward-streaming particles are trapped in the *Van Allen* radiation belts.

The shape of Earth's magnetic field (and the protection it offers from all stellar activity) is largely governed by the Sun itself. Changes in the solar wind will result in motion of the bow shock to and from the Earth. This motion flexes the magnetosheath, increasing the injection of particles into the Van Allen belts. This increase in injection rate forces particles out of the belts, along the magnetic field lines to the north and south poles, causing the *aurorae borealis* and *australis* (Northern and Southern lights).

In the Solar System, essentially every terrestrial planet has possessed a magnetic field at some point in its past, as have several moons (Schubert and Soderlund, 2011). Of the planets, only Earth retains a strong magnetic field of around 40 μT.

9.2 Magnetic Field Collapse

Mercury retains a weak intrinsic field (0.3 µT), while Venus, Mars (and the Moon) show evidence of relic fields, through magnetisation of crustal rocks.

The origin of a planet's magnetic field depends to some degree on the planet in question, but generally requires some form of *dynamo* to generate sufficient current. We can see this by reconsidering the magnetic induction equation (equation 4.73), where we now allow the fluid to have a finite conductivity:

$$\frac{\partial \mathbf{B}}{\partial t} = \nabla \times (\mathbf{v} \times \mathbf{B}) + \eta \nabla^2 \mathbf{B}, \tag{9.7}$$

where we have defined the *magnetic diffusivity*

$$\eta = \frac{1}{\sigma \mu} \tag{9.8}$$

where σ is the conductivity, and μ is the permeability of the fluid. Note that we have used the Boussinesq approximation, where density perturbations due to pressure and changes are relatively small. In an ideal MHD fluid, the conductivity is infinite and hence $\eta = 0$. However, the finite conductivity of the fluid ensures that the second term on the right-hand side of equation (9.7) acts as a diffusive term. If we set $\mathbf{v} = 0$, then the magnetic field will eventually dissipate on a time scale given by

$$t_{\text{diff}} = \frac{L^2}{\eta} \tag{9.9}$$

where L is a typical length scale in the fluid. In the case of the Earth, the dynamo motion is generated by convective currents in the mantle. The conservation of momentum under the Boussinesq approximation (in the corotating frame of angular velocity Ω) is

$$\frac{d\mathbf{v}}{dt} + \nabla . \mathbf{v} + 2\Omega \times \mathbf{v} = -\nabla \Pi + \frac{1}{\mu_0 \rho_0} (\nabla \times \mathbf{B}) \times \mathbf{B} - \alpha \Delta T \mathbf{g} + \nu \nabla^2 \mathbf{v} \tag{9.10}$$

The above describes deviations from hydrostatic equilibrium (at constant density ρ_0). The forces resulting in hydrostatic equilibrium (and the centrifugal force) are incorporated into the effective pressure gradient $\nabla \Pi$. We can therefore see that the sustenance of the magnetic field (which depends on sustaining fluid velocity) is a combination of viscosity (represented by the kinematic viscosity ν), Coriolis forces (seen on the left-hand side of equation 9.10) and buoyancy ($\alpha \Delta T \mathbf{g}$, where $\mathbf{g}$ is the gravitational acceleration). The buoyancy force results from temperature perturbations ΔT, which generate density perturbations of order

$$\rho = \rho_0 (1 - \alpha \Delta T) \tag{9.11}$$

In physical terms, the liquid outer core of the Earth generates convection cells. This is partially assisted by a compositional gradient – as the core cools, molten iron solidifies and attaches to it, reducing the density of the fluid it leaves behind. These convection cells are rolled into tubes by the Coriolis effect, which are then thought to align (roughly) with the Earth's rotation axis. Given a *seed field*, this dynamo has successfully produced a long-standing magnetic field (see Kono and Roberts, 2002, for an overview of dynamo models, and a detailed review of attempts to simulate the Earth's magnetic field).

If the core completely solidifies, this dynamo will fail, and the Earth's magnetic field will diffuse away on $t_{\text{diff}} \sim 10^4$ yr (Merrill et al., 1996; Chulliat and Olsen, 2010). This appears to have occurred to the dynamos of both Mars and the Moon. Venus is a more confusing picture, as it should still possess a molten core and yet does not have an intrinsic magnetic field. This failure has had several ascribed explanations, from Venus's failure to produce plate tectonics to allow a sufficient temperature gradient for convection cells, to the possibility that the Venusian core is not fully formed, while remaining too cool for thermally driven dynamos (see Schubert and Soderlund, 2011, and references within).

We could therefore imagine that the production of a long-term magnetic field is not a guaranteed proposition for terrestrial planets. A simple model that broadly captures the behaviour seen in the Solar System gives the magnetic moment of an exoplanet as follows (López-Morales et al., 2011):

$$M \approx 4\pi R_{\text{core}}^3 \beta \left(\frac{\rho}{\mu_0}\right)^{1/2} (F\Delta R)^{1/3} \qquad (9.12)$$

where R_{core} is the core radius, μ_0 is the permeability of free space, ΔR is the thickness of the rotating shell around the core that generates convection cells, and F is the average convective buoyancy flux. This can be expressed in terms of Terran values:

$$\frac{F}{F_\oplus} = \left(\frac{Ro}{Ro_\oplus}\right)^2 \left(\frac{D}{D_\oplus}\right)^{2/3} \left(\frac{\Omega}{\Omega_\oplus}\right)^{7/3}. \qquad (9.13)$$

The convective flux depends strongly on the ratio between the fluid's inertia and the Coriolis force, given by the Rossby number Ro, and the rotation period of the planet. This model, when applied to exoplanet data with some generous assumptions regarding their composition, suggests that magnetic moments are likely to be much greater than the Earth, provided that their rotation rates are sufficiently high. Of course, terrestrial planets orbiting low-mass stars are likely to have greatly reduced rotation rates due to tidal synchronisation, and may experience problems sustaining their fields as a result.

One possible failure channel for a civilisation, then, is that their planet's magnetic field eventually collapses, either due to core cooling or some other factor that disrupts the convection cells needed to maintain the dynamo.

9.2 Magnetic Field Collapse

It is important to note that even the Earth's magnetic field, while extant for billions of years, has experienced intervals where the field strength greatly reduces.

During a so-called *geomagnetic reversal*, the Earth's magnetic field switches polarity, where North switches with South. This transition can be easily detected at spreading seafloor faults, where the reversals can be seen as changes in crustal magnetisation in a 'bar-code' fashion as one moves away from the fault zone.

During the reversal, the field strength can significantly reduce, resulting in intervals where the Earth's atmosphere is relatively unprotected from astrophysical hazards (see next Chapter).

These intervals can be relatively prolonged by human standards. The interval between one switch and another is usually referred to as a *chron* (where chron is commonly used in geoscience to refer to a time interval separating significant events). The reversal frequency varies greatly with time – several reversals can occur in a million-year period, or a ten-million-year time interval can elapse without reversal (a so-called *superchron*).

Predicting a reversal is extremely challenging, as there are a statistically small number of reversal events on which to develop a model. Deviations from Poisson statistics suggest that the reversals exhibit correlated behaviour on very long timescales (Carbone et al., 2006).

In any case, reversals can take up to tens of thousands of years to complete, with the magnetic field being depleted to only 5–10% of its strength during the transition (Glassmeier and Vogt, 2010). During a typical chron, the *standoff distance* of Earth's magnetopause, given as the distance from the Earth's center to the substellar point of the magnetopause, is around 12 $R_\oplus$.

During the last transition (the Matuyama-Brunhes transition around 780 kyr ago), the standoff distance for a relatively quiet Sun would have moved to around 5.9 $R_\oplus$, the approximate location of geosynchronous orbit (GEO). We can imagine that this would have significant effects on our technological civilisation as satellites become bombarded with damaging radiation.

This becomes compounded during times of intense stellar activity (next chapter), when the standoff distance can be temporarily reduced to around 3 $R_\oplus$. The standoff distance remains large enough for Earth to maintain its atmospheric protection, but the topology of the field changes significantly during the transition. Most likely, this will result in the ejection of particles from the Van Allen Belt, which would then impact the upper atmosphere, causing ozone loss and increased levels of UV radiation (see section 11.3), as well as affecting other climate properties such as cloud cover.

The evidence does not currently support geomagnetic reversal causing mass extinctions (see Glassmeier and Vogt, 2010, and references within, but equally see

Courtillot and Olson, 2007). What does seem clear is that geomagnetic reversal (and decaying magnetic fields in general) is likely to exacerbate problems already present. During a reversal, Earth is even more prone to the damaging effects of stellar activity, and we can expect exoplanets orbiting more-active stars to be even more vulnerable, as we discuss next.

10
Death by Star

10.1 Stellar Activity

Stars are not steady, quiescent sources of radiation. The intense energies produced by the thermonuclear reactions at their core ensure that the Sun's gas is ionised into a highly conductive plasma (i.e., the hydrogen it contains is typically separated into its constituent protons and electrons).

This high level of ionisation means we must describe the Sun's behaviour using the laws of magnetohydrodynamics (or MHD). Along with the pressure and gravitational field of the fluid, we must also concern ourselves with the Sun's (substantial) magnetic field. Motions in this fluid generate currents (flows of electrons). According to Maxwell's equations, an underpinning component of MHD theory, these currents induce magnetic fields, which in turn act on charged particles, generating motions which generate further magnetic fields.

In the Sun, the global magnetic field arises due to the so-called $\alpha\Omega$ dynamo, which owes its genesis to Larmor and Parker. Larmor (1919) reasoned that the differential rotation of the Sun (the Sun rotates fastest at the equator, with period ~ 25 days, as opposed to ~ 35 days at the poles) was sufficient to shear a poloidal magnetic field (i.e., a field emanating from the poles). This shear would produce a field of opposite polarity in the Sun's North and South Hemispheres, neatly explaining the observed dichotomies between sunspots either side of the equator.

It was soon observed that the field would remain axisymmetric, and as such unable to sustain itself indefinitely in the face of Ohmic dissipation, i.e., energy losses from the currents generated. Parker (1955) noted that the Coriolis force would allow cyclonic motions to occur in rising fluid elements, destroying the axisymmetry and allowing the field to be maintained.

We now know the solar atmosphere has two principal zones – an inner radiative zone, where energy is mainly transported by photons executing a random walk between electrons and protons, and an outer convective zone, where the plasma

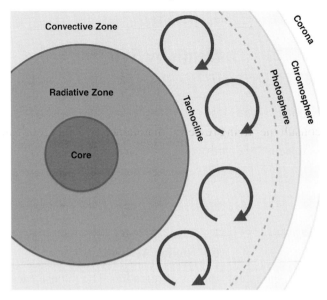

Figure 10.1 The internal structure of the Sun. Inside the core, hydrogen is fused into helium. The two principal zones of the stellar atmosphere – the radiative and convective zones – are connected at the tachocline. Photons free-stream out of the atmosphere once they reach the photosphere, passing through the hot chromosphere and beyond the coronal plasma.

principally transports energy through convection (see Figure 10.1). The convective and radiative zones are rotationally decoupled – the radiative zone rotates uniformly, and the convective zone executes differential rotation. A thin boundary layer, the *tachocline*, straddles these two zones. Significant shear occurs in the tachocline, and is the source of the majority of what we refer to as *stellar activity*.

Before we investigate this activity and its potential dangers, let us finish our study of stellar structure with investigation of the Sun's upper layers. The visible surface of the convective zone is referred to as the *photosphere* – in this region, the optical depth of the plasma (a measure of the probability that a photon will interact with a layer of material) begins to reduce below the critical value for photons to *free-stream* or escape the Sun. This region is where the *effective temperature* of stars is defined. This temperature defines the star's continuum spectrum, i.e., the regions of the spectrum that are not dominated by emission lines. In the case of the Sun, the effective temperature is approximately 5,800 K.

As photons escape the photosphere, they encounter the *chromosphere*, a 2,000–3,000 km layer of extremely low-density gas. This low-density gas is significantly hotter than the photosphere (around 20,000 K). Finally, beyond the chromosphere is the solar *corona*, a diffuse aura of plasma that extends for millions of kilometres into space. The causes of chromospheric and coronal heating are not well understood, but presumably heating by the magnetic field plays an important role. The *heliosphere* demarcates the Sun's sphere of influence on the Universe, where the solar wind meets the interstellar medium (see below).

We are focused on the aspects of stellar activity that are potentially lethal. We will therefore sidestep the many non-lethal aspects to stellar activity. The interested reader should consult the 'living reviews' maintained by Hall (2008) and Hathaway (2010) for more information on these phenomena.

10.1.1 Solar Winds

The particles of the inner corona are accelerated by the Sun's magnetic field into a range of velocities. Due to thermal collisions, these velocities are well described by a Maxwellian distribution:

$$f(v) = \left(\frac{m}{2\pi k_B T}\right)^{3/2} 4\pi v^2 e^{-\frac{mv^2}{2k_B T}} \qquad (10.1)$$

where k_B is Boltzmann's constant, and T is the local temperature. The particles in the high-energy tail of the distribution possess velocities exceeding the local escape velocity, and flow away from the Sun. Electrons entrained in the solar wind can then generate electric fields which boost the acceleration of other ions.

The solar wind is composed of a fast component, which appears to originate from coronal holes, funnel-shaped regions of open field which allow particles to escape with high speeds and lower temperatures. The slow component is significantly hotter, and is more closely related to the corona as a whole.

10.1.2 Flares and Coronal Mass Ejections

In 1859, Carrington's observations of the Sun in white light indicated that the Sun experienced a significant brightening that lasted approximately five minutes. We are fortunate that the so-called 'Carrington event' did not occur during the current era of space exploitation and exploration. Eighteen hours after Carrington's observation, the Earth's magnetic field underwent a severe disruption, causing unprecedented geomagnetic storms. These storms resulted in exceptionally strong *aurorae*. Typically such aurorae are confined to high latitudes, but during the Carrington event they were observed as far south as the Caribbean. Local accounts suggest that the aurorae were strong enough in the southern latitudes of the USA that one could read a newspaper by their light.

We now understand that what Carrington witnessed (and the Earth withstood) was a particularly strong solar flare, accompanied by a coronal mass ejection. During a flare event, the highly twisted and closely packed loops of magnetic field lines experience what is called *magnetic reconnection*. The magnetic field reorders itself to achieve a lower energy state, with nearby loops fusing and connecting together. Thanks to energy conservation, reconnection events release energy into their surroundings. In the case of the Sun, these energies can be significant (though perhaps only a billionth of the total energy emitted every second). This energy is injected into the photosphere, chromosphere and corona, causing significant heating of the surrounding plasma. This sudden energy input explains the significant brightening of the Sun during flares. Much of the increased emission from the flare comes at ultraviolet and X-ray wavelengths.

If the flare is sufficiently powerful, it can lift large quantities of plasma (and the magnetic fields threading it) from the surface. This is a Coronal Mass Ejection (CME), which often accompanies a flare (although the two are not always linked). Coronagraphs measure on average between 3 and 5 CMEs per day. Most of these events are not directed at the Earth – typically one to two CMEs reach the Earth per month. This incoming material can produce a shock wave as it strikes the Earth's magnetosphere, energising the Earth's field and triggering geomagnetic storms. This space weather can have extremely negative effects on satellites and spacecraft. The CME material includes both weakly and strongly charged ions, which will destroy electronic equipment if they bombard it in sufficient quantities. Clearly, astronauts in this environment would also suffer damage to their health if caught in this barrage. It is extremely fortunate that the timing of a particularly powerful energetic particle event was between the Apollo 16 and 17 missions. The radiation dose that would have been received by the astronauts, had they been in transit to the Moon, was almost certainly life-threatening.

Space weather can also affect ground-based systems. Geomagnetically induced currents (GICs) occur when highly charged particles in the Earth's upper magnetosphere precipitate along field lines into the upper atmosphere. This swarm of charges produces a current in the atmosphere which further induces currents in the ground. Anything that uses the ground to earth their electrical systems is at risk of surges and overcurrent. A much cited case of GICs taking down critical systems is the failure of a Hydro-Quebec transformer in March 1989 following a large geomagnetic storm.

We have yet to see what GICs might be produced by Carrington-esque events. The only major electrical equipment that appears to have suffered during 1859 were telegraph systems. If anecdotal evidence is to be believed, then GICs were definitely in play – telegraph operators reported experiencing strong electric shocks when touching their equipment, and they were able to send and receive communications despite disconnecting their systems from power.

What damage should we expect from a Carrington event to our modern civilisation? Any system composed of long conductive material placed at or under the surface will be prone to these inductive currents. Power transformers like the Hydro-Quebec transformer are prone to overload – a Carrington event would most likely cause power outages on a global scale, with damaging events concentrated at higher latitudes (as the magnetic field varies more significantly towards the magnetic poles). Steps can be taken to protect critical systems against the unwanted direct currents that are produced during GIC events (see, e.g., Bolduc, 2002) but it remains unclear how these attempts to compensate for current surges would fare in the face of an event as powerful as Carrington.

The fleet of satellites orbiting Earth would perhaps fare worst during a Carrington event. Satellites experience a form of premature ageing, especially those on high orbits such as the geostationary orbit (GEO), depending on the strength of the solar wind and the extent to which the satellite is bombarded with energetic protons. Worryingly, it seems that the satellite era has experienced a significantly quieter Sun than the 400 years preceding it (although the Carrington event holds the record for the largest flux of solar protons in the last 500 years). A Carrington-level superstorm would cause significant losses of satellites due to the bombardment of protons. Predictions for the economic cost during the current solar cycle (2007–2018) if a Carrington-level flare was to hit the Earth were placed at around $30 billion (Odenwald and Green, 2007). Disruption to location-tracking systems such as GPS are almost certain, as well as the myriad other services that rely on rapid communication across the planet's surface. In combination with GICs damaging or destroying undersea cabling, our modern society will perhaps be struck completely dumb by a powerful flare event.

The observational evidence for stellar activity is not limited to the Sun. For example, observations of low-mass stars of spectral type M show a wide range of activity signatures, particularly in the ultraviolet and X-ray wavelengths. The statistics indicate that superflares, flares with energies 10–100 times greater than solar flares, are quite common amongst M dwarf stars (Chang et al., 2015). For example, the TRAPPIST-1 system, with three 'potentially habitable' planets, experiences a Carrington-level flare once every few weeks (Vida et al., 2017).

The current predilection for searching M stars for habitable (and inhabited) planets should take note of the host stars' increased propensity for particularly damaging stellar activity. A planet orbiting within the habitable zone of an M star is likely to undergo synchronous rotation – i.e., its rotation period is close to or equal to its orbital period. This extremely slow rotation is likely to prevent the production of a planetary magnetosphere. Coupled with the fact that M star habitable zones are much closer (of order 0.1–0.2 au), we should expect that a superflare will cause significant damage to a planet's atmosphere. The increased UV radiation from a superflare will go into photolysing ozone in the upper atmosphere. While current calculations suggest the depletion of ozone will be of order a few percent only (Segura et al., 2010), repeated flare activity may compound this reduction, removing the atmosphere's capability of screening organic material from harmful radiation. Studies of the current crop of exoplanets in the habitable zone suggest that most, if not all detected so far are at serious risk of being rendered uninhabitable (Solution R.9) in the face of such prolific stellar activity (Armstrong et al., 2016).

It seems that stellar activity alone is sufficient to destroy civilisation in the spectacular way a self-perpetuated nuclear holocaust might (see section 16.3). It will certainly pose a major disruption to business as usual on planetary bodies like the Earth, and certain death for those operating in space.

10.1.3 Earth's Protection from Stellar Activity

The Earth's magnetic field is sufficient protection for the Sun's day-to-day activity. Extreme events (like the Carrington event) can penetrate this shield and cause harmful ground currents as previously described, and satellites are particularly prone.

Earthlike planets around low-mass stars must orbit much closer to their host to sustain surface liquid water. These stars are also generally more active, as the stellar magnetic dynamo grows in efficiency with the stellar rotation rate and the depth of its convective envelope. M dwarf stars can indeed be fully convective, and hence for a given rotation rate will typically be much more active than G stars. As a result, planetary habitability around M dwarfs requires strong planetary magnetic fields to sustain a magnetosphere of similar size to the Earth. The standoff distance of Earth's magnetopause is typically around $12 R_\oplus$. Placing the Earth in orbit of typical M dwarf stars decreases the standoff distance to $6 R_\oplus$ (Vidotto et al., 2013). This can reach as low as $1 R_\oplus$ if the planet is in synchronous rotation, which weaken's the planetary magnetic moment (Khodachenko et al., 2007). As the standoff distance decreases, greater quantities of the upper atmosphere can be lost, resulting in the destruction of habitable environments.

10.2 Stellar Evolution

Stars cannot live forever, and their deaths have consequences. To understand these consequences, we must consider the stellar life cycle, which begins once a collapsing protostar begins fusing hydrogen in its core. Astronomers refer to this as

10.2 Stellar Evolution

'joining the main sequence'. The *main sequence* describes all stars that are fusing hydrogen in their cores. Stars on the main sequence obey a series of very simple relations due to the relative simplicity of their internal structure. This is illustrated by the main sequence's position on the Hertzsprung Russell Diagram (or HR diagram for short). Typically, the HR diagram's axes are the star's *effective temperature* T_{eff}, and its luminosity L. The effective temperature is the temperature that satisfies the Stefan-Boltzmann Law for the star (with radius R):

$$L = 4\pi R^2 \sigma_{SB} T_{eff}^4, \tag{10.2}$$

where σ_{SB} is the Stefan-Boltzmann constant. Commonly, observational HR diagrams are constructed by computing the star's *magnitude* in a variety of bands. The *apparent magnitude* m_b of a star in a band b is computed from its flux F_b:

$$m_b = -2.5 \log_{10} \frac{F_b}{F_{b,0}} \tag{10.3}$$

where we must first define a zero point for this filter $F_{b,0}$. The *absolute magnitude* M_b of the star is defined as its apparent magnitude observed at a distance of 10 pc:

$$M_b = m - 5 \log_{10} \left(\frac{d}{1\text{pc}}\right) + 5 \tag{10.4}$$

One can then define multiple *colours* for a star by subtracting apparent magnitudes from two filter bands. The stellar effective temperature is related to its colour, and its absolute magnitude is a measure of luminosity, so it is possible to construct an HR diagram using a colour magnitude plot (see Figure 10.2).

How far along a star sits on the main sequence indicates the star's mass, with more massive stars occupying the upper-left regions.

Fusion of hydrogen into helium proceeds either by the *proton-proton chain* or by the *CNO-cycle* (where C, N and O indicate the respective elements carbon, nitrogen and oxygen). The principal proton-proton (pp) chain fuses helium directly from hydrogen (producing helium-3 as an intermediate step), while the other three pp-chain pathways rely on the presence of trace amounts of beryllium and lithium. The CNO cycle is a catalytic process, with carbon, nitrogen and oxygen being continually regenerated as hydrogen is fused into helium.

During the main sequence era, the star uses the energy released by nuclear fusion to provide thermal pressure against gravitational collapse, resulting in a stable star. As time goes on, and the star continues to fuse hydrogen into helium, the local hydrogen supply in the core begins to run out. Once the core's hydrogen is depleted, the star temporarily loses its principal power source, which has been an essential balance against the force of gravity. It is at this stage that the star leaves the main sequence.

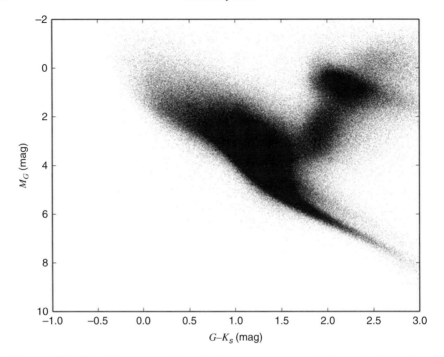

Figure 10.2 The Hertzsprung Russell Diagram, generated from the Gaia Data Release 1. Note that effective temperature (x-axis, where spectral colour is used as a proxy) increases from right to left. Most stars are located along the main sequence (central band), with more evolved stars moving to the upper right of the diagram. Credit: ESO

We can estimate the main sequence lifetime of a star by considering its energy generation mechanism (nuclear fusion) and the available fuel. If we can represent the effective temperature T_{eff} and the radius R as a function of stellar mass, we can use equation (10.2) to derive a mass-luminosity relation.

A simplistic derivation would proceed as follows. Consider first the condition for hydrostatic equilibrium, that the pressure force due to thermal energy from nuclear fusion is matched by that of gravity:

$$\frac{dP}{dr} = -\frac{Gm(r)\rho(r)}{r^2} \quad (10.5)$$

For simplicity, we will assume that the star has a uniform density ρ, and we will only look at a proportional relationship. We can immediately substitute

$$m(r) = 4\pi r^3 \rho \quad (10.6)$$

and integrate equation (10.5) from the core to the outer radius R, giving the average pressure

10.2 Stellar Evolution

$$<P> \propto \frac{GM^2}{4\pi R^4} \tag{10.7}$$

Substituting the ideal gas law:

$$P = nk_B T = \frac{\rho k_B T}{\mu m_H} \tag{10.8}$$

gives an expression for $T(R)$:

$$k_B T \propto \frac{GM \mu m_H}{R} \tag{10.9}$$

The above expression reflects that the system is in equilibrium, and that the local thermal energy is approximately equal to the local potential energy. Substituting for both R and T in equation (10.2) gives

$$L \propto M^{3.33} \tag{10.10}$$

The true relation is complicated by the density and temperature gradients in stars, which depend on which nuclear fusion reactions are occurring in the core, and whether convective layers are present. Observed main sequence relations are as follows (Duric, 2004):

$$\frac{L}{L_\odot} = \left(\frac{M}{M_\odot}\right)^4 \qquad M \geq 0.43 M_\odot \tag{10.11}$$

$$\frac{L}{L_\odot} = 0.23 \left(\frac{M}{M_\odot}\right)^{2.3} \qquad M < 0.43 M_\odot \tag{10.12}$$

The main sequence lifetime can then be gauged by the simple relation between consumption rate and fuel supply:

$$\tau_{MS} = \frac{\epsilon M}{L} \tag{10.13}$$

where the ϵ parameter reflects the fact that not all the star's mass is available for nuclear fusion. The scaling relation is therefore:

$$\frac{\tau_{MS}}{\tau_{MS,\odot}} = \left(\frac{M}{M_\odot}\right)^{-3} \qquad M \geq 0.43 M_\odot \tag{10.14}$$

$$\frac{\tau_{MS}}{\tau_{MS,\odot}} = \left(\frac{M}{M_\odot}\right)^{-1.3} \qquad M < 0.43 M_\odot \tag{10.15}$$

More-massive stars have a shorter main sequence lifetime. The Sun's main sequence lifetime $\tau_{MS,\odot} \approx 10$ Gyr. Note that at the critical value of $M = 0.43 M_\odot$, the main sequence lifetime is 125 Gyr, which corresponds to almost ten times the current age of the Universe! Conversely, at $M = 8 M_\odot$, the main sequence lifetime is a relatively brief 20 million years.

As the temperature of the core has slowly increased throughout the star's main sequence lifetime, when the core ceases hydrogen fusion, the core's shell can continue. The increased temperature in the shell ensures that the energy produced by the shell exceeds the energies produced during the main sequence phase. Much of this extra energy is absorbed by the outer envelope, resulting in its expansion.[1]

This cools the star, reducing its effective temperature. For stars below $1 M_\odot$, this results in the star moving along the *subgiant branch* to join the *red giant branch* (RGB). More-massive stars tend to move very quickly along the subgiant branch, resulting in a discontinuity in the HR diagram (known as the 'Hertzsprung Gap'; see Figure 10.3).

As the core's helium content continues to grow, its mass eventually exceeds the *Schönberg-Chandrasekhar limit*, and can no longer support itself hydrostatically, resulting in immediate core collapse. What happens next depends on the strength of the star's gravity, i.e., the star's mass. We can classify stars into three groups:

Low-Mass Stars: 0.8–2 $M_\odot$ Produce a helium core supported by *electron degeneracy pressure* – a quantum mechanical effect which forbids the electrons in the core to occupy the same quantum state (Pauli's Exclusion Principle; see section 10.5). The star then transitions to a red giant phase, before ascending the asymptotic giant branch (AGB). After the outer envelope is dissipated by winds and other processes, the end product is a white dwarf.

Intermediate-Mass Stars: 2–8 $M_\odot$ As low-mass stars, they also ascend the AGB, but these objects can fuse helium, producing a degenerate carbon-oxygen core. The end product is a carbon-oxygen white dwarf.

High-Mass Stars: > 8 $M_\odot$ continue to fuse carbon in a non-degenerate core. May continue to fuse heavier elements on the periodic table until iron is reached. Iron fusion requires energy rather than produces it, and hence an unstoppable core collapse results in a Type II supernova.

Very low-mass stars (between 0.08 $M_\odot$ and 0.3 $M_\odot$) experience a different evolution entirely, due to the presence of convection. Stars below 0.3 $M_\odot$ are fully convective, therefore they can use convection to completely mix their quantities of hydrogen and helium. In this case, the core no longer exhausts its supply of hydrogen, but replenishes it from material in the envelope, and effectively never leaves the main sequence.

---

[1] This reversal of behaviour between what is inside the shell-burning region (core) and outside it (the envelope) is often referred to in stellar evolution as the *mirror principle*. It is empirically established that wherever a shell-burning region is set up inside a star, if the region inside is contracting, the region outside is expanding, and vice versa.

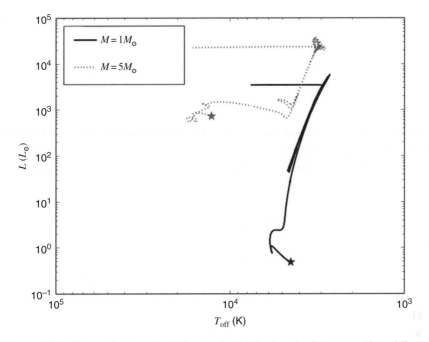

Figure 10.3 The evolution of a $1M_\odot$ star (solid line) and a $5M_\odot$ star (dotted line) on the HR diagram. We begin the tracks with both stars on the main sequence (indicated by the star symbols). The $1M_\odot$ star evolves along and off the main sequence, through the subgiant branch to the red giant branch (RGB). Once the star begins fusing helium in its core, it drops back down and joins the asymptotic giant branch (AGB), which closely tracks the RGB. After reaching the tip of the AGB (when helium burning in the core ceases), the star evolves towards the planetary nebula/white dwarf phase (not plotted). The $5M_\odot$ star jumps rapidly across from the main sequence to the RGB (through the Hertzsprung Gap), and executes a 'blue loop' before joining the AGB, as it is able to begin helium fusion into carbon before it leaves the RGB. Data generated by the MESA Isochrones and Stellar Tracks project using the MESA stellar evolution code.

Objects with masses below 0.08 $M_\odot$ are incapable of completing either the pp-chain or CNO cycle, and therefore cannot fuse hydrogen into helium. However, they are capable of carrying out the second stage of the pp-chain, where hydrogen fuses with one of its isotopes, deuterium (which contains an extra neutron):

$$^2_1\text{H} + ^1_1\text{H} \rightarrow ^3_2\text{He} + \gamma \tag{10.16}$$

Objects in this regime are referred to as *brown dwarfs*. These objects are much cooler than stars, with atmospheres dominated by molecules, dust and clouds, much like gas giant planets. Indeed, there now appears to be a continuous spectrum of objects between a few Jupiter masses (i.e., $\sim 0.001 M_\odot$), through the brown dwarf regime at tens of Jupiter masses ($\sim 0.01 M_\odot$), right through to stars $\gtrsim 0.1 M_\odot$.

The above description of stellar evolution, even as a brief summary, is incomplete, as it refers to single stars. It is believed that approximately half of the stars in the Milky Way exist in multiple systems (Duquennoy and Mayor, 1991; Raghavan et al., 2010). If the binary system is relatively compact, i.e., the stars are reasonably close to each other, then they can exchange mass. As the properties of stars depend sensitively on their mass, this mass transfer can significantly alter the trajectory of a star's evolution. A full account of pre-main sequence, main sequence and post-main sequence evolution is worth several textbooks of discussion, and very much outside the scope of this work, but we will highlight some important factors originating in stellar evolution that might affect planetary habitability.

10.3 Death by Stellar Evolution

We have looked at how stars evolve off the main sequence, and it is probably obvious to the reader that violent stellar explosions and abrupt changes to stellar structure are likely to destroy biospheres. We will consider these possibilities in the following sections, but we should be clear that habitats are likely to be destroyed even while stars remain on the main sequence.

This has of course been investigated in detail for the Sun. As the Sun fuses hydrogen into helium, its temperature (and luminosity) have been steadily increasing. We can see this by considering the mean molecular weight μ in equation (10.9) – fusing hydrogen in helium increases μ, hence increasing T (and subsequently L via equation 10.2).

This is the origin of the 'faint young Sun' paradox – the Sun's luminosity was only around 75% of its current value shortly after the Earth's formation. Conversely, it determines the future fate of life on Earth. We can approximate this luminosity evolution as follows (Schröder and Connon Smith, 2008):

$$L(t) = \left(0.7 + 0.144 \frac{t}{1 \text{ Gyr}}\right) \frac{L_{*,0}}{L_\odot} \qquad (10.17)$$

From our calculations of the habitable zone in section 4.2, it is clear that the habitable zone of the Solar System will move outward in time, and the Earth's surface will eventually become too hot for most forms of life, including human civilisation.

In a series of papers, O'Malley-James and collaborators have investigated what they dub 'the swansong biosphere' of Earth (O'Malley-James et al., 2013, 2014). As the planet's temperature increases, the rate at which silicate rocks weather increases, drawing down more carbon from the atmosphere and reducing the carbon dioxide levels. Photosynthesis requires a minimum CO_2 level – once this lower level is reached, oxygen production fails at last, and the atmosphere's oxygen content is depleted. This spells the end for most forms of multicellular life, with larger mammals and birds being the first to go, closely followed by their smaller cousins.

Fish, amphibians and reptiles are more resistant to temperature increases, but as the oceans begin to turn to steam, these vertebrates also begin to die off. Eventually, insects and other invertebrates will also perish, leaving a microbial world around 0.1 Gyr after the first photosynthesisers fail.

As time progresses, even microbes reach their limit and fail, leaving only extremophilic organisms, which eventually perish as the oceans end in a rapid phase of evaporation. The water vapour is photodissociated by UV radiation, producing a hydrogen-oxygen atmosphere that lasts for a (geologically) brief instant.

Most studies agree (see, e.g., Rushby et al., 2013) that the end of the Earth's biosphere will occur approximately 1.75 Gyr from now, some 3.5 Gyr *before* the Sun leaves the main sequence (Solution R.10). This is the end of the Earth as a habitat for life, but it could be argued that intelligent life would be able to construct new habitats, either on other planets at greater distance from the parent star, or in interplanetary space. This may buy enterprising civilisations some extra time, but such plans will only provide for an existence while the star remains on the main sequence.

10.4 Death by Red Giant

As low- and intermediate-mass stars evolve off the main sequence and initiate hydrogen shell burning, the star enters the *red giant branch* (RGB) of the HR diagram. The star begins to cool, as its primary temperature source (the core) is inactive. The envelope of the star begins to expand, thanks to the mirror principle. This expanding material can swell to an enormous size compared to the size of the main sequence progenitor, which ensures the star continues to increase in luminosity.

Once helium begins to fuse in the star's core, the star begins to increase in temperature again, and the envelope begins to contract, reducing the overall luminosity. The nature of this helium-burning phase depends on the stellar mass.

For example, once the Sun reaches the red giant phase, it is expected to swell in radius from $1R_\odot$ to around $10R_\odot$ while it ascends the RGB, and then to as much as $100R_\odot$ during its ascent of the AGB, while increasing in luminosity to as much as $30L_\odot$ (while its effective temperature lowers to around 2,000 K; see Figure 10.3).

In other words, the Sun will eventually expand to at least the orbit of Venus, and most likely to the orbit of Earth at 1 AU, with an increasing rate of mass loss due to stellar winds. As we have already discussed in the previous section, the Earth's biosphere will have been extinguished long before this occurs. If humans still exist in the Solar System during this epoch, they must already have moved to more clement planets/moons/habitats.

The habitable zone of sunlike stars during the red giant phase is of course much further away than during the main sequence. The distance to the inner and outer habitable zone edges can be estimated simply, if one knows the critical insolation flux at each edge ($S_{HZ,in}$, $S_{HZ,out}$), then the inner and outer radii of the habitable zone are given by

$$r_{HZ,in} = \sqrt{\frac{L_*}{4\pi S_{HZ,in}}} \tag{10.18}$$

$$r_{HZ,out} = \sqrt{\frac{L_*}{4\pi S_{HZ,out}}} \tag{10.19}$$

A thirty-fold increase in luminosity would result in the habitable zone boundaries being extended by around a factor of 5, i.e., the location of Jupiter and its moons in the Solar System. Of course, this is an oversimplified analysis, and the correct result requires a full radiative transfer calculation, incorporating the change in the star's temperature and spectrum. The important point is that as the star evolves, new regions of the planetary system reach temperatures suitable for terrestrial-type life.

However, the star's evolution has consequences for orbital stability. The mass of the Sun has been relatively concentrated at the centre of the Solar System during the main sequence phase. Now, this mass is being redistributed over a significantly larger volume, and a significant fraction is being lost to stellar winds. This redistribution of mass changes the gravitational field present in the Solar System.

What happens next can be determined by computer simulation. N-Body simulations of planetary systems as the star loses mass can achieve new levels of complexity – depending on the nature of the mass loss, the equations of motion for an N-Body calculation can be written in at least 22 ways!

As an example, we can select a reasonably simple set of conditions. Consider a star with a single planet, where both bodies are able to lose mass. A typical two-body calculation has the following equation of motion:

$$\frac{d^2\mathbf{r}_i}{dt^2} = -\frac{G(M_1 + M_2)}{r^3}\mathbf{r} \tag{10.20}$$

The same calculation can be applied in this case, with the proviso that M_1 and M_2 are functions of time. If we assume that the mass loss is isotropic, the resulting equation then becomes (Omarov, 1962; Hadjidemetriou, 1963):

$$\frac{d^2\mathbf{r}_i}{dt^2} = -\frac{G(M_1(t=0) + M_2(t=0))}{r^3}\mathbf{r} - \frac{1}{2(M_1(t) + M_2(t))}\frac{d(M_1 + M_2)}{dt}\mathbf{r} \tag{10.21}$$

This result is a direct application of the product rule in differentiation of the velocity $\frac{d\mathbf{r}}{dt}$. The first term represents the standard acceleration due to change in position, and the second term describes the perturbation due to change in object masses. If we consider body 1 to be losing mass and body 2's mass to be fixed, then the acceleration of body 2 slowly increases. As a result, its orbital semi-major axis also increases. The eccentricity evolution is not quite so guaranteed, but it is the case that the periapsis distance $a(1-e)$ also increases (see the review by Veras, 2016, and references within for more).

This result is easily extensible to multiple bodies (if we again assume that only a single body is losing mass isotropically):

$$\frac{d^2\mathbf{r}_i}{dt^2} = -\frac{G\sum_i M_i(t=0)}{r^3}\mathbf{r} - \frac{1}{2\sum_i M_i(t)}\frac{dM_1}{dt}\mathbf{r} \tag{10.22}$$

If mass loss is anisotropic, then the situation is further complicated (although assuming isotropy appears to be relatively safe for planetary systems within a few hundred AU of the star). From the above equation, we can make some broad inferences about the dynamical stability of post-main sequence planetary systems. Every body in the system will increase both its semi-major axis and periapsis distance. As each body's semi-major axis will evolve at different rates according to their mass and distance from the star, this is quite likely to result in orbit crossing, which will inevitably result in strong gravitational interactions. It is quite possible, even likely, that planets will be ejected from the system during this phase.

After the RGB and AGB phases, the star will have lost a substantial fraction of its mass and settled into the white dwarf phase. White dwarfs are extremely dense, containing around a solar mass of material within a radius of order that of Earth. They are relatively stable objects, typically composed of carbon and oxygen, which cool from temperatures of 10^4 K over timescales comparable with the age of the Universe, supported by electron degeneracy pressure (as opposed to the neutron degeneracy pressure that supports neutron stars).

By the time that the white dwarf phase is reached, planetary systems have been 'unpacked' into stable or unstable configurations depending on the system's initial configuration. The system can become unstable at very late times, or be 'repacked' into a stable configuration that lasts a very long time (at least as long as the age of the Universe). The habitable zone around white dwarfs evolves very slowly, although it now resides at much closer distances than during the RGB phase (Barnes and Heller, 2013).

The ultimate question then is: can intelligent civilisations survive this turmoil? The SETI literature often assumes that spacefaring civilisations would abandon stars as soon as they left the main sequence, but we should be tempted to reconsider the possibility that some species might choose to remain. While planets in the main sequence habitable zone are at serious risk of engulfment by the star in the post-main sequence phase, or ejection from the star system via dynamical instability, it is not taxing to conceive of contingency plans by which large-scale artificial habitats might ride out this crisis. Assuming an orbit highly inclined to the ecliptic of their star system would reduce the probability of collision with another planet, and the evolution of the parent star would need to be tightly monitored to ensure the habitat's orbit was correctly maintained, and the radiation environment remained suitable.

There may even be opportunities within this crisis. The energy output from the central star increases significantly, which always has advantages if harnessed appropriately. The stellar wind may provide a valuable source of hydrogen, helium and other high-energy particles. The significant heating of asteroids and other bodies by the enhanced radiation field might yield new possibilities for extracting material. Finally, the collision and destruction of planetary bodies might provide access to material otherwise inaccessible. Advanced civilisations may be able to mine the contents of terrestrial planet cores and mantles after cataclysmic collisions.

10.5 Death by Supernova

The deaths of relatively massive stars result in spectacular explosions, known as supernovae. These explosions not only result in the production of a spectacular supernova remnant, they also produce large quantitites of high-energy photons, cosmic rays and neutrinos, which can affect planets at quite some distance. Supernovae are commonly divided into two types.

10.5.1 Type Ia (White Dwarf Nova)

The true progenitors of Type Ia supernovae remain a source of heated debate. There are two camps, which both propose that a white dwarf and a companion are responsible. Both camps agree that the source of the supernova explosion involves a white dwarf's mass exceeding what is known as the *Chandrasekhar limit*. In other words, Type Ia supernovae require low-mass stars and a source of accretion material to occur.

Beyond the Chandrasekhar limit, electron degeneracy pressure can no longer protect the white dwarf against collapse. We can see this by estimating the electron degeneracy pressure under two assumptions:

10.5 Death by Supernova

1. Pauli's exclusion principle, which allows no more than one electron in each quantum state, and
2. Heisenberg's uncertainty principle, which limits our simultaneous knowledge of a particle's position x and momentum p, which can be expressed in terms of the uncertainties in each quantity

$$\Delta x \Delta p \approx \bar{h} \tag{10.23}$$

Let us assume that the pressure produced by the electron gas is given by a standard result from the kinetic theory of gases:

$$P = \frac{1}{3} n_e p v \tag{10.24}$$

We will assume for simplicity that all the electrons have the same velocity v, and we can link the number density of electrons n_e to the position simply: $\Delta x \approx n_e^{-1/3}$. Heisenberg's uncertainty principle can then give us an estimate of the typical electron momentum

$$p \approx \Delta p \approx \bar{h} n_e^{1/3} \tag{10.25}$$

In the non-relativistic case, the velocity is simply $v = p/m_e$ and hence

$$P \approx \frac{n_e p^2}{3 m_e} = \frac{\bar{h}^2 n_e^{5/3}}{3 m_e} \tag{10.26}$$

This gives a relation $P \propto \rho^{5/3}$, which can be shown to be dynamically stable, i.e., a small perturbation will result in the system returning to dynamical equilibrium. However, this is only acceptable if the electron velocity is much lower than c. As we increase the mass of the white dwarf, the radius shrinks, and hence v continues to increase until we reach the relativistic limit: $v = c$ and

$$P \approx \frac{n_e p c}{3} = \frac{\bar{h} n_e^{4/3} c}{3} \tag{10.27}$$

This equation of state $P \propto \rho^{4/3}$ is not dynamically stable, and any perturbation will result in collapse. A detailed calculation shows that this change in the equation of state, the Chandrasekhar limit, occurs at $M = 1.44 M_\odot$, and hence any white dwarf that exceeds this mass will become unstable.

It is widely agreed by astronomers that a white dwarf that exceeds this Chandrasekhar limit will explode as a Type Ia supernova (Maoz and Mannucci, 2012; Wang and Han, 2012). Disagreement springs from how the white dwarf accretes the matter required to exceed the limit.

The single degenerate (SD) hypothesis proposes that a white dwarf accretes material from a non-degenerate companion, such as a main sequence star or a giant. As this accretion pushes the mass of the white dwarf over the Chandrasekhar limit, the white dwarf explodes in a thermal runaway. The white dwarf can either obtain the material via gravitational attraction, or through the winds generated by the companion. Critics of this process point out that the accretion process requires fine tuning – too vigorous an accretion flow results in shock heating, which prevents the accretor from contracting and reaching the Chandrasekhar mass. Equally, a low accretion rate results in the material lying on the white dwarf surface. The slow accumulation of this material results in a rapid burst of hydrogen fusion on the white dwarf surface, removing the mass through radiation and outflows.

In the opposing double degenerate scenario (DD), a white dwarf binary system loses angular momentum and energy, allowing the two stars to spiral inward and merge. The final product of the merger might be a single white dwarf that far exceeds the Chandrasekhar limit and ignites, or one white dwarf is torn apart by the other thanks to tidal forces. The remains of one is then rapidly accreted by the other until it reaches the Chandrasekhar limit. Detractors note that white dwarf mergers may result in the formation of a core-collapse supernova rather than a Type Ia (see following sections). Searches for white dwarf binaries that could act as progenitor to DD supernovae have yet to deliver suitable candidates, although these searches are demonstrably incomplete and prone to selection effects.

10.5.2 Type II Supernovae (Core-Collapse)

This is the fate of stars of $8M_\odot$ and above (Smartt, 2009). These massive stars can continue to produce increasingly heavier elements through nuclear fusion. From the main sequence phase of hydrogen burning to produce helium, helium burning commences to produce primarily carbon, carbon burning to produce primarily oxygen, oxygen burning to produce primarily silicon, and finally silicon burning to produce principally iron. As each heavier element is produced (along with other elements such as neon, sodium, and magnesium), less and less energy is released per gram of reactant.

Once iron fusion begins in the core, the energy balance from nuclear fusion changes: the energy required to fuse iron is larger than the energy produced by the fusion (an endothermic fusion reaction). The balance between thermal pressure from the fusion and gravitational forces fails, and collapse begins, initially in a manner very similar to that for low-mass stars.

However, the massive stars have a much more diverse arsenal of elements in their innards. The cores of massive stars at this phase resemble an onion skin, with an outer layer of hydrogen-burning material concealing helium, with a helium-burning shell beneath, and so on (Figure 10.4).

Rather than continue fusion, iron can in fact be destroyed by photodisintegration, where photons encourage the fission of iron into helium. Further disintegration

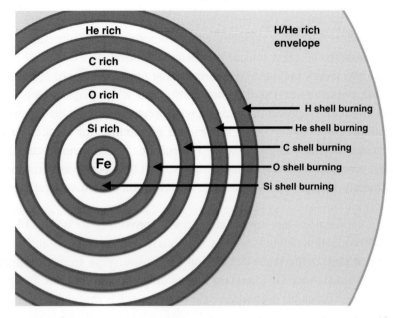

Figure 10.4 The 'onion-skin' structure of a star with a mass above $8M_\odot$. Layers of various elements are undergoing 'shell burning', fusing to form the next element upwards in the sequence H, He, C, O, Si, Fe (with other elements in relatively low abundances). Fe cannot be successfully fused as the fusion reaction becomes endothermic.

produces a dense plasma of protons and neutrons. Under the intense pressure of the core, the free electrons that were supporting the core through electron degeneracy pressure now combine with the protons, producing neutrons:

$$p^+ + e^- \rightarrow n + \nu_e \tag{10.28}$$

The core radiates a significant amount of energy away through electron-neutrinos (ν_e), and has now completely lost its electron degeneracy pressure support. The inner iron core begins to collapse, eventually exceeding the local sound speed and decoupling from the rest of the onion layers of the core. Eventually, the Pauli exclusion principle prevents the neutrons from occupying identical quantum states (neutron degeneracy pressure). The core rebounds, sending a shock wave upward from the core.

The neutrinos emitted during the collapse form a *neutrinosphere* behind the shock, with a total energy of around 10^{46} joules. Around 1–5% of this energy is transferred to the shock, generating enormous kinetic energy and causing the observed explosion. Around 10^{42} joules are released as photons, allowing the supernova to briefly radiate 100 times as much as the Sun would over its entire existence.

If the initial star mass is smaller than around 25 $M_\odot$, the neutron core stabilises to form a neutron star. They typically compress around $1M_\odot$ of neutrons into an object with a radius of a few tens of kilometers. Neutron stars owe their existence to neutron degeneracy pressure, and cannot exist above a mass of approximately $3M_\odot$. If this mass is exceeded, then the neutron star will collapse to a black hole. Stars with initial mass greater than 25 $M_\odot$ will directly produce a black hole during a core-collapse supernova. These supernovae (sometimes referred to as hypernovae) produce similarly large quantities of neutrinos as their neutron-star-forming cousins. Depending on the geometry of the explosion, hypernovae can continue to shine extremely brightly if they form an accretion disc from the stellar remnant (see section 11.2).

Would a civilisation dare to remain in close proximity to such a blast? The answer to this question depends on how well interplanetary habitats and spacecraft can parse and deflect the intense radiation and cosmic rays from the supernova. Minimum safe distances for planetary bodies are well-discussed in the literature (see section 11.3), but the minimum safe distance for a spacecraft is much less explored.

The remains of a star post-supernova have attractions for spacefaring civilisations. The neutron stars are potential sites for rather powerful Dyson spheres (Osmanov, 2018, see section 2.2.1). The frequency of a pulsar can be used as a highly accurate clock, forming the basis of a relatively simple but highly accurate Galactic navigation system (Downs, 1974; Becker et al., 2013). Their powerful emission can be used as a communication tool to synchronise directional communication between previously uncommunicating civilisations (Edmondson and Stevens, 2003). Interstellar explorers can also use the extremely powerful gravitational field in the close proximity of the neutron star to accelerate spacecraft (see section 19.3.7).

10.6 Close Stellar Encounters

The orbital evolution of stars in the Galactic potential guarantees that the distance between any two stars can evolve significantly with time. As we discussed in section 8.2, it is most likely that the cometary bodies arriving at Earth from the Oort Cloud are doing so due to the Galactic tide and perturbations from stars on close approach to the Solar System. The Galactic tide tends to dominate the secular evolution of cometary orbits, and is the principal agent of inner Solar System impactors. However, the stochasticity of stellar encounters allows the Oort Cloud to maintain an isotropic nature. Also, the Galactic tide struggles to inject comets with $a < 20,000$ AU into the inner Solar System – encounters are necessary, and such stars must come within 4×10^5 AU to drive such behaviour.

10.6 Close Stellar Encounters

The precise nature of the Galactic tide, and the probability of stellar encounters with a close approach of less than 1 pc, depend strongly on the location of the star. This is further complicated by the fact that stars can undergo a great deal of radial migration through the Galactic disc. For example, recent simulations indicate the Sun may have moved as much as 5–6 kpc in radius from its present position. Martínez-Barbosa et al. (2017) calculated the stellar encounter rate along the Sun's orbital trajectory (for close approaches less than 4×10^5 AU) to range between 20 and 60 encounters per Myr depending on the trajectory's precise nature (migration inward, outward, or neither).

These calculations can now be squared against growing surveys of stellar positions and proper motions. The Hipparcos astrometric survey identified several stars that will pass within 1 pc of the Sun (such as GL 710, which will come within 0.34 pc, or 70,130 AU of the Sun approximately 1 Myr in the future). More recently, the low-mass binary system referred to as 'Scholz's star' (WISE J072003.20-084651.2) was shown to have passed as close as 52,000 AU only 70,000 years ago. This orbit would have driven deep into the Oort Cloud, and likely caused a significant comet shower into the inner Solar System (Mamajek et al., 2015).

What dangers would a particularly close stellar encounter entail? The principal threats to habitability from stellar encounters are

1. disruption of protoplanetary discs, preventing or disrupting the planet formation process,
2. perturbation of orbits in the planetary system, which can either modify a habitable planet's orbit and render it uninhabitable, or cause close encounters between neighbouring planets. These planet-planet interactions can either result in collision or even ejection of planets from the system, and finally, as we have already stated,
3. the perturbation of cometary clouds, which can inject large numbers of potential impactors into the inner Solar System.

We can estimate the effects of a given encounter by applying the impulse approximation. For example, we can estimate the velocity change of the Sun by a close encounter using:

$$\Delta v = \frac{2GM_s}{qv} \tag{10.29}$$

where M_s is the encountering star mass, q is the distance of closest approach (sometimes called the miss distance), and v is the relative velocity between the two stars. We can then compare this velocity change to the typical orbital velocity to determine if the resulting impulse is strong enough to disrupt its orbit.

For stellar encounters with sufficiently low miss distances to disrupt planetary systems (i.e., q of a few hundred AU), we require much higher stellar densities. For example, these densities would have existed in the Sun's birth cluster, the grouping of stars that the Sun belonged to soon after its formation. During the cluster's lifetime there would have been around a 20% chance that an encounter of miss distance 200 AU would have occurred (i.e., within about 100 million years).

We can estimate the encounter rate for a given miss distance q by considering a cylinder of the same radius surrounding the path of our star as it moves through the cluster. The cylinder's length will be proportional to vt, where v is the stellar velocity and t is the time interval under consideration, and hence its volume will be $\pi q^2 vt$.

The mean number of encounters experienced in a time interval t as the star traverses a region with stellar density n is

$$N = \pi q^2 vtn \tag{10.30}$$

And the encounter timescale can be found by setting $N = 1$:

$$\tau(q) = \frac{1}{\pi q^2 vn} \tag{10.31}$$

If we know the lifetime of the cluster, and we specify the miss distance we are interested in, then we can estimate the number of encounters in this lifetime. Simulations of young open clusters indicate the mean number of encounters with $q = 1000$ AU is 4 (Malmberg et al., 2011), with 75% of stars experiencing 2 or more encounters. Most of the encounters occur within the first 10 million years of the cluster's existence. As the cluster evaporates, the density of stars decreases and the encounter rate declines.

The situation is significantly worse for globular clusters, which are older, more massive groupings of 10^5–10^6 stars. However, it has been shown that significant orbit modification and ejection of habitable planets is a rare outcome, in both open and globular clusters (Stefano and Ray, 2016).

Despite this, encounters at thousands of AU still disrupt debris belts and cometary clouds, and we should consider this hazard carefully when considering the prospects for life and intelligence in stellar clusters.

11
Death on a Galactic Scale?

Perhaps the most 'rigorous' natural destruction mechanisms originate at the very largest of scales. After all, we are interested in hard solutions to Fermi's Paradox, and hard solutions characteristically exclude larger and larger fractions of potential civilisations from existing. If single events can destroy technology and biology on scales of thousands of light-years, this would carve out regions of the Milky Way that would fall silent simultaneously, ensuring that fledgling civilisations emerging in such a volume would not see any intelligent activity – at least, not yet (see section 18.2).

11.1 Active Galactic Nuclei

Observations indicate that every galaxy contains a supermassive black hole (SMBH) at its nucleus. This is also true for the Milky Way. The radio source known as Sagittarius A* is exceptionally bright and compact, with an X-ray component that is highly variable. A combination of very long baseline radio interferometry (VLBI) and infrared observations using adaptive optics has revealed that Sag A* is orbited by ionised clouds and stars. A series of observing campaigns over twenty years provides conclusive proof that these objects orbited Sag A* at high velocities, constraining the central object to have a mass of around 4 million times that of the Sun. VLBI has further revealed that the angular size of the Sag A* source is around 40 μas (Doeleman et al., 2008), confirming that this matter resides in a physical region no more than a third of a parsec across. The compact size of the source is further confirmed by the variability of the X-ray emission, which changes on timescales of a few days, giving a structure size $r \approx c\Delta t \approx 100 AU$.

Such an arrangement of matter must collapse to form a supermassive black hole. The radio and X-ray emission that we observe in Sag A* are produced by the matter that exists around the black hole. Our studies of other galaxies have demonstrated that the Milky Way's SMBH is relatively passive. Rather than

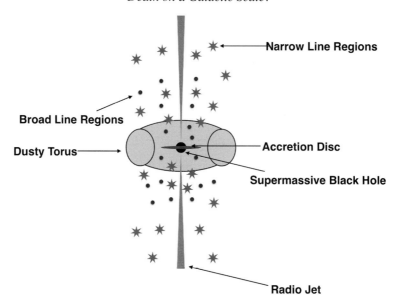

Figure 11.1 The standard unified model of active galactic nuclei (AGN). The central black hole is surrounded by an accretion disc of a few hundred AU in size, which is enshrouded in a dusty torus. This torus is punctured by powerful radio jets and winds emanating from the black hole, which illuminate broad and narrow line regions, with the breadth of the emission lines emanating from these regions a strong function of their distance from (and resulting orbital speed around) the black hole.

containing a relatively benign nucleus, these galaxies possess Active Galactic Nuclei (AGN).

Figure 11.1 depicts the classic 'unified model' of AGN. Immediately surrounding the central black hole is an accretion disc that directly feeds it. The material in the accretion disc slowly loses angular momentum, due to either gravitational instabilities or the magneto-rotational instability, and falls into the black hole. This disc is surrounded by a dusty torus that obscures the black hole (depending on the observer's orientation).

Magnetic fields at the inner edge of the accretion disc are wound extremely tightly, and launch powerful jets of ionised plasma. These jets emit strongly in the radio, and are launched along the polar directions of the black hole-disc system. The radiation from the black hole illuminates clouds of gas that emit either broad or narrow spectral lines, with the breadth of the line proportional to the relative speed of their motion around the black hole.

AGN are some of the most efficient producers of energy in the local Universe. The energy emitted from a black hole is related to the gravitational potential energy lost by material as it is accreted. This accretion luminosity is

11.1 Active Galactic Nuclei

$$L_{acc} = \epsilon \frac{G\dot{M}M}{2R} \tag{11.1}$$

where $\dot{M}$ is the accretion rate, M is the black hole mass, R is the accretion radius and ϵ is the radiative efficiency, typically around 10%. A typical black hole can emit around 10–20% of this energy as high-frequency radiation (ultraviolet, X-rays and γ-rays). Observed AGN can significantly outshine their host galaxies when active, in some cases by a factor of 10^4(!) (Krolik, 1999).

This tremendous energy output is thought to be important for the galaxy's evolution. If we assume a 10% efficiency for the energy released in feeding and growing a black hole of mass M_{BH}, this gives

$$E_{BH} = 0.1 M_{BH} c^2 \tag{11.2}$$

The binding energy of the galaxy's bulge can be approximated as $M_{bulge}\sigma^2$, where σ is the velocity dispersion of stars in the bulge. There is a relatively tight relationship between the black hole mass and bulge mass (Hring and Rix, 2004; Fabian, 2012):

$$M_{BH} \approx 1.4 \times 10^{-3} M_{bulge} \tag{11.3}$$

And as a result

$$\frac{E_{BH}}{E_{bulge}} \approx 1.4 \times 10^{-4} \frac{c^2}{\sigma^2} \tag{11.4}$$

Galaxies rarely have $\sigma > 400$ km s^{-1}. This implies $E_{BH}/E_{bulge} \gtrsim 100$. The energy released in growing supermassive black holes is several orders of magnitude larger than the binding energy of the local galaxy.

Only a fraction of this energy is likely to affect the host galaxy, and even then it will principally affect the dust and gas inventory. Injection of such enormous energy into the gas is sufficient to prevent star formation, as the significant heating of the gas would destroy any cool, dense material. The fact that the peak of star formation in the Universe is closely followed by the peak of AGN activity is unlikely to be a coincidence.

As galaxies accrete large amounts of gas, this material eventually loses sufficient angular momentum to spiral into the black hole. If this accretion rate is too large, AGN feedback suppresses the growth of the galaxy and halts star formation. This proposed regulation of star formation by a single entity at the centre of the Milky Way will have important implications for whether the 'biological clocks' of distant planetary systems become synchronised (section 18.2).

Our own galactic nucleus, Sag A*, remains inactive, but it is conceivable that it was active at some point in the past. Around 1% of a typical AGN's energy output is in the form of relatively hard X-rays. Applying typical relations for black hole mass and luminosity during the AGN phase, it has been calculated that Sag A* would have been as bright as (or brighter than) the Sun in X-rays (Chen and Amaro-Seoane, 2014). Stars orbiting the Galactic Centre closer than our distance of 8 kpc would most likely have experienced significant irradiation, and possible atmospheric losses (see section 11.3).

11.2 Gamma Ray Bursts

The most energetic events detected in the Universe can dwarf even an AGN for a brief instant. Gamma Ray Bursts (or GRBs) are flashes of intense gamma radiation, detected at a rate of several per day across the entire sky. Immediately after the gamma ray burst (which lasts somewhere between 0.01s and 1,000s), an afterglow is typically detected at a wide variety of wavelengths, from the X-rays to the ultraviolet, optical, infrared and radio.

In general, every GRB detected is co-located with a host galaxy, which becomes apparent after follow-up observations. The redshift of the host galaxies ranges from the relatively local ($z \sim 0.1$, or a luminosity distance of around 0.5 Gpc) to some of the most distant celestial objects ever detected (such as GRB090429B at $z = 9.4$, a luminosity distance of around 100 Gpc [Cucchiara et al., 2011]). Their detection at such vast cosmic range hints at the awesome energies released in a single burst. Simple estimates of the isotropic energy emission from typical GRBs is of the order 10^{45}J, which is comparable to $M_\odot c^2$, i.e., the total mass-energy of the Sun being liberated in a few minutes!

As with AGN, the variability of the burst emission also delivers clues about the typical size of the event causing the GRB. GRBs exhibit variability on millisecond scales, which suggests a diameter of a few hundred kilometres.

GRBs can be split into two categories, according to their burst duration, and their *spectral hardness*. Burst durations are typically measured using the T_{90} parameter, which measures the time at which 90% of the GRB emission has been received at Earth. The spectral hardness is the ratio of high-energy X-rays to low-energy X-rays, with the definition of high and low energy depending somewhat on the instruments being used.[1]

The two populations are: *short GRBs*, which possess burst durations of $T_{90} < 2s$, and a typically harder spectrum, and *long GRBs*, which possess longer burst durations $T_{90} > 2s$, and softer spectra. This dichotomy suggests different progenitors for each type, although the size scales of the events are approximately stellar for both long and short GRBs.

---

[1] For the ground-breaking Bursts And Transient Source Experiment (BATSE), the transition from high- to low-energy X-rays was held at 100 keV, i.e., a wavelength of 0.012 nm.

11.2 Gamma Ray Bursts

The long GRBs are thought to be *collapsars*, which are produced by particularly violent core-collapse supernovae. Consider a massive star (say 20 $M_\odot$) which is exhibiting Wolf-Rayet behaviour (essentially expunging its outer envelope through significant stellar winds). The core of the star is still approximately 10 $M_\odot$, which is sufficient to begin the core-collapse process required for a Type II supernova.

However, during the core-collapse, instead of a neutron star forming, the core continues runaway collapse until a black hole forms at the centre of the dying star. The angular momentum in the system prevents the black hole from immediately accreting all the surrounding stellar material – it can only access matter close to the rotation axis of the system, which is located along two funnels. This is still a sufficiently large amount of material to process on a relatively short timescale. As this material streams into the black hole, it leaves a low-density region. The disc quickly accretes onto the black hole, as its energy can be extracted rather efficiently via neutrino annihilation, or the intense magnetic fields generated during the collapse. The low-density funnels become the ideal location for a jet to be launched, which is what is observed by astronomers during a long GRB.

Short GRBs are widely thought to be due to the merger of two compact objects. Neutron star–neutron star (or indeed black hole–black hole) binaries are typically doomed to merge as their orbital energy is radiated away as gravitational waves (see section 21.2.2). As the objects grow closer, the gravitational radiation intensifies, resulting in an increasingly rapid inspiral, with the final moments that cause the burst taking less than a second to occur. The final object is a black hole with an accretion disc with mass $\sim 0.1 M_\odot$.

If these models are correct for GRBs, then we can make some basic predictions about their typical occurrence. Long GRBs are a consequence of massive stars, and as a result must occur near to or in active star-forming regions (as the main sequence lifetime of their progenitors is quite short). Indeed, there have been a handful of associations of long GRBs with apparent core-collapse supernovae (Galama et al., 1998; Stanek et al., 2003) but until recently this has only been possible at low redshift.

Short GRBs are not particularly constrained to occur in any specific region of galaxies. There does appear to be a correlation between low metallicity and increased GRB occurrence on the whole, but this may be a selection effect, as low-metallicity galaxies are also correlated with slightly higher star formation. This could simply be a consequence of the evolution of galaxies from gas-rich, metal-poor objects to gas-poor, metal-rich objects, which is consistent with GRBs being more frequent in the past.

11.3 Terrestrial Atmospheric Responses to Very High-Energy Events

As is clear, a supernova, AGN or GRB taking place in the vicinity of a planetary system has the capability of causing immense harm to its inhabitants. But what sort of damage might we expect? Most studies in this area have used Earth as a template. Much of this will apply to other potentially habitable planets, but we should note that much of the work in this area is a sensitive function of the oxygen content of the atmosphere. A worthwhile starting point for further reading is Melott and Thomas (2011), but note that the literature has moved forward a great deal since this.

11.3.1 UV Radiation

The response of the Earth's atmosphere to increased doses of high-energy radiation depends crucially on its ozone (O_3) content. Ozone absorbs UV radiation, splitting into oxygen molecules and atoms:

$$2O_3 + \gamma \rightleftharpoons 2O_2 + 2O \rightarrow 3O_2 \qquad (11.5)$$

This process is reversible: oxygen molecules can recombine to form ozone, and hence an atmospheric balance is established between O_2 and O_3, with the level populations of each molecule depending on the incident UV flux (in the 200 to 310 nm range). These reactions can be further catalysed by photolysis of other oxygen-bearing molecules, particularly NO and NO_2. A large number of species participate in this photodissociation network (see, e.g., Table 11 of Thomas et al., 2005a), and as such calculating the ozone depletion caused by irradiating an atmosphere requires sophisticated numerical simulations, allowing radiation to propagate through a planetary atmosphere and tracking the many photodissociation/other reactions.

The biological damage exerted by UV radiation can be significant. In particular, the absorption of UVB radiation contributes significantly to DNA and protein damage. Unicellular organisms are most prone to these effects, as they lack shielding mechanisms against such radiation. To give a concerning example, phytoplankton are responsible for around half of the world's photosynthetic activity, forming the base of the oceanic food chain. A significant UVB dose could therefore result in a significant food chain crash, with devastating consequences across the global ecosystem (Melott and Thomas, 2009).

In extreme cases, for example if a star orbits in close proximity to an AGN, extreme UV (XUV) radiation (photon energies above a few tens of eV) can result in atmospheric loss by hydrodynamic escape. Lighter atoms and molecules are driven away from the planet via a thermally driven flow, entraining the heavier atoms and resulting in significant atmospheric losses. It has been calculated that terrestrial planets orbiting within 1 kpc of our supermassive black hole, Sag A*, during its AGN phase could lose an atmospheric mass equal to that of the Earth (Balbi and Tombesi, 2017). The loss rate depends on the level of attenuation by the AGN torus. Planets orbiting well above the Galactic midplane could lose significant fractions of their atmosphere even at up to 3–4 kpc from the Galactic Centre.

11.3.2 Gamma and X-Rays

Some of the photochemistry and atmospheric loss described above can also be driven by the higher-energy gamma rays and X-rays. These are largely absorbed by the atmosphere, causing significant ionisation. This can also destroy ozone, as well as many other molecules (although not as efficiently as UV, and the timescale for their recombination is fairly swift, and hence the effect is highly transient). Absorbed X-ray and gamma ray photons can also be re-emitted as UV photons, adding to the total UV load and ozone depletion rate.

Thomas et al. (2005b) calculated the Earth's atmospheric response to a 10 s gamma ray burst around 2 kpc away. In this scenario, the resulting *fluence* (the flux integrated over a time interval) is approximately 100 kJ m^{-2}, and the ozone was depleted by approximately 35%, primarily due to the UV associated with the burst (locally, this depletion can increase to as large as 70%). After the burst subsided, oxygen recombination commenced, but took of order a decade to return to pre-burst levels.

An ozone depletion of this magnitude results in between two and three times more UV flux reaching the Earth's surface, typically far above the tolerance of most biological entities.

It also results in increased atmospheric levels of NO_2, which absorbs strongly at visible wavelengths. A transient NO_2 layer acts as a coolant, at levels likely to cause increased glaciation, and also reacts with OH to produce nitric acid:

$$NO_2 + OH \rightarrow HNO_3 \qquad (11.6)$$

which can then rain onto the surface. Its initial lowering of soil pH would be harmful to most organisms, but this would quickly subside. In fact, the nitrate deposition in the soil would mean an eventual *increase* in fertility. The initial harmful phase is unlikely to cause significant damage to a modern-day biosphere (Neuenswander and Melott, 2015), hence the nitric acid rainout is likely to be a net benefit.

11.3.3 Cosmic Rays

Cosmic rays may have effects at much greater distances than high-energy photons. The atmospheric ionisation caused by cosmic ray flux can significantly increase the lightning strike rate (Erlykin and Wolfendale, 2010), as well as boosting the rate of ozone depletion and nitrate rainout and altering the cloud formation process (Mironova et al., 2015). This ionisation is also likely to affect the Global Electric Circuit, with resulting feedbacks on the planetary magnetic field.

The cosmic ray flux from a supernova or gamma ray burst depends sensitively on the magnetic field intervening between the source and the habitable planet. A weak field fails to efficiently deflect cosmic rays, and hence the flux at TeV-PeV energies is significantly larger. Melott et al. (2017) suggest that even at distances of 50 pc, cosmic ray flux can result in moderate levels of extinction – indeed, that an elevated level of extinction on Earth around 2.6 million years ago is consistent with a supernova that exploded in the Tuc-Hor stellar group (Bambach, 2006).

11.3.4 Muons and Other Secondary Particles

As high-energy photons impact the upper atmosphere, we can expect *air showers* of high-energy particles to be produced as a form of secondary radiation.

These are principally generated in gamma ray burst events, as the photon energies of supernovae are insufficient. Indeed, long gamma ray bursts produce too soft a spectrum to generate a meaningful quantity of secondary particles.

For incoming photon energies greater than 1 GeV, we can expect at least 10^3 muons per 10^7 photons for a relatively strong, hard-spectrum gamma ray burst (Atri et al., 2014). This results in a relatively weak radiation dose for surface-bound organisms, and as such is considered a much weaker effect than high-energy photons.

However, the muon and neutron flux produced by cosmic rays can be more significant. Thomas et al. (2016) computed that for a supernova at 100 pc, where only a weak magnetic field can perturb the cosmic ray trajectories, the muon flux can exceed twenty times the regular background level. This is at least double the average radiation dose worldwide today. The deep penetration of muons means that such a dose would be administered to all but the deepest oceans.

11.3.5 The Role of Heliospheric Screening

Much of the above discussion has hinged on some level of protection afforded to Earth by its presence within the Sun's heliosphere. This magnetic protection shields us from the worst of Galactic cosmic rays (GCRs). If the heliosphere was somehow depressed and no longer encompassed the Earth, this screening effect would disappear, and the background cosmic ray flux may become substantial (even in the absence of a nearby supernova or GRB).

Such heliospheric depression can occur if the Sun passes through a molecular cloud. It has been estimated that the Solar System has entered dense molecular cloud regions some 135 times during its existence (Talbot Jr and Newman, 1977), with of order 10 events in the last 250 Myr. During these passages, if the number density of hydrogen atoms exceeds ~ 300 cm^3, this is sufficient to collapse the heliosphere. This permits the entry of an increased (or decreased) flux of GCRs, and anomalous cosmic ray events from where the molecular gas meets the boundary of the heliosphere (at the termination shock).

If this passage takes several Myr to complete, the Earth's magnetic poles may undergo a reversal (see section 9.2), and the cosmic ray flux can then induce ozone depletion of up to 40% (Pavlov, 2005a). The molecular cloud's gravitational potential can also disrupt the Oort Cloud, boosting impact flux. The cloud's dust content can reduce the Earth's received solar irradiance, resulting in snowball glaciation (Pavlov, 2005b).

Smith and Scalo (2009) find that the molecular cloud density required to collapse the astrosphere (heliosphere of other stars) scales with the mass:

$$n_{\rm crit} \approx 600 \left(\frac{M}{M_\odot} \right)^{-2} {\rm cm}^{-3} \quad (11.7)$$

This assumes the Sun's velocity relative to the ISM. Their analysis indicates that full heliosphere descreening occurs around 1–10 Gyr^{-1} for solar-type stars, and significantly less frequently for lower masses, to the point that this is only really an issue for planets orbiting Sunlike stars (or larger).

Determining cloud transitions in Earth's radioisotope record is challenging (Frisch and Mueller, 2013). There is some evidence that a cloud passage occurred some 18,000–34,000 years ago, and perhaps a passage some 20,000 years before this, but this requires careful modelling of the local interstellar medium from its current position back into the recent past, and it is widely agreed that better observations are required to reduce the uncertainties in this endeavour.

11.3.6 Mitigation against High-Energy Radiation Events

What can humanity (or ETIs) do to act against the effect of such violent high-energy events? We can categorise mitigation activity into three:

- **In situ mitigation** – directly altering or screening the emitter at the source.
- **Local mitigation** – constructing shielding at the civilisation's location, i.e., on a planetary surface.
- **Intermediate mitigation** – mitigating the radiation/particle flux in space, en route to the civilisation.

The second option is the most amenable to humanity in the present. The simplest mitigation solution to cosmic radiation of any sort is to build underground shelters, and to take cover for the duration of the radiation event. This is most practical for malignant stellar activity, whose duration is a few days. However, the radiation from supernovae/GRBs is expected to be received over much longer periods. GRBs emit gamma rays on short time scales, but the *afterglow* (over a variety of wavelengths, including UV and X-rays) can last several days, and the cosmic ray output can be enhanced for months, even years. Cosmic ray penetration is so efficient that such a shelter would need to be several kilometres deep to be effective!

Also, the underground solution only preserves selected portions of surface biomass. It does not mitigate in any way against O_3 destruction, the production of NO_2 and other NO_x species, or any other deleterious atmospheric effects. It also leaves any space-based asset prone to severely damaging effects. This places civilisations like humanity at risk of a scenario similar to that caused by Kessler syndrome (see section 13.5).

Ćirković and Vukotić (2016) propose an intermediate solution – shielding the Earth using natural debris material. Mining the Kuiper Belt for icy bodies and surrounding the Earth in a halo of said material at high orbit would increase the effective absorption coefficient of the planet. If said shield has a radius R_{sh}, and the shield must attenuate incoming radiation to a fraction ϵ, then the parent body from the Kuiper Belt required to produce the shield must have a radius

$$R = \sqrt[3]{\frac{3R_{sh}^2(-\ln \epsilon)}{4\mu}} \tag{11.8}$$

where μ is the linear attenuation coefficient, i.e., the radiation intensity through the medium (of thickness L) is

$$I = I_0 e^{-\mu L} \tag{11.9}$$

The attenuation coefficient is a function of both the material properties and the incoming radiation energy. If we assume a shield of radius $R = 2R_\oplus$, then for $\epsilon = 0.1$, we require a pure-ice body of $R = 25.4$ km for $E = 0.1$ MeV, and $R = 34.1$ km for $E = 1$ MeV (Ćirković and Vukotić, 2016). For our shield to safely encompass geostationary satellites, $k = 6.6$ and we require ice bodies more than double the radius.

Breaking up such a body and placing a swarm of material into a shell around Earth results in a non-zero porosity of the shield, depending on the particle size of the swarm. It may well even be the case that the shell is composed of *smart dust* (Farrer, 2010), effectively a large number of nanoscale particles that are manipulable or programmable. This would be able to provide real-time correction of the swarm as the high-energy event impinges on the Solar System, as well as providing telemetry to Earth on its efficacy. Of course, nanotechnology on this scale presents its own risks (see section 14.2).

11.3 Terrestrial Atmospheric Responses to Very High-Energy Events

Keeping such a shield in orbit around a host planet (and not the host star) presents energetic challenges. Maintaining a shield position requires acceleration against the natural Keplerian flow – holding such a shield nearby ready for use may become considerably expensive.

There is another, much greater challenge: prediction of the high-energy event. Currently, astronomers can monitor nearby supergiants and make very broad estimates about in what time interval they may go supernova – for example, the red supergiant Betelgeuse (at a relatively safe distance of 197 pc) is predicted to go supernova within the next 100,000 years (Dolan et al., 2016). If we are to protect ourselves against a nearby supernova, we must be able to predict its ignition within a time interval of a few years, perhaps only a few months. GRBs are effectively unpredictable, and it is unclear when Sag A* will undergo another AGN phase (if it does it all). ETIs attempting to mitigate against such events will require a much firmer grasp of high-energy astrophysics than we have today.

12

Death by Unsustainable Growth

Since the Industrial Revolution, the population of humanity has increased with an approximately exponential growth factor:

$$\frac{dN}{dt} = rN \qquad (12.1)$$

where r is the growth rate:

$$N(t) = N(t=0)e^{rt}. \qquad (12.2)$$

This exponential growth has put significant strain on the many ecosystems that comprise the Earth's biosphere. At several stages in recent history, there has been concern that Earth's human population exceeds the planet's carrying capacity.

Perhaps the most famous concerns regarding unrestricted human population growth have come from Malthus (or bear his name). In *An Essay on the Principle of Population* (published under a pseudonym in 1779), Malthus noted his concerns that a seeming universal trend in population growth and wealth creation was the growth or sustainment of the misery of the poor. This was in direct opposition to many of his contemporaries (such as Adam Smith), who assumed that wealth creation always tended to 'better the condition of the labouring poor'. In extremis, this trend would eventually beggar the realm. As Malthus (1798, *An Essay on the Principle of Population*, chapter 7) described it:

The power of population is so superior to the power of the earth to produce subsistence for man, that premature death must in some shape or other visit the human race.

This notion is referred to colloquially as a *Malthusian check*. It is important to note that Malthus's conception of what might be the cause of 'premature death' was not limited to natural forces caused by populations overshooting available resources. Indeed, Malthus was of the belief that the stresses induced on the lower classes of society as goods and resources began to vanish from their grasp would prove to be civilisation's downfall, either via war or revolution.

Malthusian checks can either be *positive*, which increase the death rate, or *preventive*, which decrease the birth rate. His skepticism regarding the ability for agriculture to be optimised without limit is characteristic of the period. It is possible that had he witnessed the scientific and technological advances in food production, that skepticism would be sorely damaged.

When Malthus wrote his essay at the end of the eighteenth century, the world's population was only around 1 billion.[1] The combine harvester (a hybrid of the reaper and the threshing machine) was a hundred years away – the full mechanisation of agriculture would have to wait for the automobile. Impressive gains in agricultural productivity also came from selective breeding of hardier strains of crops. For example, the agronomist Norman Borlaug's breeding of high-yield, disease-resistant wheat is credited by many with saving billions of lives from starvation.

Since the Second World War, human population has grown at a rate dramatically larger than any other time in recorded history. In just 40 years, the population has increased by a factor of around 2.5, from 2.5 billion in 1950 to around 6.1 billion at the beginning of the twenty-first century. Current projections for the next century from the United Nations Department of Economic and Social Affairs (2015) indicate a declining population growth rate (Figure 12.1). Earlier projections that extended to 2300 indicated a steady-state value of around 8.9 billion (UN, 2004).

This is in accord with classic models of population growth towards an equilibrium state, bounded by environmental limitations. The apparent exponential growth eventually shows the form of a 'logistic' curve (Verhulst, 1845). The governing differential equation is

$$\frac{dN}{dt} = \frac{rN(K-N)}{K} \tag{12.3}$$

where we now introduce a carrying capacity K. The typical solution is

$$N(t) = \frac{K N_0 e^{rt}}{K + N_0 (e^{rt} - 1)} \tag{12.4}$$

where at very long times

$$\lim_{t \to \infty} N(t) = K \tag{12.5}$$

A variety of factors are at play in this population evolution – a reduction in birth rates throughout the world, firstly in developed and now increasingly in developing countries, and an equalisation of life expectancy. This is accompanied by other demographic changes, including an ageing population (the median age is expected

---

[1] Historical Estimate of the US Census Bureau, www.census.gov/population/international/

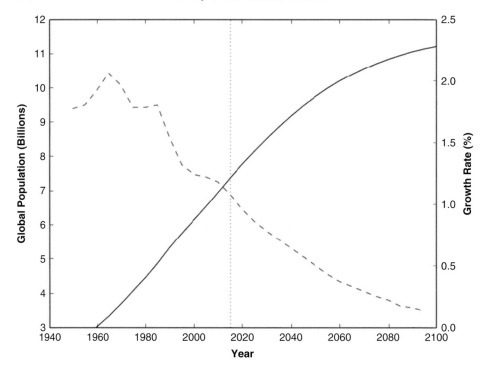

Figure 12.1 The world's population growth (solid line) and growth rate (dashed line), as measured by the United Nations Department for Economic and Social Affairs in 2015 (vertical line). Data beyond 2015 are projections.

to double from 26 years to 48 by 2300). We can rewrite the population growth equations to include an ageing population by considering the population of individuals with age $(a, a + da)$ at time t, $N(t, a)$. The population evolution is

$$dN(t, a) = \frac{\partial N}{\partial t}dt + \frac{\partial N}{\partial a}da = b(a)N(t,a)dt - d(a)N(t,a)dt \quad (12.6)$$

where b and d are birth and death rate functions which are age-dependent. The number of individuals of a given age changes according to the birth rate and to the ageing rate of individuals. We can immediately simplify this by noting that

$$\frac{da}{dt} = 1, \quad (12.7)$$

as we all age at one year per year, and the birth rate only contributes to $N(t, 0)$ (one cannot be born at the age of 1 or higher):

$$\frac{\partial N}{\partial t} + \frac{\partial N}{\partial a} = -d(a)N(t, a) \quad (12.8)$$

Given an initial population distribution $n(0, a)$, and the source term of individuals $n(t, 0)$:

$$n(t, 0) = \int b(a')n(t, a')da' \quad (12.9)$$

Solutions to these equations at a time t, given the birth time of an individual t_0, can be found using characteristics (see Murray, 2004, for more details):

$$N(t, a) = N(t_0, 0) \exp\left[-\int_0^a d(a')da'\right] \quad a > t \quad (12.10)$$

$$N(t, a) = N(t - a, 0) \exp\left[-\int_0^a d(a')da'\right] \quad a < t \quad (12.11)$$

It is easy to scoff at Malthus's predictions of woe for populations a fraction of the Earth's today, but we should be careful not to make the same mistake and put too much faith into such long-term projections. Disruptive events are at the heart of history, by their very nature unpredictable and capable of fundamentally altering the course of events.

What does this mean for Fermi's Paradox? If our ability to see ETI depends on one civilisation's ability to grow and colonise over Galactic scales, then we can resolve the Paradox by stating that either they cannot, or do not.

A common strand of thinking regarding civilisation expansion beyond the home world runs along the lines of the following: the gestation of civilisation on a single planet is either successful, in that the civilisation matures and can effectively live forever, settling other worlds and expanding its sphere of influence indefinitely; or the gestation fails, and the civilisation is extinguished before securing its existence on more than one world.

In this simple picture, the successful civilisations are free to expand throughout the Milky Way and occupy all of its habitable niches. This is in direct analogy with the terrestrial biosphere, where organisms occupy every niche they can adapt to, provided it is in some way accessible.

As civilisations continue to grow, either

1. their growth is well matched with their energy and resource demands, e.g., they practise logistic growth with some increasing carrying capacity;
2. their energy/resource demands grow beyond their ability to acquire them;
3. their energy demands strain the environment to the point of collapse (see Chapter 13).

The Sustainability Solution to Fermi's Paradox (Solution C.10, see Haqq-Misra and Baum, 2009) suggests that fundamental limits exist on the rate of civilisation growth. Civilisations that attempt to grow exponentially will most likely fall prey to scenario 2, and be extinguished.

The human race has avoided scenario 2 by technological advances that extract more energy from the environment. This improved efficiency in energy generation permits increased efficiency in resource extraction. For example, the invention of dynamite and other explosives marks an improvement in energy release, which improves the efficiency of mining operations. This runs the risk of utter exhaustion of resources, beginning scenario 3.

If the Sustainability Solution is correct, then we are unlikely to see vast interstellar empires. Successful civilisations will manage themselves to exhibit a slower population growth. Of course, slowing population growth does not mean slowing resource consumption. Individuals consume at rates nearly a factor of 2 higher than individuals 30 years ago (Fischer-Kowalski, 1997). By 2030, it is expected that consumption rates will be double that of 2005, well beyond population growth (Lutz and Giljum, 2009).

Sustainability is therefore an individual and societal pursuit. The total consumption of a civilisation depends on multiple factors, including the total population and the consumption habits of individual members. Simply reducing population is insufficient to maintain a sustainable civilisation. If individual consumption continues to rise, even a small population will eventually fall into the trap of scenario 3.

As in much of our thinking regarding alien civilisations, we inevitably overlay assumptions about our own behaviour, which remains inextricably bound by our biological constraints. Would post-biological civilisations be able to expand without limits? Ćirković (2008) argues that extremely advanced civilisations are unlikely to be expansion-driven, and are instead driven by optimisation of resource consumption. Rather than interstellar empires, the Galaxy is most likely composed of 'city-state' civilisations that manage both population and individual energy consumption. Such civilisations are likely to manipulate their environments to extract energy at extremely high efficiency. For example, extracting accretion energy from black holes yields an efficiency many times higher than that produced by stellar thermonuclear fusion. Not only do these civilisations reduce their footprint on the Milky Way, but they do not transmit signals, reducing their visibility through increasing energy efficiency and waste reduction.

As we have already stated, the global population is on course to reach a peak sometime in this century. If our gestation as a civilisation is successful, then a key measure of its success is the removal of one of the principal drivers of expansion: population pressure. Many modern views of human settlement of other planets are not a quest for increased room, but are aligned with the emotions and knowledge derived by individual achievement. The colonisation of the Moon and Mars would be a technological feat achieved through human endeavour, done for the same reasons that Mount Everest was climbed, 'because it's there'.

Of course, such views do not seem sufficient to overcome political inertia and budget constraints. The current climate suggests that *in situ* resource exploitation, and the potentially vast quantities of precious metals, will be the real impetus that propels humanity into space. We might be a species reaching peak population, but our rapacity for energy and minerals shows no signs of abating.

13
Death by Self-Induced Environmental Change

It is empirically established that human activities have increased the greenhouse effect of the Earth's atmosphere, resulting in a planet that is several degrees C warmer than it would be if humans were not present. Combined with other factors, humanity is reducing the Earth's ability to support its current biosphere.

The causes of this anthropogenic global warming (AGW) are several. For each of these causes, we can estimate the so-called anthropogenic forcing (in W m^{-2}), which broadly speaking is the increase in heating resulting from said activity. The total forcing from all human activities since the Industrial Revolution is calculated to be $2.3^{+1.0}_{-1.2}$ W m^{-2} (IPCC, 2013). When this is compared against natural forcing from volcanic eruptions, aerosol production and changes in solar irradiance, which produces a forcing consistent with zero (± 0.1 W m^{-2}), it is clear that human activities are having a substantial effect.

13.1 Atmospheric Damage

Top of the list for anthropogenic forcing (2.83 ± 0.29 W m^{-2}) is the increase in so called Well Mixed Greenhouse Gases (WMGHGs). This subset of GHGs refers to molecules that are dispersed relatively evenly throughout the Earth's atmosphere. Examples include CO_2, N_2O, CH_4, SF_6 and others.

Some greenhouse gases do not abide for long time periods in the atmosphere, but still produce significant effects, and are referred to as near-term climate forcers (NTCFs). Methane deposits (such as those released from melting permafrost) also fall into this category, as does ozone and other pollutants such as chlorofluorocarbons (CFCs) and other aerosols.

CFCs are most infamous for their damage to the ozone layer. As CFCs drift upwards, the molecules' chemical stability allows them to remain intact until they reach the stratosphere, where UV-B radiation can release the chlorine, which then undergoes the following reaction chain:

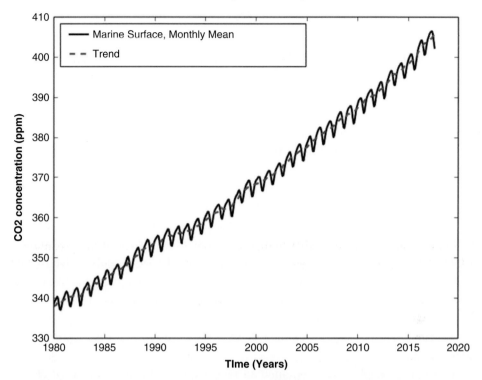

Figure 13.1 CO_2 levels (in parts per million), measured from multiple marine surface sites and subjected to a moving average (using seven adjacent points). Data Credit: Ed Dlugokencky and Pieter Tans, NOAA/ESRL (www.esrl.noaa.gov/gmd/ccgg/trends/)

$$Cl + O_3 \to ClO + O_2 \qquad (13.1)$$
$$ClO + O \to Cl + O_2 \qquad (13.2)$$
$$O_3 + O \to 2O_2 \qquad (13.3)$$

Note that chlorine acts as a catalyst in this reaction, utilising local free oxygen atoms to break O_3 into O_2 with the single Cl atom present at the end of the reaction.

As we saw in section 11.3, reducing the column density of ozone in the atmosphere increases the flux of harmful UV radiation at the Earth's surface. As the quantity of CFCs increased, the rate of ozone depletion increased in a cumulative fashion. This led to strict controls on their use, and the hole in the ozone layer is shrinking – it is expected that the ozone column will return to its 1980 levels by the middle of the twenty-first century (NOAA et al., 2010).

Despite this positive news, CO_2 levels continue to increase (Figure 13.1), driven by human activity from pre-industrial levels of around 280 parts per million (ppm) to beyond 400 ppm. This alone threatens to raise the average temperature of the

Earth by around 2–3 °C over the next century, according to a large ensemble of model projections. Without strong mitigation of CO_2 production, this will have significant effects on global sea level. Around 40% of the entire human population live within 100 km of the coast, and hence a substantial fraction of this population (around 150 million people) are threatened by sea level rises of 1 m.

As CO_2 contributes the lion's share of anthropogenic forcing, most efforts to adapt to or mitigate the damaging effects of climate change have focused on reducing its production, or removing it from the atmosphere via reforestation or carbon capture technologies.

At the most extreme end of climate change mitigation strategy is *geo-engineering*, a deliberate technological effort to alter the radiative balance of the atmosphere. A principal geo-engineering tool is to increase the aerosol content of the atmosphere, effectively emulating the effect of large volcanic eruptions, which have been shown to produce brief intervals of global cooling. Delivering 2,000–7,000 tons of sulphate aerosol to stratospheric altitudes every year would be sufficient to mitigate against warming at a rate of 0.15 K per year (Eliseev et al., 2010), but is likely to have many foreseen and unforeseen knock-on effects. Once aerosol injection ceases, the climate quickly rebounds to a warmer, pre-aerosol state (Berdahl et al., 2014).

At lower atmospheric levels, one can increase the reflective albedo of clouds by seeding them with appropriate chemical aerosols (sea water aerosols are also suitable). This, however, does produce greater levels of cloud formation in general, and increases precipitation over low-latitude regions, which can result in increased land run-off damage.

More extreme attempts to limit stellar flux onto the upper atmosphere involve placing large sun-shades in orbit. Placing such a shade (either a single object or a swarm of objects) at the L1 Lagrange point would allow a civilisation to 'tune' the flux received from the star. Calculations indicate a reduction in bolometric solar flux of around 1.7% would be sufficient to offset the total anthropogenic forcing. Neglecting solar radiation pressure, a single-piece shade at Earth's L1 point would need to be approximately 915 km in radius to achieve this (Sánchez and McInnes, 2015). If one wishes to effect specific local changes in solar irradiance, then the shade may be tilted or shifted away from L1.

While atmospheric climate change and geo-engineering tend to paint rather bleak pictures, it is this type of atmospheric evolution that is most amenable to astronomers using exoplanet data to search for intelligent life. Detecting pollutants in an exoplanet atmosphere (Lin et al., 2014), or an atmospheric composition that appears to show sources of CO_2 and aerosols that are not accounted for by natural processes, could be key signs of intelligent life attempting to mitigate its self-caused climate change. The detection of an artificial structure in an orbit semi-static relative to the planet is also an obvious sign of intelligent engineering (see section 27.3.2).

13.2 Destruction of the Land Environment

Another important change humans have made is to the surface itself. The growing human population places increasing demands on food production. Technological innovations have continued to increase caloric yields per acre, but this does not change the fact that croplands and pastures are now one of the largest biomes on the planet, occupying around 40% of the planet's surface, with some 35% of anthropogenic CO_2 produced from varying forms of land use (especially agriculture) (Ramankutty and Foley, 1999; Foley et al., 2005).

The need for highly intensive farming has resulted in a significant increase (approximately 700% in the last 40 years) in the use of fertilisers. The run-off from heavily fertilised land eventually reaches the water table, resulting in decreasing water quality.

This is coupled with an increasing need for irrigation. This thirst for freshwater has transformed the hydrological cycle, as humans siphon off increasingly large quantities and divert natural waterways. This is already beginning to impact rivers in arid/semi-arid regions, where flows are beginning to reduce and even dry up completely.

The large swathes of cropland now required to feed the human population have come largely at the cost of Earth's forests. While the trend towards massive deforestation is showing signs of slowing down, and in some regions even reversing, the first replacement forests showed significantly reduced biodiversity and resilience. Fortunately, this is now recognised as an issue, and most forest management schemes now produce significantly more biomass per unit area (in some cases an improvement of 40% in the last fifty years).

In urban areas, covering the land with concrete and tarmac changes the land's ability to drain away excess rains. This was most strikingly revealed in 2017 by the aftermath of Hurricane Harvey in Texas. Despite recent attempts to construct road systems that promoted rapid run-off of storm waters into the Gulf of Mexico, areas without this attentive construction philosophy struggled to remove the enormous volumes of water deposited by the hurricane (with estimates ranging from 30 to 60 trillion tons during the storm period).

A layer of concrete/tarmac also affects the land's ability to radiate heat. Natural surfaces usually contain vegetation or moisture-trapping soils, and can therefore effectively release heat via evapo-transpiration. Adding a layer of construction material on top, which is typically water resistant and of a low albedo, frustrates the land's ability to cool. The presence of tall buildings also results in increased heat retention, especially if these buildings are themselves generating heat (for example, by running air conditioning systems). All of this results in cities typically being slightly higher in temperature (by one to two degrees C) than rural surroundings (the so-called urban heat islands).

This rewiring of the fundamental infrastructure that underpins habitats almost always results in significant loss of biodiversity.

Despite our best efforts, nearly half of all global croplands are experiencing some form of degradation, be it soil erosion, reduced fertility or overgrazing. This drain of nutrients is combined with gradual warming to produce increasing levels of land desertification, rendering greater portions of the planet uninhabitable to humans without significant technological efforts.

Our ability to observe the surfaces of rocky exoplanets remains limited, but one could imagine a future where mapping the reflectivity of a planetary mass body (for example, through polarisation; see Berdyugina and Kuhn, 2017) could yield important clues as to whether a civilisation is affecting the land in ways similar to us.

13.3 Destruction of the Ocean Environment

The oceans are a crucial source of biodiversity on Earth. They provide an important buffer for energy as the planet continues to grow warmer, and clean water is a crucial component of any healthy ecosystem.

The increasing levels of carbon dioxide in the Earth's atmosphere have also resulted in the increasing dissolution of carbon dioxide into the Earth's oceans:

$$CO_2 + H_2O \leftrightarrow H_2CO_3 \leftrightarrow HCO_3^- + H^+ \leftrightarrow CO_3^{2-} + 2H^+ \qquad (13.4)$$

The production of hydrogen ions as a result of this process suppresses the ocean's pH. Since the Industrial Revolution, hydrogen ion concentrations have increased by around 30%, equivalent to a pH drop of about 0.1. Projections have indicated that the pH could level out as low as 0.7 below pre-industrial values.

There are few examples of such acidification in the fossil record that we can turn to for comparison. The only possible candidate is the Paleocene-Eocene Thermal Maximum (PETM), which occurred 55.8 Myr ago. The global temperature during this period rose by 5–9 degrees, which is identifiably due to a substantial release of organic carbon from sediments or clathrates. This was detected through the sudden drop in ^{13}C to ^{12}C ratio in carbonate/carbon-rich sediments (Kump et al., 2009).

The PETM does coincide with a substantial mass extinction – the largest ever recorded amongst deep-sea organisms known as *calcareous benthic foraminafera*. These are seafloor-dwelling (benthic), and possess shells of calcium carbonate (calcareous). It is unclear that this event was a direct consequence of acidification –

extinctions of lesser magnitude are common as a result of warming driven by CO_2 without significant acidification.

What is demonstrably true is that as ocean pH descends, calcium carbonate precipitation decreases. More acidic water requires a larger concentration of carbonate ions to reach saturation and precipitate. Organisms that rely on calcium carbonate to form outer shell structures struggle to precipitate, and are at greater risk of their shells dissolving. The evidence is somewhat mixed – some calcareous species tend to do better than others in warmer, more acidic waters – but larger meta-analyses indicate that the combined risk of increasing temperature, increasing acidity and decreased oxygenation are placing large swathes of the marine food chain at risk.

The greatest threat to marine ecosystems (as seen by the Intergovernmental Panel on Climate Change) is the phenomenon of *coral bleaching*.

Healthy corals depend on a symbiotic relationship with *zooxanthellae*, algae-like protozoa that live within the coral tissue. Zooxanthellae photosynthesise and provide the coral with energy and nutrients – the coral supplies CO_2 and NH_4^+ in return to facilitate photosynthesis.

Increased ocean temperatures result in the coral being unable to adequately supply the zooxanthellae. As a protective measure, the coral expels the zooxanthellae, discarding its principal energy source. The zooxanthellae are also responsible for the coral's colour – their expulsion results in coral bleaching. If the zooxanthellae cannot return to the coral quickly and resume photosynthesis, the reef structure collapses.

Coral reefs are argued to be the most biodiverse habitats on Earth. Bleached reefs cannot support this diversity – thousands of species which rely on reef nutrients and protection become vulnerable during bleaching periods. The collapse of reef biota can place the larger marine ecosystem (and other connected ecosystems) into jeopardy.

13.4 Destruction of the Biosphere

The Earth's biosphere has suffered five major mass extinction events in its history (Raup and Sepkoski, 1982; see also Table 13.1). It is argued by many that the arrival of humans (most especially industrial civilisation) is causing the sixth (Kolbert, 2014). The precise definition of when this extinction began is contested, not to mention whether it should be referred to as the Holocene extinction, or if industrial human civilisation constitutes a new geological epoch, the Anthropocene. In either case, a severe loss of biodiversity places human civilisation at risk from famine and pandemic illnesses (Solution C.12).

We can identify two significant epochs in our Holocene extinction. The first is the correlation between the disappearance of species in an area (particularly

Table 13.1 *The five major mass extinctions in Earth history. From Barnosky et al. (2011)*

Event	Estimated time	Estimated species loss
Ordovician	~ 443 Myr ago	86%
Devonian	~ 359 Myr ago	75%
Permian	~ 251 Myr ago	96%
Triassic	~ 200 Myr ago	80%
Cretaceous	~ 65 Myr ago	76%

megafauna) during the Pleistocene around 12,000 years ago, and the arrival of humans into that particular area. This is particularly evident when considering island ecologies, such as those in the Pacific (Steadman, 2003), where some 2,000 species of bird became extinct within centuries of human arrival.

The second epoch is the establishment of our modern civilisation. The widescale implementation of agriculture fundamentally altered the balance of local flora and fauna. The diversion of water supply from the ecosystem, the introduction of non-native, invasive species, and the transmission of newly infectious diseases also contributed heavily to the growing extinction. Industrial expansion accelerates the destruction of habitats, and accelerates the production of human beings to occupy the environment.

The World Wildlife Fund's Living Planet Report (2016) indicates that species are in significant decline. Between 1972 and 2012, The Living Planet Index (LPI) dropped by 58% among vertebrates, with the principal threat coming from habitat degradation.

From Table 13.1, it is clear that if humans are causing the sixth major mass extinction event, they are yet to equal the previous five in terms of species loss. It should be remembered, however, that an extinction event takes several million years to run its course – humans have had less than a thousand years to cause over half the species loss of a typical mass extinction. The prodigious drop in biodiversity in parts of the world suggests that humans are more than capable of matching nature in their ability to tear branches from the tree of life.

13.5 Destruction of the Space Environment (Kessler Syndrome)

The first destination of any newly spacefaring species is the immediate orbital surrounds of their home planet. In the case of humanity, this began in earnest with the launch of Sputnik in 1957. The orbital environment is split into three categories: Low, Medium and High Earth Orbit (LEO, MEO, HEO).

13.5 Destruction of the Space Environment (Kessler Syndrome)

Table 13.2 *The classification of the Earth's orbital environment*

Orbit	Range
Low Earth Orbit (LEO)	160–2,000 km
Medium Earth Orbit (MEO)	2,000–35,786 km
High Earth Orbit (HEO)	> 35,786 km
Geostationary Orbit (GEO)	= 35,786 km

Almost all human activity in space (with the exception of the Apollo program) has occurred at LEO – the International Space Station orbits at approximately 400 km. MEO is principally occupied by communications and navigation satellites, such as the Global Positioning System (GPS) flotilla, which orbits at the relatively crowded altitude of $\sim 20{,}200$ km.

All objects in HEO reside beyond the geostationary orbit (GEO). The orbital period at GEO (which is aligned with the Earth's equator) is equal to the Earth's rotational period. As a result, from a ground observer's perspective the satellite resides at a fixed point in the sky, with clear advantages for uses such as global communication. Activities at HEO are considerably less than at LEO and MEO. Earth's orbital environment does contain a natural component – the *meteoroids*. These pose little to no threat to space operations – the true threat is self-derived.

The current limitations of spacefaring technology ensure that every launch is accompanied by substantial amounts of *space debris*. This debris ranges in size from dust grains to paint flecks to large derelict spacecraft and satellites. According to NASA's Orbital Debris Program Office, some 21,000 objects greater than 10 cm in size are currently being tracked in LEO, with the population below 10 cm substantially higher. Most debris produced at launch tends to be deposited with no supplemental velocity – hence these objects tend to follow the initial launch trajectory, which often orbits with high eccentricity and inclination. However, these orbits do intersect with the orbits of Earth's artificial satellite population, resulting in impacts which tend to produce further debris.

The vast majority of the low-size debris population is so-called *fragmentation debris*. This is produced during spacecraft deterioration, and in the most abundance during spacecraft break-up and impacts. The first satellite–satellite collision occurred in 1961, resulting in a 400% increase in fragmentation debris (Johnson et al., 2008). Most notably, a substantial source of fragmentation debris was the deliberate destruction of the Fengyun 1C satellite by the People's Republic of China, which created approximately 2,000 debris fragments.

As with collisions of 'natural debris', debris–debris collisions tend to result in an increased count of debris fragments. Since the late 1970s, it has been understood that man-made debris could pose an existential risk to space operations. Kessler and Cour-Palais (1978) worked from the then-population of satellites to extrapolate the debris production rate over the next 30 years. Impact rates on spacecraft at any location, I, can be calculated if one knows the local density of debris ρ, the mean relative velocity v_{rel}, and the cross-sectional area σ:

$$I = \rho v_{rel} \sigma \qquad (13.5)$$

Each impact increases ρ without substantially altering v_{rel} or σ. We should therefore expect the impact rate (and hence the density of objects) to continue growing at an exponential rate:

$$\frac{dI}{dt} = \frac{d\rho}{dt} v_{rel} \sigma = \frac{d \ln \rho}{dt} I \qquad (13.6)$$

Kessler and Cour-Palais (1978) predicted that by the year 2000, ρ would have increased beyond the critical value for generating a *collisional cascade*. As new collisions occur, these begin to increase $\frac{d \ln \rho}{dt}$, which in turn increases $\frac{dI}{dt}$, resulting in a rapid positive feedback, with ρ and I reaching such large values that LEO is rendered completely unnavigable.

This has not come to pass – LEO remains navigable, partially due to a slight overprediction of debris produced by individual launches. The spectre of a collisional cascade (often referred to as *Kessler syndrome*) still looms over human space exploration, as debris counts continue to rise. Without a corresponding dedicated effort to reduce these counts, either through mitigating strategies to reduce the production of debris during launches, or through removal of debris fragments from LEO, we cannot guarantee the protection of the current flotilla of satellites, leaving our highly satellite-dependent society at deep risk.

What strategies can be deployed to remove space debris? Almost all debris removal techniques rely on using the Earth's atmosphere as a waste disposal system. Most debris is sufficiently small that atmospheric entry would result in its complete destruction, with no appreciable polluting effects. Atmospheric entry requires the debris fragments to be decelerated so that their orbits begin to intersect with lower atmospheric altitudes. Once a critical altitude is reached, atmospheric drag is sufficiently strong that the debris undergoes runaway deceleration and ultimately destruction.

There are multiple proposed techniques for decelerating debris. Some mechanical methods include capturing the debris using either a net or harpoon, and applying a modest level of reverse thrust. These are most effective for larger fragments, and especially intact satellites (Forshaw et al., 2015). Attaching sails to the debris is also a possibility if the orbit is sufficiently low for weak atmospheric drag.

13.5 Destruction of the Space Environment (Kessler Syndrome)

The Japanese space agency JAXA's Kounotori Integrated Tether Experiment (KITE) will trail a long conductive cable. As a current is passed through the cable, and the cable traverses the Earth's magnetic field, the cable experiences a magnetic drag force that will de-orbit the spacecraft.

Orbiting and ground-based lasers can decelerate the debris through a variety of means. For small debris fragments, the radiation pressure produced by the laser can provide drag. A more powerful laser can act on larger debris fragments through ablation. As the laser ablates the debris, the resulting recoil generated by the escaping material produces drag and encourages de-orbit.

A more lateral solution is to ensure that launches and general space-based activity no longer generate debris. These approaches advocate lower-energy launch mechanisms that do not rely on powerful combustion. The most famous is the *space elevator* (see Aravind, 2007). Originally conceived by Tsiolkovsky, the elevator consists of an extremely durable cable extended from a point near the Earth's equator, up to an anchor point located at GEO (most conceptions of the anchor point envision an asteroid parked in GEO).

'Climber' cars can then be attached to the cable and lifted to LEO, MEO and even GEO by a variety of propulsion methods. Most notably, the cars can be driven to GEO without the need for chemical rockets or nuclear explosions – indeed, a great deal of energy can be saved by having coupled cars, one ascending and one descending.

Space elevators would solve a great number of problems relating to entering (and leaving) Earth orbit, substantially reducing the cost of delivering payload out of the Earth's atmosphere. The technical challenges involved in deploying a cable tens of thousands of kilometres long are enormous, not to mention the material science required to produce a cable of sufficient tensile strength and flexibility in the first place. The gravitational force (and centrifugal force) felt by the cable will vary significantly along its length. As cars climb the cable, the Coriolis force will move the car (and cable) horizontally also, providing further strain on the cable material. The relatively slow traversal of the biologically hazardous Van Allen Belt on the route to GEO is also a potential concern for crewed space travel.

Whatever the means, a spacefaring civilisation (or at least, a civilisation that utilises its local orbital environment as we do) must develop a non-polluting solution to space travel, whether that is via the construction of a space elevator, a maglev launch loop, rail gun, or some other form of non-rocket acceleration. If it cannot perform pollution-free spacecraft launches (or fully clean up its pollution), then it will eventually succumb to Kessler syndrome, with potentially drastic consequences for future space use, with likely civilisation-ending effects (Solution C.13).

14

Self-Destruction at the Nanoscale

14.1 Genetic Engineering

Humans have unwittingly engineered the genomes of many species through artificial selection (for example, in the quest for productive yields in agricultural crops). As our knowledge of genetic manipulation continues to improve, the prospect of directly editing genes opens up a whole new vista of biological innovation and engineering.

Recently, genetic engineers have discovered a powerful new set of gene editing tools, hiding in plain sight in microbial DNA. CRISPR (Clustered Regularly Interspaced Short Palindromic Repeats) were discovered by Ishino et al. (1987). CRISPR are DNA sequences which contain short, repetitive sub-sequences. These subsequences contain palindromic repeats (which read the same in both directions), interspaced with spacer DNA.

CRISPR is a record vital to the immune systems of bacteria and archaea. When a bacterial cell encounters a virus, a protein samples the viral DNA and records it in the bacteria's own DNA, at the CRISPR locus. Future invasions of the cell by the virus can then be recognised by a Cas enzyme, guiding it to attack invaders (Horvath and Barrangou, 2010).

It was soon realised that a two-component system could be used to very accurately edit DNA. The CRISPR-Cas9 system consists of a Cas9 enzyme, which acts as a pair of molecular scissors, and a small RNA molecule that guides the scissors to where to make the cut. Judicious selection of RNA allows Cas9 to make the cut at essentially any location on the DNA strand. Once the DNA is cut, standard DNA repair tools in the cell step in to seal the breach. During these repairs, new DNA can be inserted, allowing the genome to be modified, making CRISPR tools capable of removing any gene and replacing it with another.

This system not only allows much more precise DNA editing than previously possible, but it also allows biologists to take direct control of gene expression.

14.1 Genetic Engineering

If a mutated Cas9 enzyme is used, the DNA is no longer sliced, but the enzyme remains near the site it was guided to, and prevents transcription. In effect, the gene is unaltered, but switched off. The gene can then be activated by tethering the enzyme to proteins responsible for gene expression.

Epigenetic effects can also be studied. The epigenome is composed of an ever-changing film of proteins and molecules that adhere to DNA, affecting gene expression. Clever use of Cas9 allows these epigenetic components to be activated and deactivated at will.

These modifications are not heritable – an organism's descendants will not share the genetic changes. This can be subverted by using *gene drive* techniques. CRISPR/Cas systems can not only insert a desired gene, but they can insert genes that activate the CRISPR/Cas system itself. This allows us to modify the chromosomes of organisms, deleting unwanted genes from evolution (or adding new ones). When these organisms reproduce, the CRISPR system activates, ensuring that a desired trait propagates far further in a population than natural selection would permit. These gene drive methods have been successfully used to reduce the fertility of mosquito populations that act as vectors for malaria (Hammond et al., 2015).

In short, we can now truly foresee a future where heritable gene editing is cheap, and relatively simple technology. Indeed, some argue that in many respects that future is already here. Scientists aim to use CRISPR methods to study diseases, identify genes associated with a range of disorders, and to further our general understanding of genomics (and epigenomics, the study of how genomes and gene expression are modified by external agents).

Like all technologies, there can be immoral uses. Diseases could be modified for extra virulence and pathological effect. Malevolent modification of human genomes could result in genetic disorders being inserted into victims. Animals could be modified to be stronger and more aggressive with a reduced fear response.

The law of unforeseen consequences looms large here. The genome is ferociously complex, and much of its structure is yet to be understood. Only 2% codes for proteins, with the function of the other 98% remaining unclear (some of this codes for RNA, some of this merely 'enhances' gene expression, but with puzzling levels of success). Modifying or deleting genes can affect other gene expression.

Evolution via mutation, drift, recombination and selection is already extremely complex. Artificial selection (principally via indirect methods such as selective breeding) has been a heavily used tool by humanity. It seems reasonable that ETIs will have deployed similar tools. However, the ability to interfere directly in genetic structure and gene expression may be the final act of any civilisation, as it extinguishes itself in a fit of poor engineering.

14.2 Nanotechnology, Nanomachines and 'Grey Goo'

The continuing miniaturisation of technology delivers enormous benefits to computing and materials science, but there are associated risks. *Nanoscience* refers to the science of phenomena at scales of around 100 nm or less. Nanotechnology refers to the design of systems that manipulate matter at these scales. The concepts underpinning modern nanotechnology can be traced back at least as far as Feynman's 1959 talk *There's Room at the Bottom*. The idea of being able to manipulate individual atoms was definitely alluring (Feynman imagined the Encyclopaedia Britannica written on the head of a pin) and it was soon realised that nanoscale physics had valuable attributes that could be exploited.

The initial driving force behind nanotechnology was the production of integrated circuits of smaller and smaller size (electron beam lithography would etch circuits onto silicon chips on 40–70 nm scales in the 1970s).

The study of *nanomaterials* has yielded important advances in a range of fields. The carbon nanotube is only a few nanometres in width, but with extremely high tensile strength, some 100 times stronger than steel for about one-sixth of the mass. Their thermal and electrical conductivity properties rival (and exceed) copper.

Nanoparticles have particularly promising applications in the fields of medicine, from the antimicrobial properties of silver nanoparticles to the accurate targeting of drugs at specific sites in the body. The development of nanofilms has similar advantages (as well as the remarkable 'self-cleaning' windows that utilise superhydrophobic nanofilms).

In the now classic *Engines of Creation*, Eric Drexler imagined the apex of nanotechnology – a *molecular assembler* that could rearrange atoms individually to produce nanomachines. Such a molecular assembler can in principle reproduce itself. If this reproduction cycle was in some way corrupted, the assembler could begin producing indefinite copies. Exponential growth of assemblers would rapidly result in the conversion of large fractions of a planet's mass into assemblers. This is referred to as the 'grey goo' scenario. Freitas (2000) further sub-categorises this into grey goo (land based), grey lichen (chemolithotrophs), grey plankton (ocean based) and grey dust (airborne).

Assuming that molecular assemblers are possible, what can be done to mitigate against this risk? Careful engineering can ensure that there is a rate-limiting resource in replication. If the assembler can use commonly found materials to replicate, then its growth is unlimited. If the assembler requires some rare element or compound, then replication will cease when that element is exhausted. Like biological replicators, one might imagine that mutation or copying errors may find templates that bypass this element, and exponential growth recommences unabated.

Freitas (2000) imagined a range of defence mechanisms specifically designed to dismantle assemblers, being triggered by the byproducts of assembly (e.g., waste heat). Producing a fleet of 'goodbots' to dismantle or remove corrupted assemblers is one such strategy.

Is a grey goo scenario likely on Earth? In 2004, the Royal Society published a report on the then consensus – that molecular assembly itself is unlikely to be produceable technology in the foreseeable future, due to fundamental physical constraints on locating individual atoms.

On the other hand, molecular assemblers exist in nature. Organisms depend on molecular assembly for a range of fundamental processes. Molecular machines exploiting these biological processes are very possible, and currently being produced in research labs. Constructing a molecular assembler from these machine components remains challenging.

Perhaps the best way to mitigate this risk is to simply not build autonomous assemblers. Phoenix and Drexler (2004) suggest that centrally controlled assemblers are more efficient at manufacturing, and hence the desire for assemblers to be able to replicate themselves is at odds with industrial goals.

Of course, a molecular assembler could be weaponised. We may need to consider grey goo as an associated risk with future conflicts.

14.3 Estimates of Existential Risk from Nanoscale Manipulation

The nature of the nanoscale risk is quite different from the risks posed by (say) nuclear weapons. The nuclear arsenals of the Earth remain under the control of a very small number of individuals. As genetic engineering and nanotechnology mature as fields of research, the ability to manipulate the nanoscale becomes available to an increasing number of individuals.

Any one of these individuals could accidentally (or purposefully) trigger a catastrophic event that destroys a civilisation. If there are E individuals with access to potentially destructive technology, and the probability that an individual destroys civilisation in any given year is P, then we can define the probability that the civilisation exists after time t and can communicate, C, as a set of independent Bernoulli trials, i.e.,

$$P(C|t, E, P) = (1 - P)^{Et}. \tag{14.1}$$

Therefore, we can compute the number of civilisations existing between $t = 0, t_0$ as

$$N(t_0) = B \int_0^{t_0} P(C|t', E, P) dt' \tag{14.2}$$

where we have assumed a constant birth rate B. This can be easily integrated to show (Sotos, 2017):

$$N(t_0) = B \frac{S^{t_0} - 1}{\ln S}, \quad S = (1 - P)^E \qquad (14.3)$$

At large t, this tends to the steady-state solution

$$N(t) \to \frac{B}{EP} \qquad (14.4)$$

i.e., there is an inverse relationship between the number of individuals in a given civilisation, and the number of civilisations remaining (Figure 14.1).

The generality of this solution to Fermi's Paradox – that the democracy of technology is the cause of its downfall – is particularly powerful. All we have assumed is that technology can be destructive, and that it eventually becomes accessible to a large enough number of individuals that the product EP becomes sufficiently large.

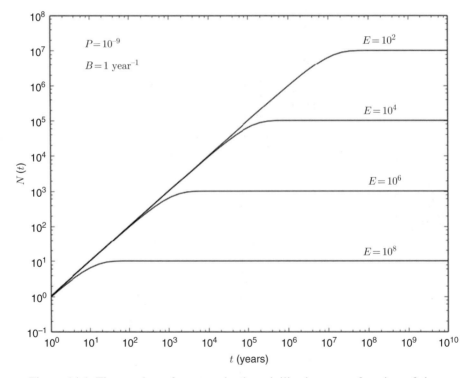

Figure 14.1 The number of communicating civilisations as a function of time, given E individuals causing self-destruction with a probability of P per individual per year. The civilisation birth rate B is fixed at one per year.

This is a sufficiently general solution that we may consider it to be hard – our definition of intelligence hinges on technology, and almost all advanced technology has a destructive use. We can already use the Great Silence to rule out the upper curves in Figure 14.1, which would yield civilisation counts that would be visible on Earth (see also section 19.2). This suggests that if this is the solution to Fermi's Paradox, then roughly $EP > 10^{-3}$.

15

Artificial Intelligence and the Singularity

A common source of existential dread in humans comes from our attempts to produce computers and machines that process the world with similar (or better) ability than ourselves. This attempt to create *artificial intelligence* (AI) evokes classic science fiction tropes such as the intelligent cyborgs of *Terminator*, or the machine-tyrant of *Colossus*.

Computers already carry out certain kinds of data processing far in advance of human ability, for example, the recall of large tranches of text, the rapid evaluation of mathematical operations, and in recent years the playing of rule-based strategy games such as Chess and Go. On the other hand, humans are traditionally better at accomplishing tasks without prior knowledge, such as navigating a complex evolving spatial environment.

We can define AI as follows (Rich, 1983):

The study of how to make computers do things at which, at the moment, people are better.

AI can then be broken down further into two categories, *weak AI* and *strong AI*, each with subfields of research. Weak AI systems are non-sentient, and can only be applied to specific problems or tasks (e.g., playing chess or Go). Strong AI systems are proposed to be sentient, applicable to general-purpose problems. If we consider a human brain as yielding a set of outputs O from a set of inputs I by some abstract function F, i.e., $O = F(I)$, then a strong AI system will be able to approximate F to a high level of accuracy for a wide variety of different input sets.

A common fear of AI is a proposed event referred to as the Singularity. First described as 'the intelligent explosion' by Good (1966), and popularised by Ray Kurzweil in the early 1990s, the Singularity refers to the accelerating progress of humanity, measured roughly as the time interval between technological innovations. Rather than pursuing a linear growth of innovation, human civilisation's achievements appear to follow an exponential growth paradigm.

If we extend this exponential growth to the development of AI and other systems that store and represent information, we should therefore expect that once AI reaches a threshold level, it will be able to participate in the design of its successors. It is at this point the so-called Singularity begins.

An AI A, of human-like intelligence, modifies the design of AI B to improve its efficiency and performance. AI B is capable of making more advanced modifications to AI C, and so forth. This produces an exponential growth of intelligence until AI systems are designed to specifications far in excess of human comprehension. This production of so-called *superintelligence*, an entity that rapidly and profoundly outstrips the best human mind in any and every task, is a natural concern.

If the Singularity occurs, machines slip away from the control of humanity forever, and in some dark scenarios spell humanity's end. We must therefore ask: *Can the Singularity occur? And what can biological intelligences do to protect themselves?*

This is clearly a vast field of research, and we can only summarise parts of it. We will focus on some basic definitions and categorisations of AI that exist today and that may exist in the future.

15.1 The Turing Machine

Let us define in the most general terms the functions of a computer. Computers are a class of Turing machine. One imagines such a machine (which we shall label T) receiving a series of input symbols I (given the era in which Turing defined his machine, we consider these symbols to be placed sequentially on tape). The machine T scans each symbol on the tape in sequence. According to a set of instructions (the program R), T will then either modify the symbol it reads, move the tape left or right, or halt the computation.

This description, while extremely simple, is sufficient to describe any algorithm or function that a computer can execute (this is a consequence of Gödel's completeness theorem for first-order predicate logic). Its simplicity also means it is amenable to theorems that circumscribe its capabilities. We can describe a *Universal Turing Machine U* as a Turing machine which can emulate any other conceivable Turing Machine T. Turing's genius was to realise that U could easily be constructed by allowing U to read a description of any Turing machine T from tape, before executing T on the subsequent inputs.

In more plain language, a computer can emulate any other computer provided it has received a description of how the other computer works (the program R, which may itself require input data D). This is almost axiomatic to modern computer users today, but it is difficult to overestimate how important this insight proved to be to the development of computer science. If we apply our human brain analogy from earlier, where the brain acts on inputs I with some function F to deliver output O, this would suggest that if human intelligence were a Turing machine, it could be distilled into a program, i.e., $R \equiv F$.

The Universal Turing machine is perhaps more powerful in the abstract than it is in silicon and plastic. Turing and his successors were able to deduce important constraints on the computability of certain types of problem, in much the same way that physicists are able to place constraints on energy consumption through thermodynamics.

Gödel's incompleteness theorem demonstrates that in higher-order logic, there are true statements whose truth is unprovable. In a connected discovery, Turing was able to show that there is no program R that determines if an arbitrary program R' (with input D) will run in an infinite loop (known as the halting problem). Contemporaneously, Church proved a similar set of axioms for arithmetic, and hence this is now referred to as the Turing-Church thesis.

15.2 A Brief History of AI Techniques

We can identify several areas of technological innovation in the history of AI, which sheds light on not only our understanding of machine intelligence, but human intelligence also.

AI methods can be classified as either *numeric* or *symbolic*. From these classifications, we can derive three broad approaches in developing AI. While no single approach has achieved strong AI, hybrid approaches are making continued advances, allowing AI systems to penetrate deeper and deeper into the data-rich culture that humanity has developed.

15.2.1 Logic-Driven AI: Symbolic Approaches

Before the computer was a physical reality, symbolic approaches were essential, and the work of Gödel and Turing defines this early period. The development of symbolic processing languages like LISP and PROLOG in the 20 years following Turing coincided with a period of perhaps wistful imagination. The completeness theorem for first-order predicate logic (PL1) states that every true statement renderable in such logic is provable, and the converse is also the case (we can always define a calculus such that only true statements are provable). Statements in PL1 render variables into formulae. For example:

$$\forall x \text{ likes}(x, cake) \tag{15.1}$$

or in English, 'For all x, x likes cake' or more simply 'Everyone likes cake'.

15.2 A Brief History of AI Techniques

A large number of AI programs were initially developed to prove statements in PL1, using the so-called 'resolution calculus' (referred to as 'automatic theorem provers'). For example, Simon and Newell's 'Logic Theorist' (LT), one of the first theorem provers, provided independent proofs of various theorems in Russell and Whitehead's *Principia Mathematica*. Notably, the prover began at the end, using the theorem as a starting point and searching for relevant axioms and operators to develop the proof.

Some AI practitioners held to the belief that the crafting of automatic theorem provers was the penultimate step in true AI – the final step being the correct phrasing of all problems in PL1. McCarthy (1960) proposed such an ideal program, which he named the 'advice taker'. Such a program would be able to infer conclusions from a set of premises. If the program could

[deduce for itself] a sufficiently wide class of immediate consequences of anything it is told and what it already knows

then it could be said that such a program had 'common sense'. The historic Dartmouth conference in 1956, widely regarded as the birthplace of AI as a discipline, introduced these new concepts to the world, including the special-purpose programming language LISP, specially designed for symbolic processing.

The dream of Dartmouth was hampered by what is often referred to as 'explosion of the search space'. The highly combinatorial nature of the inferential steps involved means that an automatic theorem prover can solve a theorem in finite time, but 'finite' typically means 'impractical'.

This can be addressed by better algorithms for searching the solution space. For example, most solution spaces can be organised in *decision trees*, and the algorithm used to navigate the decision tree (which can grow exponentially large) can greatly reduce the computation time and allow theorem provers to function in a range of environments.

Human cognition also suffers from this combinatorial explosion. Given the finite resources of the human brain (and often extremely strong time constraints on decision making), humans apply *heuristics* to arrive at a solution. By their definition, heuristics are simple and rapidly calculated, and if appropriately chosen can provide quick shortcuts to a solution that is more or less optimal.

A classic example of a heuristic is in the A^* algorithm for finding paths. Path-finding algorithms are commonly applied to problems such as traffic control and routing of utilities in cities. Let us define a set of vertices $V = \{v_1, \ldots, v_n\}$, and edges connecting the vertices $E = \{e_1, \ldots, e_n\}$, with the vertices defined by their position, and the edges defined by the two vertices they connect (and the subsequent distances between the vertices, which we will record as d_k for edge e_k). We can draw a path between vertices v_i and v_j by using a subset of E. Optimally, the subset will minimise the sum of all edges in the subset

$$D = \sum_{e_k \in E} d_k. \tag{15.2}$$

The A^* algorithm finds the shortest path from e_i to e_j, vertex by vertex, by calculating the following cost function at each:

$$C(v_k) = g(v_k) + h(v_k) \tag{15.3}$$

Here, g is the total cost function calculated from all previous vertices to v_k, and h is the heuristic. It is standard to set h as the distance between v_k and the target v_j (Figure 15.1). In English, the heuristic is 'pick the next vertex so that my distance to the end is the smallest' and is one that most humans would recognise as they navigate the world. Heuristics are not in general guaranteed to find optimal solutions, and in some cases heuristics can actually increase the computation time for a problem rather than reduce it. In humans, heuristics are the roots of several types of *cognitive bias*.

Also, we are likely to wish to solve problems at higher orders of logic. For example, the following induction statement is itself PL2, as it renders both variables and formulae:

$$\forall p\; p(n) \rightarrow p(n+1) \tag{15.4}$$

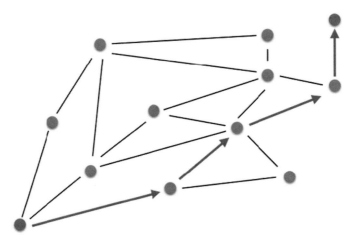

Figure 15.1 The A^* algorithm in action. To find the shortest path between the leftmost and rightmost vertex on this graph (indicated by arrows), the algorithm calculates a cost function for all possible vertices to move to the next vertex, using the distance between that vertex and the destination as a heuristic (equation 15.3).

PL2 systems are incomplete, that is, there exist true statements in PL2 that are not provable. Some PL2 statements can be reconstituted into PL1, but not all.

Predicate logic in general is not well disposed to the incorporation of *uncertainty*. Being able to reason while in only partial possession of the facts is a key aspect of animal intelligence in general, and any machine intelligence is likely to suffer the same information droughts that are presented to all species. For example, *expert systems* employed to render diagnoses of medical conditions must be able to present judgements based on a subset of symptoms presented by a patient. Even with access to the complete genome of the subject, and a detailed medical history, any diagnosis must be expressed as a probability, not a certainty.

15.2.2 Biologically Motivated Approaches

If we are to build a human-like intelligence, can we simply program a brain? This question motivated the development of *neural networks*, a technique that became possible thanks to the growth of computer power to model a large number of programming elements. Neural networks were at the vanguard of a new movement in AI, now known as *connectionism*.

The human brain contains approximately 10^{11} neurons. Each neuron consists of an axon, with dendrites which form connections with neighbouring neurons. Each neuron can store and release charge along the connections. The releasing of charge (known as 'firing') occurs once the voltage of a neuron exceeds a threshold, and hence the firing of any one neuron is correlated with the firing of the other neurons connected to it. As each neuron fires, it increases the voltage of its connected partners until they can fire.

The 'clock speed' of a given neuron is no more than 1 kHz, some million times lower than a modern computer. The brain's neural network is massively parallel, allowing it to perform extremely complex cognitive tasks in real time.

The enormous number of connections (each neuron typically has $10^3 - 10^4$, resulting in a total of around $10^{14} - 10^{15}$ connections) and their three-dimensional mapping into layers of soft tissue forbids us from simply constructing a complete digital copy of the human brain. This is completely aside from the fact that neuron connections are in a constant state of change, forming when needed and dissolving when they fall out of use. It is this adaptation which identifies the site of human learning as the connection (the synapse), not the neuron. The electrons are not entirely free to travel from one neuron to the other, as there is typically a charge gap between the two. Instead, chemical substances lying along the gap (*neurotransmitters*) can be ionised by an applied voltage, bridging the gap. The conductivity of a synapse indicates its use: high-conductivity synapses are commonly used, and the entire network is affected by the supply and concentration of various forms of neurotransmitter.

Formally, a single neuron can be described by a sequence of n input values $\{x_j\}$, each describing the signal received from an individual synapse and a set of weights $\{w_{ij}\}$. The output signal w_i is given by the activation function f which acts as a function of the total weight:

$$x_i = f\left(\sum_{j=1}^{n} w_{ij} x_j\right) \quad (15.5)$$

If we propose a simple Heaviside step function for f:

$$f_\Theta(x) = 0 \text{ if } x < \Theta, \quad (15.6)$$
$$1 \text{ otherwise} \quad (15.7)$$

then this is referred to as a *perceptron* with threshold Θ. This neatly captures the essential behaviour of binary neurons.[1]

Like the human neural network, an artificial neural network can be trained, by allowing the weights w of every neuron to evolve. The most typical neural network model used in machine learning is the three-layer backpropagation network (Figure 15.2). The network consists of three layers of neurons, an input layer of n_1 neurons, a hidden layer with n_2 neurons and an output layer of n_3 neurons. Every neuron in the input layer connects to the hidden layer, and every neuron in the

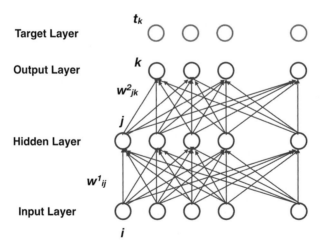

Figure 15.2 A three-layer backpropagation network. Each neuron i in the input layer is connected to the hidden layer, with a connection weight w^1_{ij}. Each neuron in the hidden layer is then connected to a neuron in the output layer k by weight w^2_{jk}. The outputs x_k are then compared to the target outputs t_k.

---

[1] If we wish to model neurons with non-binary activation, then f can be replaced with a sigmoid function.

hidden layer connects to the outer layer. The output value of the output layer $\{x_k\}$ is then compared with the target $\{t_k\}$.

Initially all the weights w_{ij} are given random values. The network is then trained on a set of data, using a series of *forward propagation*. The inputs are propagated from the input layer, through the hidden layer to the output layer. Each neuron computes its x by equation (15.5), where n refers to the number of neurons in the layer below it.

The outputs in the output layer are then compared with the target, and the weights of the network are modified, beginning first with the hidden layer (*backward propagation*), so that the outputs more closely resemble the targets given as a training set.

The weights are modified according to the quadratic error function between the output layer and the target data:

$$E(\mathbf{w}) = \frac{1}{2} \sum_{k \in output} \left(t_k^2 - x_k^2\right)^2. \tag{15.8}$$

The weights of the network are adjusted at each backward propagation step to minimise E:

$$\Delta w_{ij} = -\eta \frac{\partial E}{\partial w_{ij}}, \tag{15.9}$$

where we introduce a constant parameter η as the *learning rate*. This is typically referred to as *gradient descent*, as one can imagine E being a potential surface of which the network is attempting to find minima. To avoid overfitting, one can perform the gradient descent with a random selection of neurons disconnected.

This basic model has had many successful applications in a variety of regimes, especially in pattern recognition, and other problems relating to robotics. Gradient descent can become challenging when there are a large number of local minima in E, or there are many thousands of weights and a significant tranche of training data. *Convolutional neural networks* are a common variant, with a convolutional filtering layer to reduce the inputs. This design is more fitting for analysing visual (i.e., pixel) data.

So-called *deep learning* neural networks use more than one hidden layer. This appears to allow better abstraction of a dataset into fewer distinct units, but obviously vastly increases the computational load to perform backpropagation.

15.2.3 Data-Driven Approaches

Machine learning, originally a subfield of AI, has become immensely popular in the field of *data science*. Indeed, it is now a common misnomer to interchange AI and machine learning as labels.

At its core, machine learning is a discipline where computers use existing data to make predictions or responses to future data. Generally speaking, machine learning algorithms construct an internal model of the data, and use this model as a predictor. Most machine learning algorithms lean heavily on statistical measures and optimisation.

We can classify most branches of machine learning into one of three categories. In *supervised* machine learning, the target outputs are clearly understood, and the system uses a *training set* to adjust its weights to correctly map from input to output. Neural networks are commonly used in this mode, as are *support vector machines*, which classify datasets into categories.

Unsupervised learning has no training set, and typically no prior target outputs. The machine is given a set of inputs and is expected to find structure inside the data, e.g., by classification or pattern matching. Neural networks can be used in an unsupervised mode – an example is the *self-organising map*, which rearranges multi-dimensional data points into a 2D grid such that similar data points are mapped onto adjacent grid points (Kohonen, 1982). Another is the generative adversarial network (GAN), which pits two neural networks against each other (Goodfellow et al., 2014). One network generates candidates, the other evaluates, with each attempting to 'defeat' the other in a zero-sum game, where the generator attempts to 'fool' the evaluator into thinking a generated item is in fact an authentic data point.

Reinforcement learning also has no training set, but the machine does have a concept of what is a 'good' outcome. An action a modifies the state s of the environment, and gives a reward $r(s, a)$, where the sign of the reward indicates positive or negative reinforcement. This is commonly used in robotics, where there are clear indicators of when a manual task is performed successfully or not.

15.3 The Turing Test

A critical problem for most, if not all, of the above approaches is the issue of *transparency*. This is true both for weak and strong AI. For example, a neural network can be trained to execute a task – driving a car – with high efficiency and accuracy. One could ask: Where inside the neural network are the machine's instructions for dealing with children on the road?

The answer of this question is at some level encoded on the weights of the neurons. Extracting the answer is far from trivial, and becomes exponentially difficult as the number of neurons in the network increases. Such machine learning products can easily become *black boxes* – processes that accept inputs and deliver outputs, with the user being unable to discern how the process functions.

One could argue that this is a likely feature of human-like AI – after all, the brain is in some respects a black box. We are fortunate in that we can interrogate biological neural networks and receive linguistic responses with some form of explanation. Of course, these responses are imperfect, but they shed some insight into our brain's operations. If a strong AI were to be developed, it is quite possible that we would not fully understand its machinations, no more than we would understand a human interlocutor.

If this is so, how are we to understand if a machine contains a strong AI? Turing's test of the quality of AI was simple. A human conducts a conversation with an anonymous entity. If the human cannot determine that the responses come from a machine, then the machine is deemed to have passed the Turing test.

This test implies an underlying philosophy: anything that appears intelligent is intelligent. At dinner, I am unable to determine the process by which the person across the table from me constructs responses to my statements. Anything that appears to be a strong AI is in fact a strong AI.

Searle's Chinese Room argument is a convincing retort to this. Consider a computer program that is able to conduct a conversation in a language you do not understand (Searle uses Chinese) that passes the Turing test. Chinese characters are inputted to the system, and appropriate Chinese characters are given as output.

Now imagine sitting with a hard copy of the source code for this program. If you are placed in a room with two doors, where Chinese characters arrive from one door, it is possible in principle for one to use the program to generate a response, and send it out the other door. In effect, you are a computer, executing a program. *But do you understand Chinese?* No!

One might argue that this is mere semantics, and that the human in the Chinese Room plays only a mechanical role in the intelligence of the machine, as one might consider a region of the brain in isolation, itself incapable of consciousness alone, but a crucial component of the conscious mind as a whole. In effect, this argument is a subset of centuries of arguments regarding the granularity of human consciousness.

15.4 Is the Singularity Inevitable?

Kurzweil's vision of the Singularity is highly utopian – it signifies a major evolutionary transition, from the biological to the post-biological, with humans being able to either take complete control of their biological entity with nanotechnology and gene therapies, or to shed them and 'upload' their consciousness to a non-biological substrate.

Unsurprisingly, Kurzweil and his contemporaries share many detractors. First of all, for the Singularity to occur, artificial intelligence must be able to rise above a certain intelligence threshold, and we will show arguments at the end of this section that suggest this is impossible.

Much criticism is made of Singularity proponents' tendency to consider the timeline of technological progress on logarithmic scales. Critics note that presenting the time interval between significant events on log-log plots is likely to be biased in favour of a straight-line fit.

Perhaps more damningly, the concept of unending exponential growth has been touted in several different arenas of existential risk, and the trend has been shown not to hold (e.g., Malthusian theory for population growth, see Chapter 12).

Given similar datasets to Kurzweil's on other topics (world population, economic growth), one can fit exponential growth models to the data and claim a 'singularity' where one does not otherwise exist. This is especially easy to do if the latter half of the dataset is missing, which is true for the technological Singularity, as it lies in our future.

We can see from Bringsjord (2012) that the 'rational basis' for the Singularity is *logically brittle*. A natural statement S is said to be logically brittle if, once rendered into formal logic, it is either provably false or logically unsound. Let us consider

S = 'The Singularity, barring external circumstances, is inevitable'.

We can render S into the following logical inference:

Premise 1 There will be human-level AI (created by humans)
Premise 2 If there is AI, it will create AI^+
Premise 3 If there is AI^+, it will create AI^{++}
Therefore There will be AI^{++}.

Why is this argument brittle? We can investigate this by considering the classes of machine that can be produced by humans (and by machines). We can define three classes:

- M_1: Push-down automata – more simplistic than the Turing machine, it can only read the input tape and the top element of a stack of data, as opposed to Turing machines which can read the entire stack
- M_2: Standard Turing machines
- M_3: Infinite-time Turing machines, which are able to calculate using the same method as M_2, but at infinitely high speed (see Hamkins and Lewis, 2000)

It is easy enough to demonstrate by looking at the current status of machine learning, and other realms of AI, that human-developed AI is either at or below M_2. Turing-Church theory demonstrates that a machine of class M_2 cannot produce a machine of class M_3.

We can see this by returning to the halting problem. Infinite-time Turing machines are in principle capable of solving the halting problem. If an M_2 machine can build an M_3 machine, in principle this would constitute a standard Turing machine that can compute and solve the halting problem, which is not possible.

This can be more generalisable (via complexity theory) to the statement that 'M_i is unable to produce M_j, where $i < j$', and there are hence definable roadblocks on the road to AI^{++}.

If we define human intelligence as M_p where $p > 2$, then by the above argument, AI cannot be raised beyond M_2, and therefore human-level AI is not possible, thereby falsifying premise 1 and showing that S is unsound. Singularity pessimists hold this as strong evidence that the Singularity will never happen.

On the other hand, Singularity optimists can rightly argue that this in fact defines a clear condition for human-level AI to be possible: humans must be able to develop machines of class M_3.

15.5 Controls on AI

If the Singularity does occur, and superintelligent AI emerge, how can humans (and other intelligences) protect themselves?

Bostrom (2014) classifies the protective mechanisms available to biological intelligence into two categories: *capability control* and *motivation selection*. In the first, fail-safes are built into the AI before they become superintelligent. Such fail-safes can be physical (disconnection from the environment, quarantine from information systems, etc.), algorithmic (trip-wires that deactivate the AI if it exhibits dangerous behaviour) or simply stunting the AI's performance by restricting it to run on limited software or hardware.

In the second class, humans relinquish direct control of the AI's actions and capability, but instead attempt to imbue it with appropriate moral or ethical instructions, and trust its ability to make good decisions based on its own internal rules. Again, these protective mechanisms may be direct programming (such as Asimov's Three Laws of Robotics), behavioural training, programming the AI with rules for generating its own set of norms (*indirect normativity*) or augmenting the AI by adding a human component.

Alfonseca et al. (2016) argue that the halting problem itself places severe constraints on our ability to control superintelligent AI. The halting problem states that a function acting on machine T with input I, *Halt*(T,I) is uncomputable. We can then define a function *HarmHumans()* that harms humans in finite time, and write a further algorithm *HaltHarm*(T,I), which first executes T(I), then executes *HarmHumans()*.

This algorithm only harms humans if T executes without halting. If we can tell that this algorithm will harm humans, then we can determine that $T(I)$ will not halt, and hence we find a contradiction, proving that we cannot compute whether a machine will harm humans or not. We therefore cannot use an AI to determine if another AI is safe!

Following the 2017 Asilomar conference on AI, experts in the field became signatories to 23 guiding principles for future AI research.[2] These principles address the day-to-day political issues of AI research, as well as the ethics and values that researchers should abide by, and some guidelines for assessing the risk of one's research. These are highly worthy aims for any attempt to develop machine intelligence – one can only hope that they are followed.

15.6 Is Strong AI Likely?

Given around seventy years of development, we cannot yet point to a convincing example of strong AI. We can point to several very powerful examples of weak AI (such as IBM's Watson), but these examples are not likely to evolve from their current limitations into a general artificial intelligence.

In a recent survey, experts gave a 50% probability that AI systems will eventually outperform human workers within the next 50 years at all tasks surveyed, with an effectively fully automated workforce within 120 years (Grace et al., 2017). This implies that AI development will itself become the province of AI, and a Singularity-like event could occur, although the survey respondents ascribed a low probability of this occurring (around 10%).

Those who believe strong AI to be unlikely often quote Fermi's Paradox: *if strong AI is possible, where are they?* We see no signs of strong alien AI, and hence they are either impossible or very short-lived.

We can consider several possible reasons as to why we see no sign of strong AI, human-made or otherwise:

1. Intelligence is not a one-dimensional property. As much as humans find the concept appealing, one cannot satisfactorily arrange animals in order from smartest to dumbest. Intelligence is instead a very high-dimensional combinatorial space. Weak AIs assume very small subsets of this combinatorial space (and now easily defeat humans in these regions). A strong AI would not necessarily be 'smarter' than humans, but merely operate in a different subset of intelligence space.
2. From reason 1, we can perhaps conclude that humans do not satisfy the conditions for strong AI either. We also only inhabit a small subset of the intelligence

[2] https://futureoflife.org/ai-principles/

space. Our brains were evolved to permit our genetic legacy to endure, not to achieve fully rational cognition. A series of trade-offs has occurred to shape our minds to ensure their survival. It is quite possible an analogous set of trade-offs will be made for AIs to function as required in human society. Will these trade-offs prevent strong AI?

3. Human-like intelligence is utterly wedded to its biological substrate. To produce a human-like AI, one might argue we must simulate an entire human, including its digestive and endocrine systems, to truly replicate a human mind. Of course, this does not rule out disembodied intelligence, but it remains unclear how such an AI would behave. Can a disembodied AI truly pass the Turing test?

Together, these arguments suggest that strong AI as we have described it is unlikely. Instead, we may find ourselves producing 'good-enough' AI – more general and flexible than weak AI, but not able to reproduce the level of flexibility and creativity of humans.

Even if strong AI is impossible, weak AI will continue to be installed into a multitude of institutions and products. Existential risks are still posed even if we are not confronted with a malicious intelligence. AI, like many tools developed by humans, is dual use. There are many ways that weak AI systems can be subverted by unfriendly actors. Exploiting vulnerabilities in digital security, interfering with black box models to alter the AI's predictions, or repurposing of airplane pilot AIs by terrorists are just a handful of threats posed by the increasing use of AI.

We must therefore remember: it may not be a superintelligence, but a hacked weak AI control system – the equivalent of a picked lock – that ends humanity (and other ETIs).

16

War

In 1986, the United Nations Educational Scientific and Cultural Organization (UNESCO) released what is now known as the Seville Statement on Violence. In it, the undersigned scholars, from fields ranging from anthropology to neuroscience and psychology, were moved to make five propositions on the human propensity for violence and war.

1. It is scientifically incorrect to say we have inherited a tendency to make war from our animal ancestors.
2. It is scientifically incorrect to say war or any other violent behaviour is genetically programmed into our human nature.
3. It is scientifically incorrect to say that in the course of human evolution there has been a selection for aggressive behaviour more than other kinds of behaviour.
4. It is scientifically incorrect to say that humans have a 'violent brain'.
5. It is scientifically incorrect to say that war is caused by 'instinct', or any single motivation.

These are five bold propositions to make on the causes of war, violence and aggression in human civilisation. The factual accuracy of these statements has been roundly questioned in the intervening three decades, which we will address briefly, but they are an excellent place to begin our discussion of a potential solution to Fermi's Paradox – that there can be no civilisation without war, and with war there will eventually be no civilisation.

To address this solution – that civilisations destroy themselves through their own aggressive impulses and/or a political need to make catastrophic war – we must consider these five statements carefully, and decide whether war is inevitable in human civilisation, and indeed whether it is inevitable in any technological civilisation.

16.1 The Origins of Human Warfare

What are the roots of warfare? We should first make a careful definition of warfare, as opposed to aggression, which is evident in many non-human species. A predator must show aggression if she wishes to eat her prey. In many cases, a male must use violence to defend his harem of breeding partners against competing males. A mother will answer an attack on her young with extreme prejudice. These are examples of instinctive response, in varying degrees essential to the life cycle of the organism. Defending one's young or breeding partner is obviously an adaptive trait, and often crucial if one's genetic material is to propagate.

A passable definition of war might be 'collective killing for a collective purpose beyond consumption'. Hunting and predatory behaviour exists throughout the biosphere, and is heavily circumscribed by instinct and breeding – predators carve out territories to avoid encountering rivals, and if conflict occurs, it is brief and often marked by submission of one party before escalation.

Warfare clearly goes beyond hunting. Humans kill their own species with no intention to eat them, and not to protect their own reproductive partners. The genetic imperative of warfare is significantly diminished in comparison to the above examples. Systematically murdering competition may yield an advantage in breeding, but the cost of making war on the environment and fellow individuals may well nullify said advantage.

The development of increasingly advanced weaponry also dissociates us from our victims. Animal aggression is triggered by specific environmental stimuli, and can be suppressed by other stimuli. For example, primates can be stimulated into aggressive behaviour by appropriate manipulation of the limbic system with radio receivers, but will not attack dominant individuals even when stimulated thus (Delgado, 1969). The disgust response at seeing a fellow member of one's species grievously hurt also acts as an inhibitor to further aggression. Hurling a spear from a distance weakens the disgust response – firing a missile from a computer terminal further so.

The origin of war as collective behaviour lies well down the family tree. Groups of chimpanzees have been observed to conduct campaigns of violence against other groups that bear all the hallmarks of modern conflict. The *casus belli* for such conflicts is typically resources. We should also note that insects from the *Hymenoptera* order (ants, bees, wasps) are also capable of waging what we would probably call war. When two ant colonies compete for resources, worker ants from both sides bite, sting and die in the hundreds of thousands. These are typically eusocial societies, so the genetic imperatives for worker ants going to war are extremely different (see page 124).

One could imagine that the earliest human warfare was conducted for similar reasons. A hunter-gatherer band happening on an agricultural settlement may have decided that killing the farmers and taking their crop and livestock was the easiest way to feed themselves. Hunting large animals as a group, using tools and weapons to level the playing field, may have quelled their fear response, and taught them the rudiments of military strategy.

It is important to note some distinguishing features of chimpanzee aggression compared to that of early human societies. In particular, there is a marked trend in in-group violence: chimp groups are thought to display non-lethal aggression between members around 200 times more frequently than, say, Australian Aboriginal groups of similar size (Wrangham et al., 2006). This suggests warfare is a hijacking of internally aggressive behaviour by the social group, channelling violent behaviour outward. As human social networks grew denser and more complex, the concept of 'us vs them' allowed such dangerous impulses to be focused on external threats that could only be perceived by brains capable of sufficient abstraction.

This abstraction has allowed war to perpetuate even as humans have successfully vitiated their violent impulses. The continuing decline in violent crime and aggressive behaviour throughout the last few hundred years has been remarkable, and yet the casualties in the two World Wars reached distressingly high levels. The need for individual aggression has been superseded by the highly developed concepts of the nation-state, duty, honour and the belief that one's moral code and doctrine requires defending, even if such defence is the cause of individual trauma and horror.

This would seem to confirm the validity of statement 5. There are multiple causes of modern wars, and 'instinct' plays an extremely diminutive role. Military commanders cannot formulate strategy and tactics on instinct alone. One can point to many cases where soldiers cannot operate on instinct – for example, infantry personnel ascending a trench wall into a hail of machine gun fire must do so against every instinct afforded an animal with a strong drive for self-preservation.

16.2 Is Violence Innate to Humans?

Can we identify the source of violence in the human brain? There are indications that humans with a predisposition to violence possess markedly different brain structures to their 'normal' counterparts. These changes are most commonly associated with the prefrontal cortex, which regulates emotional behaviour. A study of 41 murderers that pleaded not guilty due to insanity were shown to have significantly reduced activity in the prefrontal cortex, as well as increased activity in the limbic system, which is associated with aggression itself. But are these genetically driven changes?

There is a good deal of debate regarding violence as an adaptive trait in early humans. As we have already noted, there are circumstances where aggression and killing may confer an advantage in primordial human societies, and there are many adherents to this 'homicide adaptation theory' (Buss, 2005; Durrant, 2009). Equally, killing may well be a by-product of achieving other goals, which are themselves adaptive. The human urge to achieve dominance or reproductive success, in what should be a non-lethal competition, can sometimes end in death, an accident against the wishes or desires of both parties. Given that most killing in peacetime is carried out by men of reproductive age in regions of high social inequality, both theories offer explanations to this observation. Equally, this observation is also strongly suggestive that nurture and environment play key roles, reinforcing aggressive behaviour if and when it is successful in furthering an individual's goals.

While we may search for biological origins of aggression, and congratulate ourselves on the apparent decline in individual violence in society, we cannot ignore the fact that even if violence has a biological origin which might one day be mitigated by intervention, violence is also a tool of societal manipulation. Feudal monarchies, totalitarian dictatorships and tribal chiefdoms have used aggression to further their own ends. Violence without aggression is very real, and represents a disconnect between our biological drives and the demands of our ambient culture.

16.3 Nuclear War and Weapons of Mass Destruction

The ultimate example of politicised violence, absent personal aggression, is the use of nuclear weapons and other weapons of mass destruction. As I write these words, the Doomsday Clock, which measures the threat of global nuclear annihilation, has moved 30 seconds closer to midnight, where midnight symbolises the beginning of Armageddon.[1] The Bulletin of the Atomic Scientists instigated the setting of the Clock in 1947, with the hands set at 7 minutes to midnight. Every year since, the Doomsday Clock has been set closer to and further from midnight. As additional threats to humanity have emerged, the Clock now takes into account several global risk factors, such as climate change and biosecurity risk.

In 2017, the Clock was set at two and a half minutes to midnight, reflecting the enhanced risk to our planet, particularly from nuclear attack. As of 2013, the Earth possesses 10,213 nuclear warheads, distributed between several nation-states, with a total destructive power of order 6 million kilotonnes (kt). Some 90% of the world's arsenal remains divided between the United States of America and the Russian Federation, with both states continuing to modernise their weaponry. Perhaps

---

[1] http://thebulletin.org/doomsday-dashboard

most concerningly, Pakistan and India continue to dispute the territory of Kashmir, and both states are card-carrying members of the nuclear club.

Between 2010 and 2015, some 58 'security events' have occurred, putting nuclear material at risk through loss, theft or other violations and lapses in procedures that contaminate the material. These events leave open the possibility that either rogue states or non-state actors such as terrorist organisations could gain access to radioactive material, causing significant destruction with low-yield 'dirty bombs', and potentially triggering World War III itself.

Einstein is famous for having said

> I know not what weapons World War III will be fought with, but the fourth World War will be fought with sticks and stones.

This seems to have been a common trope at the time, but whether it was Einstein who coined this phrase, or General Bradley, or an anonymous GI stationed in Germany during the postwar transition, its meaning is clear. If the nations of Earth chose to engage in total war, as they did between 1939 and 1945, and nuclear weapons are used, then human civilisation as we understand it is likely to end.

A nuclear explosion has two principal sources of radiation – directly from the nuclear explosion, and indirectly from radioactive fallout. This refers to the radioactive particles thrown into the upper atmosphere by the explosion, which then descend slowly to the ground. If the entire nuclear arsenal were to be deployed evenly across the entire Earth's surface, this is roughly equivalent to 25 kt of explosive per square kilometre (Stevens et al., 2016), i.e., more than one Hiroshima bomb for every square kilometre. If we consider the radiation from the fallout alone, then nuclear tests conducted in Samagucha indicate that this would result in a radiation dose rate of approximately 1 milliGray/hour, which eventually decays to normal background levels over a period of a hundred days. A whole body dose of around 2–6 Grays is generally fatal, meaning that the fallout is likely to cause a fatal dose if the subject is exposed to these levels for a few weeks. Given that these estimates are based on a single test site, and do not take atmospheric currents into account, the entire terrestrial biosphere may be effectively extinguished in less than a month.

It would therefore seem that a strong solution to Fermi's Paradox is that 'civilisations cannot survive the invention of atomic power and nuclear weapons'. A worthwhile counter-argument to this is the fact that human civilisation is not yet extinguished. The Doomsday Clock is now 70 years old. The carnage visited upon Hiroshima and Nagasaki has stayed the hands of two superpowers through what some have called 'the Long Peace'. Despite fighting proxy wars and engaging in vigorous subterfuge, the USA and USSR did not launch a nuclear attack against each other, despite some moments where it seemed likely. We have transitioned from trembling under a sword of Damocles, to being able to look through the steel. Nuclear weapons are a threat to our existence, but they have also been a bulwark against war. As much as the doctrine of Mutually Assured Destruction may seem indeed MAD, it has stayed the hand of people that would be otherwise willing to use it.

However, past performance is no guarantee of future competence.

16.4 Is War Inevitable amongst Intelligent Civilisations?

We must sadly answer this by giving no definitive answer. We might prevaricate and say that aggression is a common feature of metazoan organisms. Sentient creatures whose ancestors were predatory may develop those predatory instincts into warlike conflicts, but such a phenomenon is most likely conditional on external factors, especially the availability of resources.

We might also say that conquest can be adaptive under the right circumstances. It seems clear from human history that the military solution to hunger was discovered quite early, and a literal arms race began between attackers and defenders. Again, this might only establish competition over resources – if the highly ritualised nature of resource competition between animals was somehow imprinted into a sentient species, then the total war that humanity experienced in the twentieth century might not occur at all, and the alien equivalent of war is merely a sport where the winner takes all, and the loser's wounds are bloodless.

17

Societal Collapse

The preceding chapters have outlined various stressors or pressures that can exert themselves upon a civilisation. A combination of these factors can result in a breakdown of the social and political contracts between individuals, and the disintegration of society.

Human history, in many ways, is the history of the genesis and death of societies. We should note that the death of a society is not guaranteed to be the death of the individuals that form this society. Much is made of the rise and fall of the Roman Empire, which saw a Mediterranean society grow from a single city to a state that stretched from Baghdad to the Atlantic coasts of Europe (in 117 CE, under Emperor Trajan), then tear itself apart through a raft of civil wars. The Western Roman Empire is considered by most historians to have effectively ended when Odoacer deposed the Western Emperor Romulus (although the emperor's hold on distant territory had long been non-existent, and Roman territories would continue to raise emperors for decades afterwards).

However, the dissolution of the Western Empire did not destroy the Roman people, only their way of life. It is clear that many, especially the invading Goths, benefited enormously from what some would have considered an aggressive reorganisation rather than total destruction. In outlying territories, the demise of central bureaucracy did not require the demise of local bureaucracy, and Roman culture was deeply ingrained in the new states born from the Empire's demise.

From our point of view, we want to know what makes a society collapse so fundamentally that its constituent people are no longer capable of civilisation. We have very few examples of this, as humans still exist on the Earth. However, we can look to societies which have failed locally, and comment on how these failures could become a global phenomeon.

17.1 Common Causes of Societal Collapse

There are many schools of thought on the causes of societal collapse. There are dozens of ascribed causes for the collapse of the Western Roman Empire, and this is one among hundreds of extinct societies to which we have archaeological access.

Diamond (2013) notes five principal factors that form his framework of societal collapse:

- Self-inflicted environmental damage,
- Climate change,
- Hostilities with other societies,
- Trading relations with other societies,
- Cultural attitudes.

The first three factors are covered to varying degrees by the previous chapters of this part of the book. Indeed, we could generalise this framework as

- Self-inflicted stressors,
- External stressors,
- Relations with neighbouring societies,
- Cultural attitudes.

We have listed many self-inflicted stressors in previous chapters (genetic engineering, nanotechnology), and external stressors indicate those factors over which civilisations have limited to no control (e.g., asteroid impacts or supervolcanism).

Joseph Tainter's seminal work *The Collapse of Complex Societies* interrogates the ability of societies to solve problems, a fundamental skill needed to weather the effects of the above factors, and adapt to an ever-changing environment. Tainter's view is informed by economics and complexity theory – as a society attempts to solve new problems, it designs institutions and bureaucracies to achieve this task.

This inevitably increases the complexity of a given society – i.e., it is composed of a greater number of parts, a greater number of differentiated types of part, and greater integration between parts of the same type and different types. More complex societies demand greater energy resources, in a strong analogy with the negentropy demands of organisms in biology. Complex societies rely on their complexity as a problem-solving tool, resulting in a positive feedback loop.

Tainter argues that each increase in complexity offers diminishing economic returns for the society. In particular, the flow of information becomes less coherent as it traverses a complex organisation. Decision making suddenly has unforeseen consequences, which add unexpected and possibly heavy costs. In general, as societies continue to grow in complexity, diminishing returns result in depleted resources, generating sufficient intersocial tension that drives the society to collapse.

Complexity therefore becomes a proximate cause of poor decision making, which itself is a cause of societal failure. Diamond proposes four categories of poor decision making:

- Failure to anticipate an imminent problem,
- Failure to perceive a current problem,
- Failure to attempt to find a solution,
- Failure to find a solution.

The above four failures depend on a society's previous experience with problem solving, the institutional memory of a society's dealings with previous problems, and the vigour of its institutions in responding to and solving problems.

The institutions a society builds are heavily influenced by its cultural values. Complex societies are predicated on the flow of information, and the education of its citizens becomes increasingly vital if this information flow is to be correctly manipulated. However, it can be stated with reasonable mathematical rigour that the productivity ascribable to new acquired knowledge is initially very high, but certainly begins to decline as the sum of acquired knowledge becomes very large (Tainter, 2000). As new challenges impact a society, the ability to generate solutions to the problem becomes prohibitively expensive.

Paraphrasing Toynbee in his (now largely ignored) thesis *A Study of History*, civilisations eventually find a problem they cannot solve. This statement could be construed as a solution to Fermi's Paradox. Whether this solution is hard or soft remains a matter for debate.

The Need for Social Capital Even in a society free from internal or external stressors, it is entirely possible that a society can collapse merely because its binding elements begin to fail. If the fabric of society itself has a finite lifespan, then this ultimately means a finite lifespan for civilisations themselves. Human societies are founded on mutual trust and co-operation, formalised by a variety of social and political institutions. We can quantify this via 'social capital', a concept that describes the intrinsic value of social networks.

The precise definition of social capital varies from author to author – Adler and Kwon (2002) present a range of definitions. Common to all these definitions is a recognition that goodwill generated between humans is the substance of social capital, which generates varying levels of influence and solidarity between individuals in the social network.

This capital can itself be divided into two principal types: internal or 'bonded' social capital, and external or 'bridging' social capital. Bonded social capital is derived from networks of people who are in some way similar or homogeneous:

17.1 Common Causes of Societal Collapse

examples include close friends and relatives. Bridging social capital crosses divides in society, whether they are socio-economic, political, or otherwise. Examples of networks with bridging social capital might be playing for a local sports team, groups of hobbyists or enthusiasts, or fellow employees in a company.

The multiple definitions of social capital vary depending on the focus (internal/external), but they do agree on an important property that distinguishes social capital from other forms (such as economic, human or environmental capital). Like a sum of money subject to inflation, social capital needs continual maintenance. Relationships that are not tended will wither, and network contacts are unlikely to assist a colleague if they do not feel like the connection is valued.

Studies have shown that increased social capital is correlated with individual health benefits, better-functioning governments and more-productive economies (Helliwell and Putnam, 2004). Conversely, poor social capital correlates with increased rates of illness (especially illnesses caused by mental or emotional stress), increased crime rates (Sampson, 2012) and higher levels of institutional corruption (Putnam, 2000).

Social capital also feeds back into environmental capital, and the notion of sustainability. For one generation to care about the circumstances of the next requires social capital to link the two. A world rich in social capital is a world invested in its future inhabitants, and as such a world that would attempt to build and maintain environmental capital.

It is important to note that bridging social capital – the foundation of institutions – is key to societal success. Bonded social capital plays a role in maintaining individual health and well-being (e.g., having grandparents available for childcare, or a spouse to look after their sick partner), bridging social capital brings down barriers and dilutes the effects of segregation. A society with a preponderance of bonded social capital and a lack of bridging capital encourages an 'us versus them' mentality that can erode implicit trust in strangers and institutions, and encourage us to prioritise ourselves against others, to the detriment of all.

There are no guarantees that an economically successful society is rich in social capital. Societies that rely on immigration to drive economic growth may in turn damage sources of social capital, in particular the sources of bonded social capital (belonging, social solidarity) (Putnam, 2007; Ariely, 2014). The opposite may even be true – individuals in a society overabundant in bonded social capital may fail to properly scrutinise their network connections' decision-making skills, due to an implicit trust engendered by social or ethnic ties. Such failures have been identified, for example, in pricing of stocks, where ethnically homogeneous markets have been shown to be more prone to overpricing, and subsequent 'price bubbles' that can be financially damaging.

We could propose a solution to Fermi's Paradox along the lines of the following: 'Civilisations have a short lifespan because they fail to reliably generate healthy social capital'. Civilisations that follow this failure mode disintegrate because the implicit trust between strangers dissipates. Even a society rich in bonded social capital may fail through what Putnam calls the 'hunker down' effect. Family units close ranks around each other, and look after their own. The concept of paying taxes to assist others becomes reprehensible, and the social contract dissolves.

17.2 Intellectual Collapse

It is a common, almost hidden assumption amongst SETI scientists that any intelligent civilisation will contain curious, exploring individuals that are unsatisfied with their current level of knowledge about the Universe. It is partially a reflection of the scientists' own predilections – a hope that if another civilisation is found, they are a kindred spirit.

Like all assumptions, we must examine it carefully. Is there any guarantee that an alien civilisation will remain intellectually curious? We could imagine a civilisation that has achieved a perfect understanding of the Universe and its constituents (or it has decided that its understanding is sufficient). Such a civilisation could construct a high-fidelity simulation of the Universe as a more satisfying, pleasing alternative to our reality (see Chapter 24).

If ETIs retreat into leisure and forsake exploration, scientific research and communication, then their detectability is bound to decrease. The only conceivable signature such inward-looking creatures would produce would be the waste heat from their entertainment devices.

This could be thought of as the thin end of the wedge. A civilisation that becomes overly engrossed in leisure pursuits could do so to the detriment of a civilisation's other, more crucial functions: the storage and transmission of knowledge, the maintenance of environmental and social capital, and the survival of its citizens.

We can only speculate on what would happen if humans underwent such an intellectual collapse. A true mental regression is unlikely to occur – the *homo sapiens* brain has existed in its current form for millennia, and a few decades of idling is not going to change its configuration. If the transmission of knowledge failed – for example, if the skill of literacy was lost, and oral transmission ceased – most of civilisation would fall apart rapidly as essential equipment failed and was not maintained. Launching spacecraft and sending transmissions would certainly cease, and our visibility as a species would be reduced to atmospheric signatures.

Like other social solutions to Fermi's Paradox, this is soft. We are demanding all civilisations collapse or recede into their own fantasies because they simply stop trying – what does this say about our own civilisation? Or are we simply projecting our own fears about the trajectory of our species?

Part IV

Uncommunicative Solutions

In this final category, we consider the class of solutions that reduce f_c in some way. This is philosophically risky territory, for two reasons. First, this class commonly relies on assumptions regarding extraterrestrial motive which are fraught with anthropocentric bias. Secondly, many solutions require a kind of anti-epistemology – the so-called 'proof' is in the absence, not the presence. This criticism is typically levelled against SETI in all its forms, and it is a criticism with merit. As we go forward, we must take care to present our logical arguments carefully, and avoid these pitfalls wherever we can.

As with the C solutions, we can define two broad subclasses. In the first, civilisations are truly uncommunicative – they either do not broadcast, or broadcast in a fashion that makes them undetectable. The second class (Chapter 23 onwards) is the realm of conspiracy – civilisations *are* communicating (or have communicated) with Earth, or have actively interfered in Earth's development. As we will see, solutions from the second subclass rarely stand up to scientific scrutiny.

18

Intelligent Life Is New

At first glance, this grouping of solutions, which is temporal in nature, sits rather uncomfortably in any categorisation using the time-agnostic Drake Equation. In its simplest form, it reduces the value of f_c because any intelligent civilisations in the Milky Way are not yet capable of interstellar communication (or exploration via probes). But why might we expect that there are no ancient intelligences, despite there being apparently quite ancient stars and planets?

18.1 Carter's Argument against SETI

Using a line of argument reminiscent of the Doomsday Hypotheses we discussed in Chapter 7, Carter and McCrea (1983) considered the apparent coincidence between the evolutionary timescale required for the genesis of intelligent life on Earth:

$$t_i \approx 0.4 \times 10^{10} \text{yr} \tag{18.1}$$

and the main sequence lifetime of the Sun:

$$t_{MS,\odot} \approx 10^{10} \text{yr} \tag{18.2}$$

Is this truly a coincidence? At first glance, these two timescales should be uncorrelated, and there are three possibilities:

1. $t_i \gg t_{MS}$
2. $t_i == t_{MS}$
3. $t_i \ll t_{MS}$

If the two timescales are uncorrelated, the second possibility should be *a priori* the least likely. Carter also dismisses the third possibility, as it appears to be at odds with our observation that $t_i \sim t_{MS}$. We are then forced to typically accept the first possibility, which is that the typical timescale required to produce intelligent life is extremely long compared to the constraints imposed by astrophysical considerations.

This might seem at odds with the fact that we are present to make these deductions. Carter invokes the *anthropic principle* in this case. The universe looks this way to us due to the selection effects that allow our existence. In general, $t_i \gg t_{MS}$, but for us to be able to observe the Universe and find the Great Silence, we must be one of the few fortunate species where in fact $t_i \approx t_{MS}$.

This argument therefore seems to indicate that ETIs are extremely unlikely. Most civilisations fail to develop in advance of their host stars leaving the main sequence, which would most likely result in the destruction of non-technological civilisations. As a result, only a very small number of ETIs develop sufficiently rapidly to avoid this end.

One way to refute this argument is to argue that t_i and t_{MS} are in fact correlated, or that in a general sense astrophysical and biological evolution are more deeply interlinked. Global regulation mechanisms are one way in which this could be possible.

18.2 Global Regulation Mechanisms and the Phase Transition Solution

In our study of the Catastrophist solutions, we came across galactic-scale natural disasters that could spell doom for civilisations within thousands of light-years of the event (Chapter 11). Habitable planets that exist within range of these catastrophic events may be subject to a 'resetting' of their biological clocks by these astrophysical events. These resetting events influence the biological timescale, by forcing regions of the Milky Way to recommence biological evolution at the same moment (Annis, 1999; Vukotic and Ćirković, 2007; Vukotic and Cirkovic, 2008).

This is not proposing a direct link between the biological timescale and the *local* astrophysical timescale. What it does propose is a link between neighbouring biological timescales, and that biological timescales are forced by astrophysical constraints beyond the local environment.

Multiple mechanisms have been proposed, with several criteria that need to be satisfied. Firstly, the events must occur sufficiently regularly at early times that a quasi-equilibrium state can be achieved. Secondly, the event must be able to affect a sufficiently large spatial extent; and lastly, there should be some sort of secular evolution of the event rate.

The most obvious mechanism is gamma ray bursts. These extremely powerful explosions can be effective at distances in excess of a kiloparsec, and would have occurred much more frequently in the past than the present, due to gamma ray bursts being preferentially the fruit of low metallicity progenitors (Piran and Jimenez, 2014).

18.2 Global Regulation Mechanisms and the Phase Transition Solution

Other candidates may include the Galactic tide, which in the case of the Solar System perturbs the Oort Cloud and increases the impact rate on the inner terrestrial planets. In a similar vein, the passage of the Sun through a Galactic spiral arm can also perturb the Oort Cloud, and has also been claimed by some to affect the Earth's climate through an increase in local optical depth between the Earth and Sun (although the strength of this latter effect is vigorously argued).

Whatever the cause, if these global regulation mechanisms are functional, they can synchronise biospheres over large regions of the Galaxy. As the rate of resetting events begins to decrease, the equilibrium state of astrobiological systems begins to shift. During the early, high-event-rate phase, intelligent life fails to appear as resetting events prevent life from achieving this phase. As the event rate drops below a critical value, habitats are given sufficient time to 'grow' a technological civilisation, and ETIs begin to appear in the Milky Way (Figure 18.1).

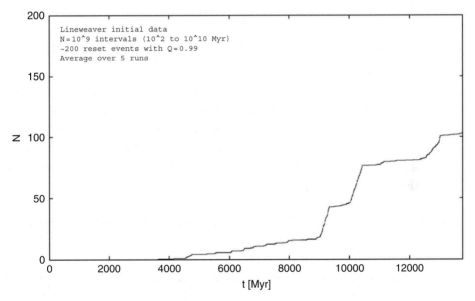

Figure 18.1 An example of global regulation mechanisms inducing a phase transition in a very simplified model of the Milky Way, as originally published in Vukotic and Ćirković (2007). Multiple sites are assigned a timescale for producing an intelligent civilisation (selected uniformly between 10^8 and 10^{16} years). Given an appropriate distribution of ages, and a probability Q that a resetting event will destroy a given site's chance to grow a civilisation, one can produce a phase transition in the number of intelligent civilisations N. The number of sites with an intelligent civilisation is suppressed at early times due to a high frequency of resetting events. As the frequency of resetting events decays, more civilisations can eventually appear in this simple model.

This has given rise to the notion of a 'phase transition' solution to Fermi's Paradox – the Milky Way is emerging from a state in which the number of ETIs is low, to a new state where the Milky Way is populated with a large number of ETIs, all at a similar stage in technological development thanks to the resetting effects of an increasingly weak global regulation mechanism. Fermi's Paradox is then resolved by saying that we are amongst the first (or indeed, we are the first).

19
Exploration Is Imperfect

If we assume that ETIs will discover Earth through their exploration of the Milky Way via spacecraft, then a viable solution to Fermi's Paradox is that ETIs have, for one of many reasons, failed to or have yet to explore the Earth. In this chapter we will consider humanity's thinking on the uncrewed exploration of the Milky Way, and consider how it may fall short of a full mapping.

19.1 Percolation Theory Models

It is best to begin considering the concept of imperfect exploration with the seminal work of Geoffrey Landis (1998), using percolation theory to model interstellar settlement.

A concept well understood in condensed matter physics (amongst other fields), percolation theory describes the growth of an entity across a number of lattice sites. For an entity to percolate from one site to another adjacent site, a probability P is prescribed. Percolation models set up a number of occupied lattice sites, and then test each site for percolation to an adjacent unoccupied site. If a site fails to occupy an adjacent site, it ceases its motion.

A primary concern of percolation modelling is to determine a critical probability P_c, which is typically a function of the connectivity of lattice sites, or more simply the geometry of the system being modelled. If the percolation probability $P > P_c$, then the entity can percolate over all lattice sites and fill the system. If the percolation probability $P < P_c$, then percolation will eventually cease, and regions of the system will remain untouched. However, even with $P > P_c$, a small number of sites can remain unoccupied.

We can think of interstellar colonisation in this form. Star systems are our lattice sites, and inhabitants of these star systems can colonise nearby stars with a fixed probability P. Landis's original work considers a simple two-dimensional cubic lattice system. This is obviously not representative of true stellar cartography, but

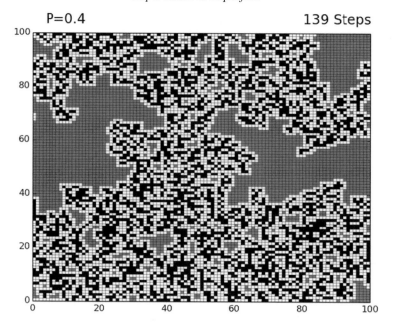

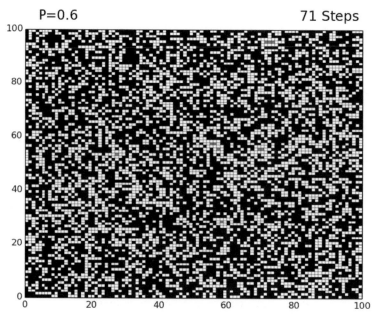

Figure 19.1 Examples of Landis's percolation model (in 2D). Black squares indicate active colonisation sites, white squares indicate sites that have ceased colonising, and grey squares indicate sites that have yet to be explored. Top: $P = 0.4$, which is slightly below the critical value for 2D square lattices. The filamentary structure of the colonisation sites can be seen, as well as the large regions of unexplored sites. Bottom: $P = 0.6$, which is slightly above the critical value, and the entire domain is explored.

it does have the advantage that P_c is well understood for a cubic lattice system, and it can be shown that $P_c = 0.311$ (see, e.g., Wang et al. 2013). If $P = 0.333$, i.e., slightly greater than the critical value, then occupied sites can be found across most of the domain, but the distribution of occupied sites is fractal, and significant voids also exist. This leads to a rather simple solution to Fermi's Paradox – we exist in one of these voids, and the void is sufficiently large to either preclude contact, or to require significant light travel time for it to occur.

In this simple model, the probability that we reside at an 'unoccupied lattice site' is proportional to PN_{adj}, where N_{adj} is the number of adjacent sites connected to any given site. But the model is simplistic in several aspects. Obviously the connectivity of each star system is likely to vary (as the distances between star systems change with time), but so also is its probability of colonising another site, so we may expect P and N_{adj} to evolve over the course of a percolation simulation.

Wiley (2011) noted that several refinements could be made to Landis's model. Firstly, we might expect colonies to die after some period, allowing new colonisation fronts to continue moving through recently vacated systems. Secondly, a colony that decides to stop colonising other systems might change its mind. This 'mutation' of colony motive seems perfectly acceptable given the changing goals and attributes of human culture over the last few thousand years.

The most thorough treatment of percolation on 3D cubic lattices belongs to Hair and Hedman (2013). Landis's work assumed that connectivity between lattice sites could only occur at the faces of the cubes, giving $N = 6$. Hair and Hedman proposed that all adjacent cubes be considered as lattice sites, giving $N = 26$. In a slightly different approach, they define a quantity similar to P, and a quantity m which corresponds to the maximum number of sites colonisable from one site, with $m \in [1, N]$. They also find that critical limits exist for complete exploration of the lattice. If either P or m is too low, then the exploration is finite, and does not extend to the edges of the lattice. If m is sufficiently large, then the lattice is completely explored provided that $P \sim 0.5$. The growth rate (measured by the maximum distance between vertices) is independent of m, but is sub-exponential if $m < N$.

19.2 Exploration of Realistic Star Fields

More recently, the simulations of Bjørk (2007) considered a compartmentalised approach to probe exploration. Their work proposed a probe/sub-probe scenario. Mother probes are dispatched to a region of the Galaxy, and on their arrival dispatch several sub-probes to explore the local environment. Once this exploration is complete, the probes return to the mother probe and move to a new region.

268 *Exploration Is Imperfect*

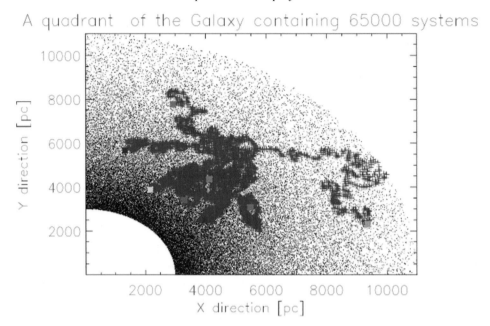

Figure 19.2 A snapshot of probe exploration in the simulations of Bjørk (2007) (their Figure 3, reproduced with permission).

Attempting to simulate the entire Galaxy at once is computationally too demanding. Even sifting the stellar population to remove potentially hazardous massive stars and short-lived binary systems (and only considering stars inside a broad annular Galactic Habitable Zone between 3 and 11 kpc) still leaves approximately 10^{10} stars that probes may wish to visit to search for habitable (or inhabited) planets. Bjørk began by simulating boxes containing 40,000 stars (Figure 19.2), where each box represents a portion of the Galactic disc. In this region, it is assumed that one mother probe has been dispatched, and the sub-probes explore the 40,000 stars. These initial simulations give scaling relations for the exploration time as a function of position from the Galactic Centre, with the box exploration time

$$t_{box} = ae^{br}, \tag{19.1}$$

where a and b are fitted parameters, and r is the box's distance from the Galactic Centre. The box's position determines the local density of stars. Given that the Galactic stellar density decays exponentially with r, the form of t_{box} is intuitively sensible.

These scaling relations can then be applied to a quadrant of the Galactic disc, which is modelled as a set of 64,000 boxes, with t_{box} calculated from the above. Hence a larger, coarser simulation can then be run which shows how a set of mother probes explores a significant fraction of the Milky Way. Bjørk (2007) was forced to conclude that the probe/sub-probe approach as simulated here was not a tenable

19.2 Exploration of Realistic Star Fields

approach for exploring the Galaxy in a sensibly short time period. Even with 8 probes containing 8 sub-probes each, around 400,000 stars can be explored in 292 Myr, which would result in a total Galactic exploration time of many Gyr.

In much the same sense as in the percolation theory example, we can imagine the star field as a network of *nodes* (stars), with the nodes connected by *edges* (the probes' chosen routes through the star field). Cotta and Morales (2009) noted that as this network description is complete, the selection of routes for the sub-probes is an instance of the notorious 'travelling salesman' problem, which attempts to reduce the total length of a route composed of edges connecting all nodes in a network, where each edge is used once. This problem has been identified as 'NP-hard', where the NP stands for nondeterministic polynomial.

In general, a polynomial problem (or P problem) is defined as one with at least one solution, where the algorithm has a number of steps N, and the time taken for the algorithm to complete is bounded by N to some power n. Usually, this is denoted $O(N^n)$. For example, the calculation of pairwise forces between N particles (such as the gravitational force), if calculated explicitly for each particle pair, scales as $O(N^2)$. Approximations to this method (such as using kd-trees to reduce the computation of particle pairs at large distances) can improve this calculation scale to $O(N \log N)$. NP-hard problems cannot (in general) be solved in polynomial time. They can be *verified* in polynomial time, i.e., if presented with the solution we can confirm its validity in polynomial time.

The travelling salesman problem does possess an algorithm to achieve an exact solution – the brute force approach of testing all permutations of the salesman's path – but the solution scales as $O(N!)$, where N is the number of stops on the path. This scaling quickly becomes impractical even for a small number of stops. Much better scaling can be achieved in most circumstances by using simple heuristics (as we discussed on page 237).

In Bjørk's (2007) work, sub-probes were assigned a set of stars that occupied a given range in the vertical direction, to avoid overlap with their sub-probe companions. Once a target was visited, sub-probes selected their next target according to the distance of all targets, selecting the nearest target. Simple heuristics such as this nearest neighbour heuristic (NNH) have clear computational benefits, but they do not guarantee that a group of sub-probes will co-operate to produce optimised routes for the group as a whole.

Cotta and Morales (2009) consider two improvements to the nearest neighbour heuristic (NNH). These begin by constructing the entire route using the NNH, and then make slight modifications of connections on the total route, testing if these reduce the total travel time. The *1-opt* approach looks at the last star on a sub-probe's route, and attempts to assign it to another sub-probe, to see if this can result in a total travel time reduction. The second approach (known as *2-opt*) takes any two edges on a route that are not currently adjacent, and reconnects them. If this modified route is better than the previous, it is kept. This procedure is performed for all possible non-adjacent connections, and so this result is optimal relative to any change of two non-adjacent edges.

With these heuristics in place, Cotta and Morales generated new scaling relations for sub-probes exploring boxes, and then applied this to a Galactic quadrant in a very similar manner to Bjørk (2007) (with the exception of using grid cells at Galactic scales rather than particles). The NNH heuristic is used unmodified for the probes – only the sub-probes attempt to optimise their exploration.

Using a grid cell approach along with multiple runs allows some interesting probability calculations to be made. The probability p that a given cell (i, j) (where the indices refer to polar angle ϕ and distance r from the centre respectively) is visited by probes launched from cell (i', j') at time t is given simply by a relative frequency that this event occurs, as established from multiple runs. The rotational symmetry of these calculations means

$$P(i, j \to i', j') = P(i+k, j \to i'+k, j) \qquad (19.2)$$

where k is integer. This rotational symmetry allows the addition of the Galactic rotation into the calculation, by simply adding a value of k that corresponds to the total rotation over time t.

With the means of calculating the probability that any grid cell in the Milky Way will have fully explored the grid cell containing the Solar System, Cotta and Morales can then compute the complementary probability that no probes have explored the Solar System's grid cell. This requires an assumption of the number of exploring parties in the Milky Way at any one time, and some estimates of the probes' expected lifetime, and how long residual evidence of the Solar System's exploration is expected to last (see Chapter 25).

This work concluded that if more than $10^2 - 10^3$ civilisations were currently exploring the Milky Way using this approach, it would be statistically very likely that we would see evidence of this in the Solar System.

Cartin (2013) went on to explore the consequences of ETIs exploring the Solar neighbourhood. By running smaller-scale simulations and using known positions of stars within ten parsecs of Earth, they were able to show that our star system can still exist within a void of unexplored space with relatively high probability, depending on the maximum travel time and general resilience of the exploration fleet, as well as the fleet's total complement of probes.

19.3 Human Designs for Interstellar Probes

Clearly, the exploration timescale derived in these simulations is a function of the maximum probe velocity. It is quite common practice in this area to assume the maximum velocity is less than or equal to 10% of the speed of light. But is this a feasible figure?

We should consider some of humanity's designs for propulsion systems, and explore this top speed.

19.3.1 Chemical Rockets

To date, all launches of spacecraft have required the use of liquid oxygen and liquid hydrogen rockets. To leave the Earth's gravitational field, rockets must accelerate to speeds near the escape velocity:[1]

$$v_{esc} = \frac{2GM_\oplus}{R_\oplus} = 11.2\,\text{km}\,\text{s}^{-1} \tag{19.3}$$

The energy requirements to enter outer space force us to use energy-dense fuels. As matter is propelled from the rocket, we can estimate the acceleration using Tsiolkovsky's rocket equation:

$$\frac{dv}{dt} = \frac{v}{m}\frac{dm}{dt} \tag{19.4}$$

The change in velocity (Δv) between the beginning and end of the rocket's 'burn' is

$$\Delta v = v_e \ln\left(\frac{m_i}{m_f}\right). \tag{19.5}$$

v_e is the so-called 'exhaust velocity', the speed at which propellant is expelled from the craft, and (m_i, m_f) are the initial and final (total) mass of the spacecraft.

If we wish $\Delta v = 0.1c$, then how much propellant will be burned? If we assume an exhaust velocity of $v_e = 2.5\,\text{km}\,\text{s}^{-1}$, which is typical of the first stage of the Apollo program's Saturn V rocket, then

$$\frac{m_i}{m_f} = e^{\Delta v / v_e} \approx e^{12,000} \;(!) \tag{19.6}$$

Clearly chemical rockets are not capable of reaching suitable interstellar speeds. In fact, the ability to launch material into orbit using this technology becomes increasingly difficult for more massive terrestrial planets, and becomes effectively impossible above $M_p = 10 M_\oplus$ (Hippke, 2018).

19.3.2 External Nuclear Pulse Propulsion

Both nuclear fission and nuclear fusion have been proposed as potential external propulsion methods. Freeman Dyson's Project Orion design relied on a series of

---

[1] Spacecraft can achieve escape at speeds below the escape velocity with constant acceleration.

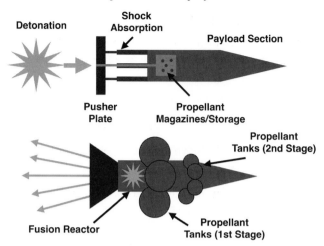

Figure 19.3 The external and internal nuclear propulsion systems (top and bottom). These simplified diagrams illustrate the Orion and Daedalus designs, respectively.

nuclear fission pulses. Effectively a series of small atomic bombs, these explosions would occur exterior to the spacecraft, with the shockwave from the bombs impinging on a pusher plate, which forces the spacecraft forward (Figure 19.3). This design has two key advantages over internal nuclear propulsion. The first is that much of the limitation for nuclear propulsion comes from the melting temperature of the engine components. Having the 'engine' external to the spaceship allows the craft to operate at lower temperatures for the same energy input.

The second major advantage is that the 'fuel' is already present in the world's arsenal of nuclear weapons. In the original vision for this craft, Dyson imagined a 'swords into ploughshares' moment as nations disposed of their bombs by detonating them behind interstellar craft. The feasibility of such a romantic vision was deeply damaged by test ban treaties forbidding the detonation of nuclear weapons in space.

19.3.3 Internal Nuclear Pulse Propulsion

In this scheme, a controlled thermonuclear fusion reaction is used to generate plasma, which is confined by lasers. The exhaust nozzle is threaded by strong electromagnetic fields that force the plasma to be expelled as propellant (see Figure 19.3). This concept underlies the Daedalus interstellar project, which is outlined in detail by Martin and Bond in a series of papers in the Journal of the British Interplanetary Society (e.g., Bond, 1978). The fuel for Daedalus was proposed to be micropellets composed of a mix of ^{3}He and ^{2}H (deuterium).[2]

---

[2] The more easily ignitable fuel mix of deuterium and tritium (^{3}H) was discarded due to its overproduction of thermal neutrons, which cause long-term damage to the combustion chamber walls.

The rocket equation is slightly modified for internal nuclear propulsion. Our spacecraft begins with total mass M and velocity v. The nuclear fuel dM_n is converted into energy, with some of this energy being used to drive electrical power, and some potentially being transferred to a nuclear-inert fuel dM_i to be used as exhaust (known as *working fluid*). Each component has its own exhaust velocity $v_{e,n}$ and $v_{e,i}$ respectively. Conservation of momentum gives:

$$Mv = (M - dM_n - dM_i)(v + \Delta v) + dM_n(v - v_{e,n}) + dM_i(v - v_{e,i}) \quad (19.7)$$

The exhaust velocity of each component can be estimated via energy conservation. The kinetic energy produced by the nuclear reaction dKE_n depends on the conversion rate of matter into energy δ_n, and the efficiency of energy input into the exhaust ϵ_n:

$$dKE_n = \delta_n \epsilon_n dM_n c^2 = \frac{1}{2} dM_n v_{e,n}^2 \quad (19.8)$$

and hence

$$v_{e,n} = \sqrt{2\delta_n \epsilon_n} c \quad (19.9)$$

The kinetic energy of the inert material dKE_i depends on the efficiency of nuclear energy to electric energy conversion ϵ_e, and the efficiency of electric energy to exhaust kinetic energy ϵ_i:

$$dKE_n = \delta_n \epsilon_e \epsilon_i dM_n c^2 = \frac{1}{2} dM_i v_{e,i}^2 \quad (19.10)$$

The matter conversion rate for fission systems such as Orion is around $\delta_n = 0.00075$ (Shepherd, 1952). The exhaust velocity of Orion was never fully characterised thanks to legal obstacles, but it was estimated at around 200 km s^{-1}.

Dyson calculated that for a mass ratio of 4 (see later), an interstellar cruise with speeds of around 0.0035–$0.035c$ could be achieved. The matter conversion rate for fusion systems such as Daedalus is around 0.004, giving an exhaust velocity of around $0.03c$ (Matloff, 2006).

19.3.4 Antimatter Drive

By definition, antimatter drives are the most efficient form of nuclear propulsion. The annihilation of protons and antiprotons results (after several decay stages) in the production of γ-ray photons and neutrinos. These photons can either be expelled as propellant – again using powerful magnetic fields, of order 50 Tesla for a Daedalus-style design (Morgan, 1982) – or be used to heat a working fluid, which can then be expelled as a source of thrust.

If the annihilation of protons and antiprotons were perfect, all of the matter would be converted into exhaust energy. In practice, around 40% of the proton-antiproton rest mass energy is converted into pions. These pions are best suited to heating working fluid (with a small fraction of the γ-ray flux also contributing).

What is the ideal mass ratio for this antimatter drive? Let us consider the mass of annihilating material M_{an}, which is used to heat a working fluid M_w, which from the rocket equation

$$M_w = M_{\text{vehicle}} \left(e^{\Delta v/v_e} - 1\right) \tag{19.11}$$

The annihilating mass converts to energy with an efficiency ϵ and heats the working fluid:

$$\epsilon M_{an} c^2 = \frac{1}{2} M_w v_e^2 \tag{19.12}$$

This gives an expression for the annihilating mass

$$M_{an} = \frac{K}{x^2} \left(e^x - 1\right) \tag{19.13}$$

where $x = \Delta v/v_e$, and K is a collection of constants (for a given mission profile). Minimising M_{an} gives $x = 1.59$, and hence $M_w/M_{\text{vehicle}} = 3.9$, which holds true for any rocket (Crawford, 1990).

19.3.5 The Interstellar Ramjet

Despite achieving a much higher exhaust velocity than conventional chemical rockets, the rocket equation demands that the Daedalus mission plan contain a significant amount of fuel as payload. If this fuel could be collected en route, the mission duration could be significantly reduced. The *interstellar ramjet* concept proposed by Bussard (1960) uses the interstellar medium to feed the reactor. As the craft traverses between stars, protons from the interstellar medium are fed into the reactor to be used as reaction fluid.

Consider a ramjet vehicle moving with velocity v_0 and rest mass m_s, which ingests and expels a rest mass dm of interstellar medium (which is initially stationary). A fraction α of this mass fuels a nuclear reaction with energy release $\alpha dm c^2$, with the rest $((1 - \alpha) dm)$ being expelled from the vehicle at velocity v_e.

The nuclear energy is converted into kinetic energy with efficiency η (with the rest being lost as thermal radiation, which emits transverse to the vehicle's path). We can compare the energies before (the rest mass energy of the ISM gas and the relativistic energy of the craft) and after (the new relativistic energy of the craft, the gas accelerated to speed v_e and the thermal energy emitted):

$$E_{\text{before}} = \gamma m_s c^2 + c^2 dm \tag{19.14}$$

$$E_{\text{after}} = (\gamma + d\gamma) m_s c^2 + \gamma_e (1 - \alpha) c^2 dm + (1 - \eta) \alpha dm c^2 \tag{19.15}$$

Our vehicle's speed has increased. Equivalently the vehicle's Lorentz factor has increased from γ to $\gamma + d\gamma$, where

$$\gamma^2 = \frac{1}{1 - \frac{v^2}{c^2}} = \frac{1}{1 - \beta^2}. \tag{19.16}$$

It can be shown that the gas velocity v_e must satisfy

$$\frac{v_e^2}{c^2} < 2\alpha\eta \frac{1 - \alpha(1 - \eta/2)}{(1 - \alpha(1 - \eta))^2} \tag{19.17}$$

If the gas velocity is too large (relative to the lab frame, not the craft), then the vehicle begins to decelerate. Too much kinetic energy is injected into the interstellar gas, and not enough is injected into the vehicle to deliver thrust. The overall efficiency of the ramjet (energy added to ship as a fraction of energy released) is

$$\epsilon = 1 - \frac{(E_{\text{burned}} - E_{\text{rest mass}}) + E_{\text{lost}}}{E_{\text{released}}} \tag{19.18}$$

Dividing out the common factor of dmc^2 gives

$$\epsilon = 1 - \frac{\gamma_e(1 - \alpha) - (1 - \alpha) + \alpha(1 - \eta)}{\alpha} \tag{19.19}$$

Approximately, this is (Bussard, 1960)

$$\epsilon \approx \eta - \frac{1 - \alpha}{2\alpha} \frac{v_{e0}}{c^2} \tag{19.20}$$

For low velocities, the efficiency is simply the thermal efficiency of the energy conversion. The pp-chain reaction delivers $\alpha = 0.0071$, and hence the velocity of the expelled ISM gas must be kept to a minimum to avoid decelerating the craft.

Although the ramjet does not require large amounts of fuel initially, the craft must be accelerated to the stage where the ramjet begins to operate. Variations in gas density will also affect the feeding rate of the ramjet, which affects v_e and the efficiency of the engine.

19.3.6 Light Sails

The need for chemical and nuclear propulsion technologies to carry their fuel places severe constraints on the available payload of any interstellar craft (not to mention its maximum velocity).

Photon sails neatly avoid this problem by relying on radiation pressure as a source of thrust (Tsander, 1961). If we are aiming to travel between star systems, then our Sun and the target star are abundant sources of photons. For a uniform stellar disc, ignoring limb-darkening, the photon pressure at a distance r is (McInnes and Brown, 1990):

$$P(r) = \frac{L_*}{3\pi c R_*^2} \left(1 - \left(1 - \left(\frac{R_*^2}{r}\right)^2\right)^{3/2}\right) \quad (19.21)$$

and the maximum force on a sail of area A is simply $P(r)A$. In practice, the sail normal $\hat{\mathbf{n}}$ will not be perfectly aligned with the separation vector between the sail and the star, $\mathbf{r}$. In this case

$$F(\mathbf{r}) = P(|\mathbf{r}|)A\hat{\mathbf{r}}.\hat{\mathbf{n}}. \quad (19.22)$$

The maximum kinetic energy produced during initial acceleration can be found by calculating the work done on the sail:

$$KE = \int_0^r F(\mathbf{r})dr \quad (19.23)$$

The sail undergoes a diminishing acceleration as the stellar flux falls away according to the inverse square law, resulting in a cruise velocity that depends principally on the sail area and mass (Figure 19.4). The acceleration (and survival of the sail) depends primarily on the reflectivity of the craft. If too many photons are absorbed rather than deflected, the sail will overheat and potentially be destroyed. Forward (1989) made a careful study of finite absorbency sails (so-called 'grey sails') and confirmed that the optimal course was to maximise the reflectance of the sail.

Rather than relying on stellar photons for acceleration, it was also proposed that powerful lasers provide the initial accelerative pulse (Marx, 1966). This has several advantages over broad spectrum solar sails, chief among these being that humans control the 'wind' the sail takes. The initial acceleration phase is initiated at the press of a button. The narrow bandwidth of the laser permits optimisation of the sail design – for example, the use of dielectric films rather than metallic sheets – to reflect the chosen wavelengths with high efficiency, resulting in more efficient conversion of photons into thrust.

The principal issues with laser sails in particular are (a) the diffraction limit of the laser, which sets a minimum size for the sail, and (b) stability of sail flight inside the laser beam. Manipulation of the beam structure (and careful design of the sail shape) can mitigate against these problems (Manchester and Loeb, 2017).

The 'heavy sail' concept was most famously captured by Forward (1984). A one metric ton lightsail is illuminated by a laser focused by a 1000 km diameter Fresnel zone lens. The laserlight provides a constant acceleration of 0.36 m s^{-2} over a three-year period, attaining a cruise velocity of $0.11c$. When the sail reaches its destination, a double mirror system allows the sail to reverse the laser light received from Earth and induce a deceleration (see Figure 19.5).

19.3 Human Designs for Interstellar Probes

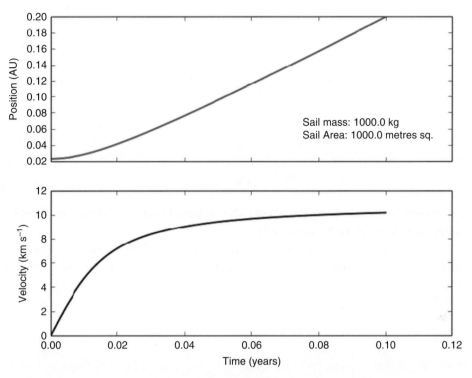

Figure 19.4 The position and velocity evolution of a 1 ton solar sail (area 1,000 m²) as it accelerates away from the Sun, from an initial distance of 5 solar radii.

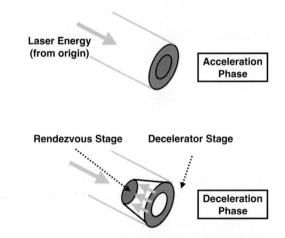

Figure 19.5 Schematic of Forward's (1984) two-sail concept for acceleration and deceleration phases. Laser light from the sail's origin (the Sun, thick arrows) impinges on both sails constructively during the acceleration phase, and destructively during deceleration, where the net momentum transfer on the rendezvous sail comes from photons deflected by the deceleration sail.

19.3.7 Gravitational Assists

There are ways to accelerate probes to near-relativistic speeds that do not require significant energy input from engines, if one is initially rather patient. The approach is analogous to that of the Voyager probes' strategy for accelerating out of the Solar System. These probes utilised *slingshot trajectories* around the planets, which relies on the planet's gravitational potential well to modify the probe's velocity. To satisfy momentum conservation, the planet's momentum is very slightly decreased. The decrease is essentially negligible due to the extreme difference in mass between the giant planet and the probe.

This procedure can be scaled up from interplanetary travel to interstellar travel, where stars are now utilised instead of planets (Gurzadyan, 1996; Surdin, 1986, see Figure 19.6).

Consider a probe moving towards star 1 with velocity $\mathbf{u_i}$ (in star 1's reference frame). After the gravitational assist, it will move towards star 2 with a velocity $\mathbf{u_f}$ in star 1's frame. The magnitude of the probe's velocity before and after the maneouvre is constant:

$$|u_i| = |u_f|. \tag{19.24}$$

In the lab reference frame (which we take to be the frame where the Galactic Centre is at rest), the probe's speed is changed by the following (see bottom panel of Figure 19.6):

$$\Delta v = 2|u_i| \sin \frac{\delta}{2} \tag{19.25}$$

The maximum Δv achievable by this trajectory depends on the closest approach the craft can achieve before becoming captured by the star:

$$\Delta v_{max} = \frac{u_{esc}^2}{\frac{u_{esc}^2}{2u_i} + u_i}, \tag{19.26}$$

where the escape velocity

$$u_{esc} = \sqrt{\frac{2GM_*}{R_*}}, \tag{19.27}$$

and M_*, R_* take their typical meanings as stellar mass and stellar radius. Clearly, potential velocity boosts are greater around more compact objects (Dyson, 1963). Neutron stars have extremely high escape velocities near their surface – black holes possess the maximum possible escape velocity at the event horizon. These trajectories are clearly non-Newtonian, so the above analysis is invalid, but the conceptual principles remain similar.

19.3 Human Designs for Interstellar Probes

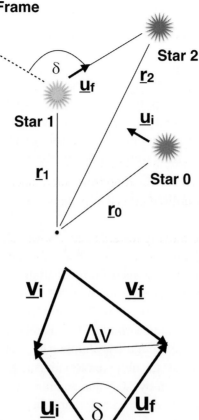

Figure 19.6 The slingshot maneouvre, or gravitational assist. Top: The gravitational assist in the rest frame of star 1. Beginning from star 0, the probe achieves a velocity $\mathbf{u}_i$ in star 1's rest frame, and leaves star 1 towards star 2 with velocity $\mathbf{u}_f$, with the same magnitude. Bottom: Comparing lab frame velocities $(\mathbf{v}_i, \mathbf{v}_f)$ to the velocity in the frame of star 1.

Simulations that use these dynamical shortcuts (Forgan et al., 2013; Nicholson and Forgan, 2013) show that even if the craft can only sustain a maximum velocity of $10^{-5}c$ under powered flight (i.e., Voyager's current speed), slingshots can eventually accelerate the probes to around $0.05c$. If a self-replicating probe fleet elected to use slingshot trajectories (see section 20.1), the Milky Way could be explored in around 10 Myr!

19.3.8 Relativistic Effects

We have largely ignored the effects of special relativity on interstellar travel. Chief among them is time dilation. During flight, system clocks onboard the craft would record the so-called *proper time interval*, τ_0. An observer at rest with an identical clock would record a time interval

$$t = \gamma \tau_0 \tag{19.28}$$

where γ is the Lorentz factor. For a flight at $0.1c$, $\gamma \approx 1.005$, and hence a passenger may record a total journey time of 100 years, whereas observers on Earth will record a journey time of 100.5 years (i.e., there will be a six-month discrepancy between the craft's clock and mission control's clock).

The measured return signal from a craft will also be redshifted. If the craft moves at $\beta = v/c$, then the observed frequency v_o will differ from the transmitted frequency v_t according to the relativistic redshift

$$\frac{v_o}{v_t} = \sqrt{\frac{1-\beta}{1+\beta}}, \tag{19.29}$$

which takes the above form due to time dilation. For our flight at $\beta = 0.1$, this results in an observed frequency around 9.5% lower than transmitted.

Most bizarrely, if our craft is travelling at relativistic speeds, then we must also take account of relativistic aberration. This is related to the phenomenon of *relativistic beaming*, where an emitter's luminosity seems large if it is moving with relativistic speed towards the observer. Relativistic beaming is commonly seen, for example, in the jets driven by AGN and gamma ray bursts.

Let us assume a reference frame where the emitters are stationary, and the observer moves at relativistic speeds. The observer must measure light rays from directly ahead and perpendicular to the observer to travel at the same speed. Under Lorentz transformation, this adds extra components of velocity to light rays that are not dead ahead. As a result, an observer looking ahead on our relativistic craft will see a star field that encompasses more than the field of view should permit. At highly relativistic speeds, the observer should be able to see stars that are 'behind them' (see Figure 19.7).

If we are in the observer's frame, and the emitter moves with speed v at an angle θ_t relative to the vector that connects the emitter and observer when the light is emitted, then we can compute the same angle as measured by the observer θ_o

$$\cos \theta_o = \frac{\cos \theta_t - \beta}{1 - \beta \cos \theta_t} \tag{19.30}$$

This is clearly advantageous if we wish to observe on a wider field of view. At highly relativistic speeds, light emitted a few degrees astern of the craft will appear only a few degrees off the bow! On the other hand, this also requires us to be able to correct for this relativistic aberration when reconstructing a 'true' image taken by our spacecraft's instrumentation.

19.3 Human Designs for Interstellar Probes

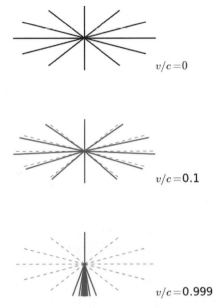

Figure 19.7 Illustrating relativistic aberration. Light is emitted by sources at rest along radial lines of sight towards our spacecraft at rest (top figure). Emitters from slightly behind these positions will still be visible from an observer travelling on the spacecraft at $0.1c$ (middle). At highly relativistic velocities, the observer's field of view becomes close to 360° (bottom).

A less-discussed problem when considering fleets of probes is the issue of communication. Signalling at lightspeed incurs a significant time delay over Galactic distances, and as a result a given probe's memory of which stars have been visited will be necessarily incomplete, as signals from other probes take their time to arrive. Most simulations have sidestepped this problem by assuming instantaneous communication.

One solution to this issue is the establishment of data 'dropboxes' at each visited star system. When a probe visits a star system, it creates a dropbox and fills it with data regarding the stars it has visited. If the star system is subsequently visited by another probe, then it 'syncs' with the dropbox, downloading the star visitation data and uploading its own visits. This is a relatively efficient system of data transmission, as it does not require interstellar broadcasts, and local dropboxes can rely on stellar radiation as a power source.

20
Probe Exploration Is Dangerous

Exploration of the Milky Way via probes requires the probes to possess a relatively high level of autonomy, self-repair and (most likely) self-replication capability. These are key attributes of biological organisms, and the probe population is left to its own devices for (probably) millions of years. Depending on the probe's ability to reduce replication errors in each generation, it is not unreasonable to assume that the probe population would be subject to the equivalent of genetic drift and natural selection.

A whole family of solutions to Fermi's Paradox suggest that the unintended consequences of these evolutionary effects are too dangerous for ETIs to permit the construction of self-replicating probes (see, e.g., Sagan and Newman, 1983). We will now explore these arguments in more detail.

20.1 The Self-Replicating Probe

The most stringent version of Fermi's Paradox revolves around the self-replicating probe, a device that can produce copies of itself as it explores the Galaxy. This results in an exponentially increasing number of probes exploring the Galaxy on a relatively short timescale.

The concept of self-replicating machine has its logical foundations with Von Neumann (1966) (hence their being commonly referred to as *Von Neumann machines*). Initial attempts at studying self-replication were entirely digital and focused on *cellular automata* – grids of cells with on/off states defined by rules relating to their neighbour cells.

A classic example of the cellular automaton is Conway's Game of Life, which defines whether a cell is either active or inactive by the following rule-set, which is applied to the cells repeatedly:

1. An active cell with fewer than two active neighbours becomes inactive.
2. An active cell with three active neighbours persists to the next repetition.

3. An active cell with greater than three active neighbours becomes inactive.
4. An inactive cell with three active neighbours becomes active.

It was quickly shown that cellular automata could be designed such that self-replication and self-repair behaviour will appear. Von Neumann designed such a cellular automaton with 29 possible states. Cells in this automaton, once excited, then proceed through a subset of these 29 states depending on their neighbourhood cells. This 29-state automaton eventually produces self-replicating behaviour.

Self-replicating systems are therefore possible in theory. This is hardly surprising, because biological systems are self-replicating systems. What was less clear was whether self-replicating mechanical systems were possible.

Beginning with Von Neumann's suggestion of self-configuration/self-assembly from caches of available parts, there have been many proposals for self-assembling or self-replicating machines or swarms of machines (see Jones and Straub, 2017, and references within). Many have attempted to ape the self-replication of biological systems, with some even relying on evolutionary methods to generate successful iterations (cf. Pan, 2007).

The advent of 3D printing as a viable technology has revolutionised our thinking on how machines can self-replicate. The self-replication problem can eventually be boiled down to: *can a 3D printer print a copy of itself?* The answer to this question, thanks to technologies such as RepRap, appears to be yes (Jones et al., 2011). Of course, the real question should be *can a 3D printer create a fully assembled, functioning copy of itself?* The current answer to this question seems to be no, as RepRap's printer copies currently require human assembly. However, robotic assembly of parts is a relatively straightforward task compared to replication, and so we must conclude that self-replicating machines are very possible (Ellery, 2016), and in fact likely to appear in human industry in the near future. We should therefore consider self-replicating machines as a viable technology for interstellar travel, provided that the raw materials for making copies exist *en route*.

The most detailed first description of the mechanics of a self-replicating interstellar probe comes from Freitas (1980). Taking the Daedalus interstellar probe blueprints as a starting point, Freitas speculated on designs for humanity's own self-replicating craft.

Freitas's design pre-dates modern construction technology, in particular 3D printing. In his scheme, the probe (dubbed REPRO) is a scaled-up Daedalus, with a mass approximately two times larger. Upon arrival at a star system, REPRO orbits a gas giant with moons (Freitas is assuming a Jovian system). REPRO then deploys SEED to the moon to establish a (much larger) FACTORY, which in turn constructs a copy of REPRO. Freitas argues that building REPRO directly from SEED is suboptimal, and the intermediate FACTORY stage allows the system to vastly reduce the construction timescale for REPRO. Also, once FACTORY has been constructed, SEED is free to build another FACTORY, doubling the production rate of REPRO probes.

The replication process is carried out by an ecosystem of robotic entities, with varying levels of artificial sentience and autonomy. These entities are the first elements that SEED builds, which allows the system to mine the lunar surface, process the ore, assemble the components and verify the build is carried out according to design (while also extracting fuel from the upper atmosphere of the giant planet).

20.2 Can Probes Self-Replicate without Evolving?

Consider the two elements necessary for a probe to make a copy of itself. It requires raw materials (hardware), and it requires instructions as to how to assemble them (software). If both elements are present, it can either construct the copy, or build tools that carry out the task of construction.

How would the software be stored? In human systems, DNA and mRNA contain the necessary data to code for proteins and produce the necessary structures. The propagation of human genetic information requires reproduction. Humans possess two *alleles* (copies/variants) of a gene, at the same genetic locus on a chromosome.

During reproduction, the allele pairs from each parent are mixed amongst their children. The resulting genotype is either homozygous (matching alleles) or heterozygous (differing alleles). Which allele ends up being expressed in the organism depends on which is dominant (or recessive).

In the New Horizons probe which recently visited Pluto, the data is stored on two solid-state drives. The data is stored as a stream of zeros and ones. These bits of data can be corrupted or changed by radiation impacting the system. Most spacecraft are shielded to some degree against this event, but high-energy particles such as cosmic rays are extremely difficult to mitigate against. If zeros are switched to ones (and vice versa), then the instructions are rewritten, in a manner analogous to mutation in organisms (transition/transversion, see page 115). In most cases, a switch of a digit would result in a program that cannot be executed. In some cases, it may result in code being executed with a different set of variables, with a different output as a result.

If a self-replicating probe executes copying with flawed instructions, we can expect the result to be fatally flawed in most cases, in the same manner that many mutations are fatal to biological organisms. However, given a large enough population, some mutated probes will be viable systems, and their own software is likely to inherit the same flaws as the parent. If the local selection pressure is weak, this results in genetic drift of the probe population.

But what if a mutated probe is 'better' than the original design? A 'successful' mistake that increases the relative fitness of the probe can then be propagated through a probe population, as high-fitness mutations propagate through biological populations. The combination of genetic drift, gene flow and natural selection (acting on timescales longer than the interval between copying events) would then result in a variety of forms, each adapted to varying environments and challenges, just as happens in biological systems (see section 5.3.6).

This logic relies on the premise that replication errors cannot be rectified by the parent during the copying process. Error checking is a fundamental part of modern computing and signal processing. A well-designed replicator will possess a variety of fail-safes and checking algorithms during a copy.

In signal processing, the average *bit error rate* can be calculated by dividing the number of bits transmitted in error by the total number of bits. Given that information is copied and transmitted by Earth's Deep Space Network to probes in the Solar System with bit error rates of 10^{-6}, i.e., one erroneous bit per million bits (Taylor, 2014), and that most transmission systems are designed to be redundant against erroneous bits, it might seem quite possible that the level of error could be extremely small indeed, removing the possibility that probes could evolve 'naturally'.

However, we should note that replication of the human genome constitutes an extremely low bit error rate of 10^{-10} per site (with similar low rates for other species, such as *E. coli*). If biological systems can possess such a low error rate and still evolve, what hope is there for mechanical and electronic systems?

Also, we have already noted that probes are likely to be highly autonomous, and possess software approaching (or exceeding) our definition of artificial intelligence. A flexible goal-seeking system may alter its own programming as it receives new data regarding itself and its environment. Therefore, rather than evolving via natural selection, it may well be the case that probes evolve via intelligent (auto)design!

If probes do evolve and change in ways unforeseen by their creators, then there are some dangerous scenarios that we can consider.

20.3 Predators and Prey

During the Ediacaran Period (around 575 million years ago), the growing abundance of animals allowed a fundamental shift in how biological organisms obtained energy from the environment – the consumption of other animals. While predation is not unique to large organisms, the sudden boost in energy available from the practice triggered a significant increase in organism size, and fundamentally shifted the tempo of life.

We can imagine a similar shift might occur in an population of self-replicating machines. Instead of being forced to replicate from whole cloth, a machine can scavenge ready-made parts from another, and significantly decrease the time and energy cost of replication. This 'predatory' strain of self-replicating probe will begin to outgrow other strains that require more costly construction requirements.

If mutated programming encourages machines to seek out other machines to consume, it is quite possible that these machines will be conferred the same evolutionary advantage as the first animal predators on Earth.

Once this behaviour is established, we can think of the probes in the same way that we think of animal predator–prey populations. In population dynamics, a typical starting point is the Lotka-Volterra system of equations describing the number of predators P and the number of prey R:

$$\frac{dR}{dt} = \alpha R - \beta R P \tag{20.1}$$

$$\frac{dP}{dt} = \delta P R - \gamma P \tag{20.2}$$

α is the growth rate of prey, β is the death rate of prey due to predators, δ is the growth rate of predators due to consuming prey, and γ is the death rate of predators due to natural causes. This system assumes that the predators only eat prey (with infinite appetite), prey have an ample food supply, (i.e., will grow unchecked in the absence of predators), and both populations grow exponentially.

This non-linear system (coupled through the RP cross-terms) either results in predators destroying the prey population ($R \rightarrow 0$), or produces a periodic solution for P and R, with the values of both constrained to a fixed trajectory in $R - P$ phase space (see Figure 20.1).

We can define the conditions of this steady state by rearranging the coupled equations into a single form:

$$\frac{d}{dt}(\delta R + \beta P - \gamma \log R - \alpha \log P) = 0 \tag{20.3}$$

We can then take a Hamiltonian approach, where the Hamiltonian

$$H(R, P) = \delta R + \beta P - \gamma \log R - \alpha \log P \tag{20.4}$$

As the second derivatives of H are all positive, H is a convex matrix. It therefore has a well-defined unique minimum, which we determine from $\nabla H = 0$:

$$\frac{\partial H}{\partial R} = \delta - \frac{\gamma}{R} = 0 \tag{20.5}$$

$$\frac{\partial H}{\partial P} = \beta - \frac{\alpha}{P} = 0 \tag{20.6}$$

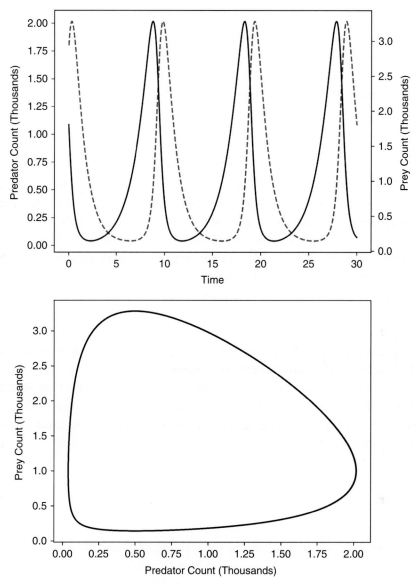

Figure 20.1 The predator–prey dynamics of a Lotka-Volterra system. The top panel shows the total population counts with time, for a Lotka-Volterra system where $\alpha = 1/3$, $\beta = 0.666$, and $\delta = \gamma = 1$. The evolution of both populations is periodic, and traces a closed locus in $R - P$ phase space (bottom panel).

We can then solve for

$$(R, P) = \left(\frac{\gamma}{\delta}, \frac{\alpha}{\beta}\right) \tag{20.7}$$

This identifies the beginning point of the system's trajectory in R-P space, with each curve being a contour of constant H.

In our case, we should consider modifying the 'vanilla' Lotka-Volterra system to include mutation of prey into predators, and ideally some consideration of competition for asteroid resources. Pure exponential growth is an unphysical model if the resources are finite (see Chapter 12). If we instead require that the prey practise logistic growth with some carrying capacity K:

$$\frac{dR}{dt} = \alpha R \left(1 - \frac{R}{K}\right) - \beta RP \qquad (20.8)$$

$$\frac{dP}{dt} = \delta PR - \gamma P \qquad (20.9)$$

This fundamentally alters the system's stability. Depending on the system parameters $(\alpha, \beta, \delta, \gamma, K)$, one can either destroy the prey population, or produce a solution with initial periodicity, quickly settling toward an equilibrium solution with constant (R, P) (Figure 20.2). If the predators also practise logistic growth (which may occur if they, for example, find a secondary food supply), then this approach to an equilibrium can be especially rapid.

Indeed, scavenging probes may not restrict their appetites to other probes. Rather than being forced to refine asteroid ore, if a probe happens upon any object containing pre-processed material, its programming may consider it an advantage to simply dismantle the object and use its components.

Predators may even predate on each other. We can think of this type of interaction using the Lanchester equations, originally developed to describe the losses of armies in battle. If we imagine two strains of predators, A and B, then the Lanchester system is given by

$$\frac{dA}{dt} = -\beta B \qquad (20.10)$$

$$\frac{dB}{dT} = -\alpha A \qquad (20.11)$$

where α and β now describe the predatory ability of A and B respectively. In its original use, α describes the rate at which individuals in army A fire bullets at army B, and vice versa for β. This combat model assumes that either numerical superiority or greater firepower are the only factors in the success of an attacking population. From it, we note Lanchester's Law, that a force's power is proportional to the square of its population.[1]

In an extreme case, a predatory strain of self-replicating probes could become a marauding band, harvesting any technology it encounters to convert into further probes. We can think of this as a macroscopic 'grey goo' scenario (see section

---

[1] In practice, military strategists tend to work with the salvo combat model, which more accurately reflects the discrete nature of munitions and warfare generally.

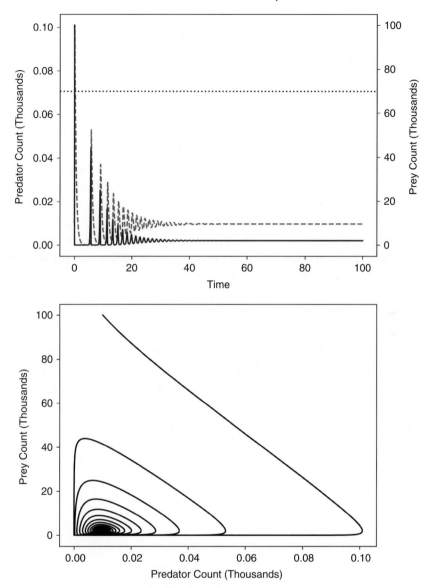

Figure 20.2 As Figure 20.1, but for a prey population practising logistic population growth. The prey-carrying capacity K is shown by the horizontal dashed line. The predator–prey dynamics of a Lotka-Volterra system. The top panel shows the total population counts with time, for a Lotka-Volterra system where $\alpha = 10$, $\beta = 0.1$, and $\delta = 1, \gamma = 2$. The evolution of both populations is periodic but quickly settles towards constant values, tracing a spiral in $R - P$ phase space (bottom panel).

14.2). Wiley (2011) points out some important differences between animal and probe predators, namely that the prey do not necessarily remain at their point of origin, but are likely to be dispersing radially at maximum speed, without stopping to rest. Predators chasing such prey would sweep out a 'predator shadow' of unexplored territory behind them as they encounter prey moving toward them.

In summary, the effects of predator/prey dynamics on Fermi's Paradox are unclear. A solution could be that predators will eventually reduce the number of probes in the Milky Way, but the number of predators may tail off (or at least plateau) as prey disappears. Are we looking at the Milky Way during an instant of predator dominance? At the moment, we have insufficient data to approach this question.

21
The Aliens Are Quiet

21.1 They Communicate at a Different Frequency

In earlier chapters, we discussed the motivations for SETI scientists to adopt specific frequencies for their searches. Most of radio SETI has focused on frequencies near that of atomic hydrogen emission, specifically the HI emission line. Optical SETI searches have looked at a range of frequencies from the visible into the near infrared, being constrained partially by available photon-counting technology, and by the relatively opaque Milky Way at visible wavelengths.

We have considered likely transmission frequencies under two constraints: that of the interstellar medium, and that of the Earth's atmosphere. All ETIs are bound by the first constraint, but it is not obvious that they are bound by the second. Depending on the structure and composition of their planet, their atmospheric conditions may vary from ours. The principal absorbers of radio waves in Earth's atmosphere are water vapour and oxygen molecules – varying the relative abundances of these species will increase or reduce the absorption in bands above around 10 GHz.

Messerschmitt (2015) notes that interstellar communication is likely to be optimised in a different way to terrestrial communication. Terrestrial communication is *spectrally optimised* whereas interstellar communication is *energetically optimised*.

More rigorously, optimal interstellar communication maximises the information rate for a given input signal power (or minimises the signal power for a given information rate). There is a further desideratum that the receiver's energy requirements to process any signal be minimal.

We can see this by considering the information-carrying limitations of a signal. Spectrally efficient transmission maximises the spectral efficiency S:

$$s = \frac{R}{B} \qquad (21.1)$$

where R is the information rate in bits per second, and B is the bandwidth in Hz. Energy-efficient transmission maximises the power efficiency (bits per joule):

$$p = \frac{R}{\bar{P}} \qquad (21.2)$$

where $\bar{P}$ is the average radiated power (usually obtained by simply dividing the total energy consumed in the transmission, and dividing by the signal duration). It is also quite common to see power efficiency described by the energy-per-bit $E_b = 1/p$.

In a perfect universe, a transmission would maximise both the spectral and power efficiencies. This is not possible – a signal with high power efficiency will have low spectral efficiency, and vice versa. This is due to the so-called *Shannon limit* on channel capacity, C. If our information rate $R \leq C$, then, mathematically speaking, perfect communication is possible. Information rates above the channel capacity cannot be communicated reliably.

The simplest example of a channel model is the Additive White Gaussian Noise model (AWGN). We take a 'pure' transmitted signal X, and add noise N (which is a Gaussian random process with zero mean), which results in a received signal Y:

$$Y = X + N \qquad (21.3)$$

The channel capacity in this case is

$$C = B \log_2(1 + SNR) = B \log_2\left(1 + \frac{P}{N_0 B}\right) \qquad (21.4)$$

In this equation N_0 is the noise amplitude per unit frequency, and hence $N_0 B$ is the total noise in the signal bandwidth. In this case, it can be shown that the Shannon limit $R < C$ can be re-expressed as an upper bound on power efficiency (relative to spectral efficiency):

$$\frac{p}{p_{\text{max}}} < \frac{s \log 2}{2^s - 1}, \qquad (21.5)$$

where p_{max} is the maximum possible power efficiency:

$$p_{\text{max}} = \frac{1}{N_0 \log 2}. \qquad (21.6)$$

Increasing the power efficiency results in a sharp decrease in the permitted spectral efficiency. Terrestrial communications have typically prized spectral efficiency due to:

a) the limited transparent frequencies in the Earth's atmosphere,
b) the commercial and legal limitations on what frequencies can be used for broadcast, and
c) the potential for interference in broadband signals as a result.

Interstellar communications may not be limited by atmospheric concerns at all, as is the case for space-based telescopes orbiting the Earth. Given that interstellar transmissions require large amounts of energy regardless of their signal properties, it seems reasonable that ETIs would attempt to maximise their power efficiency and obtain the greatest bit-per-joule rate possible.

Further limitations on both power and spectral efficiency come from the so-called *interstellar coherence hole*. This is a combination of time/bandwidth constraints imposed on signals attempting to traverse the interstellar medium. The frequency constraints come from wavelength dispersion by free electron plasmas, and scattering, which disfavour high bandwidth signals. Time constraints depend on the motion of the transmitter/receiver pair relative to the interstellar medium, which changes both the total path length of the signal (acceleration), as well as the line of sight of the signal, ensuring it encounters different regions of space and variations of scattering and dispersion as a result (scintillation). As a result, shorter-duration signals are less susceptible.

We can therefore define the interstellar coherence hole as a region in parameter space where signals will pass from transmitter to receiver coherently with $\Delta t < \Delta t_c$ and $B < B_c$, where Δt_c and B_c are the naturally imposed upper limits on time interval and bandwidth.

We can see now why interstellar communication will generally adopt energy-efficient modes. While narrow bandwidths suggest a spectrally efficient signal, the time constraints ensure that the bits per joule must remain high for a signal to be completed in short time. The received signal strength will be significantly lower than the signal strength at transmission due to the inverse square law, but the range of frequencies at our disposal is large.

There are several strategies available to reduce energy consumption when transmitting:

Reduce the information rate This is the motivation behind so-called 'beacons'. These are 'on/off' signals, transmitted in a single channel. To maximise the energy efficiency, these beacons typically emit pulses of radiation, where the pulse duration and pulse intervals can be adjusted to optimise against one of several constraining factors (distance, systems contacted, energy costs, etc.). This is the extreme limit in information rate.

Increase the carrier wave frequency This is perhaps the strongest argument for optical SETI over radio SETI. Increasing the frequency of the transmission allows more-intensive modulation of the signal, which can increase the information rate (with everything else being fixed). Typically, the required transmit power drops with increasing carrier frequency $P \propto \nu^{-2}$.

Select a less-distant target If a civilisation is severely restricted in its energy input, this is likely to be an adopted strategy. Closer targets demand less energy to broadcast a signal that can still be successfully detected ($P \propto D^2$).

Accept transmission errors Signals that are weak may be indistinguishable from noise at certain instants depending on the intervening medium and the tools available to the receiver. A signal that is repeatedly broadcast will allow receivers to reconstruct the entire transmission from multiple error-prone versions, so this can be an acceptable strategy if one is willing to repeat oneself. We will return to the issue of successfully decoding transmissions in later chapters.

A crucial concept in interstellar communication is the energy bundle. In effect, this is the quantum of any transmission – a pulse of radiation with a bundle energy and a specific waveform. We can concoct a highly flexible transmission by chaining together an ensemble of energy bundles. Naively, one might assume that an energy bundle is equivalent to one bit of information transmitted (and in some encoding strategies, it is), but as we will see, this is not true in general. As we have already stated, there are two fundamental parameters that govern the information rate of any transmission – the power, and the spectral bandwidth.

We give some examples of spectrally efficient and energy-efficient encoding schemes (see Figure 21.1). The most conceptually simple spectrally efficient scheme is referred to as On-Off Keying (OOK). This is quite simply transmitting on a narrow bandwidth in a binary fashion, where pulses equal 1, and gaps equal zero. This can be considered to be the next step up from a beacon transmission, where the pulses are timed such that a series of bits are transmitted.

Multilevel Keying (MLK) can increase the information rate by allowing the signal strength to vary. If the signal can be one of M levels of energy (including zero), then each energy bundle can transmit a total of $\log_2 M$ bits of information. OOK is of course multilevel keying with $M = 2$.

Spectrally efficient coding schemes can be impressively information-dense. For example, frequency-shift keying (FSK) utilises multiple frequency bands, as opposed to multiple energy levels. This is extremely energy-efficient compared to multilevel keying due to the need for receives to discriminate energy levels against a noise background. For safety, the average energy per bundle needs to increase with M as M^2 in MLK, which can become prohibitive for large M.

By contrast, FSK demands a single energy level for each bundle, but that the frequency of the bundle shifts to one of M bands. Again, this delivers $\log_2 M$ bits per bundle, without increasing energy demands.[1]

---

[1] Using a wide total bandwidth may seem to put transmissions at risk of scintillation and other attenuation effects from the surrounding ISM, but the quantisation of the signal into energy bundles of narrow bandwidth prevents this from becoming a serious issue.

21.1 They Communicate at a Different Frequency

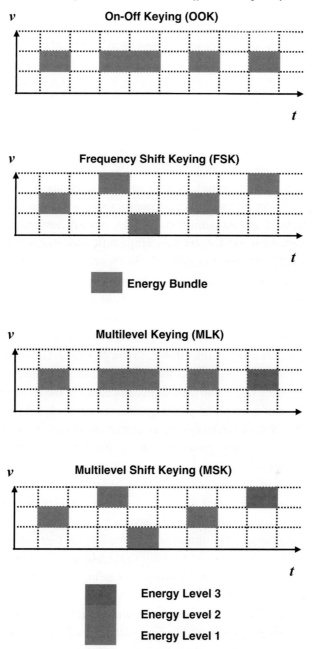

Figure 21.1 Spectrally and energy-efficient encoding schemes. Each graph represents energy bundles (boxes) transmitted at a specific frequency v at a time t. On-Off Keying (OOK, first) and Frequency Shift Keying (FSK, second), which use a single fixed bundle energy. Bottom: Multilevel Keying (MLK, third), which maintains spectral efficiency by broadcasting a range of M bundle energies. In this case $M = 3$, giving $log_2 3 = 1.58$ bits of information per bundle; and Multilevel Shift Keying (MSK, fourth), which combines the advantages (and disadvantages) of both MLK and FSK.

21.2 They Communicate Using a Different Method

It is entirely possible that interstellar transmissions do not use the electromagnetic spectrum at all. Our current understanding of the physical Universe suggests several possible communication methods that are theoretically possible, but extremely challenging for humans to receive or decode given current technological capability.

21.2.1 Neutrinos

We have conducted our discussion of practical, observational SETI entirely in the electromagnetic spectrum. Detecting photons of various energies has been the standard approach of astronomers for centuries. Our recent advances in particle physics have shown us that photons are not the only messenger particles available to sufficiently advanced civilisations.

An enticing candidate is the neutrino. Neutrinos are extremely low-mass, weakly interacting particles, produced not only naturally in stars, supernovae and AGN, but also by humanity as the byproducts of particle accelerators. Neutrinos are leptons, and come in three flavours, corresponding to the other three leptons of the Standard Model of particle physics (electron e, muon μ, tau τ). Neutrinos can *oscillate* between each of the three flavours, as their flavour/mass is actually a superposition of three possible eigenstates. The probability of changing from one state to the next varies, as the phase of each eigenstate depends on each state's mass.

The local (solar) flux of neutrinos at any instant is very high (of order $10^9 - 10^{10}$ particles cm^{-2} s^{-1}), but most pass through the Earth without interaction. This is because neutrinos only couple to other particles through either the weak force or the gravitational force. The weak force acts only at extremely short range, and the gravitational force is some 10^{38} times weaker than the other fundamental forces (and the neutrino is itself extremely low mass).

As such, despite being a fermion, the scattering cross section of neutrinos with other fermions is extremely small, making their detection extremely challenging. The majority of the Sun's neutrino emission passes through the Earth without being absorbed.

Neutrino detectors cannot directly detect the particle. Instead, they detect the byproducts of interactions between neutrinos and other fermions, such as protons and electrons. Modern detectors typically employ large tanks of very highly purified water (buried at high depths to remove unwanted signals from cosmic rays). Neutrinos travelling through the water occasionally interact with an electron, causing it to be accelerated beyond the speed of light in the medium (which is approximately 75% of light speed in a vacuum). This results in Cherenkov radiation, which is then sensed by an array of detectors placed around the tank's edge.

Neutrino astronomy is in its infancy – the only astrophysical sources currently detected are the Sun and recent supernova SN 1987A, using the Super-Kamiokande and Irvine-Michigan-Brookhaven observatories in Japan and the USA (Hirata et al., 1987). The most recent attempts to find significant signals beyond the solar neutrino fluxes have yet to yield a convincing detection from a specific object (IceCube, 2017).

Neutrinos are also produced by human activities. Antineutrinos are commonly produced in nuclear reactors. As neutrons decay into a proton and electron, an electron-antineutrino ($\bar{\nu}_e$) is released. A typical fission reactor will emit around 5% of its output energy as antineutrinos which are not harnessed and escape to (presumably) interstellar distances.

Particle accelerators can also produce neutrinos. Protons, when accelerated at a target, produce charged pions and kaons (π/K), which can then decay into muon/muon-neutrino (ν_μ) pairs. Thanks to the relativistic velocities reached by the protons before particle production, the neutrino emission is highly beamed.

If ETIs were able to generate and direct neutrino beams, these would offer several advantages compared to photon beams:

1. The universe emits far fewer neutrinos than photons. Even a relatively weak neutrino beam, given an appropriate particle energy, propagates in a region of low noise and hence has a favourable S/N ratio.
2. The low interaction cross section of neutrinos means their signals arrive without attenuation, even if they traverse particularly dense regions of the interstellar medium.
3. Neutrino beams arrive at the detector with virtually no scatter, resulting in no positional 'jitter', which improves a receiver's ability to locate the signal. The lack of temporal jitter also allows for higher-fidelity encoding.

Neutrinos are generated by natural sources over a relatively wide range of energies, just like photons. Analogously, the cross section of neutrinos changes with energy, although the relationship between cross section and energy is relatively simple, compared to the opacity laws that govern photons.

Like ETIs transmitting photons, ETIs transmitting neutrinos also have important decisions about which energy to transmit at. Most natural sources produce neutrinos at the 10 MeV scale (for example, supernovae tend to produce 40 MeV neutrinos).

Increasing the neutrino energy moves the signal out of the range of natural sources and also boosts the cross section. In particular, electron anti-neutrinos with energies of 6.3 PeV interact with electrons producing a *Glashow Resonance*. This creates a *W* boson which can then decay back into lepton and neutrino pairs. The fact this resonance occurs at a unique energy is quite analogous to the photon-based Water Hole in the microwave band.

If ETIs were able to produce such a neutrino beam, for example, by producing high-energy pions (π^-) which decay into $\bar{\nu}_\mu$, some fraction of the neutrino beam would eventually oscillate into $\bar{\nu}_e$. Information could be injected into the beam by (for example) switching the beam between π^- and π^+, resulting in an intermittent stream of $\bar{\nu}_e$, with the potential for two-level signal keying ($\bar{\nu}_\mu, \bar{\nu}_e$).

The energies required to make PeV neutrinos is clearly immense. The Large Hadron collider at CERN produces neutrinos with energies less than 100 GeV, around ten thousand times too weak (Bailey et al., 1999). It has been argued that muon colliders are a more effective neutrino source – indeed, the neutrino radiation field could become so intense in the vicinity of the source as to be a biohazard (Silagadze, 2008; Learned et al., 2009).

While humans are yet to achieve such extraordinary feats of particle acceleration, instrumentation such as the IceCube observatory should easily discriminate neutrino beams with energies in excess of 10 TeV from the background (where the expected flux from background sources is less than $1\, \text{yr}^{-1}\, \text{km}^{-2}$). A 100 TeV beam is calculated to be easily detectable even at 20 light-years distance (higher-energy beams are increasingly collimated, making their detection less and less probable).

The most extreme proposal for neutrino SETI concerns advanced civilisations' efforts to study the Planck scale, at which quantum theory and general relativity can no longer be considered separately, and must be unified into a quantum gravity theory. This requires a particle accelerator of YeV energies (10^{24} eV, or 1 billion PeV), or the entire rest mass energy contained within $100 M_\odot$ of matter! Such particle accelerators would necessarily be much larger than a planet. As well as producing enormous amounts of synchrotron radiation as particles are accelerated in the presence of a magnetic field, YeV neutrinos are almost certainly produced.

Such neutrinos are in principle easier to detect due to their boosted cross section, but we expect to see a far lower number count, and it is likely that modern neutrino detectors are simply not big enough to detect YeV particles – even detectors covering the Earth's entire surface area are likely to be insufficient (Lacki, 2015).

On a more encouraging note, neutrino SETI can also be conducted in serendipitous or 'piggyback' mode. Neutrino observatories are not 'pointed' at a particular astrophysical source – instead they receive flux from the entire 4π solid angle of the sky (including the Earth beneath them). SETI studies can occur exactly in concert with other experiments focused on neutrino detection, such as the problem of further constraining neutrino masses, or the study of supernovae.

21.2.2 Gravitational Waves

As of 2015, the Advanced LIGO system has directly detected the presence of gravitational waves. Gravitational waves are oscillations in the fabric of space-time, a

21.2 They Communicate Using a Different Method

wave solution of Einstein's equations of general relativity. In GR, the propagation of the gravitational force has a finite velocity – the speed of light. In the same way that perturbations in the electromagnetic field (photons) are constrained to travel at c, perturbations in the space-time metric (gravitational waves) must also travel at light speed.

We can see how wave solutions emerge from the field equations by considering their linearised form. We do this by assuming that the space-time metric $g_{\mu\nu}$ is close to a flat (Minkowski) metric $\eta_{\mu\nu}$, with a small perturbation $h_{\mu\nu}$:

$$g_{\mu\nu} = \eta_{\mu\nu} + h_{\mu\nu} \tag{21.7}$$

We then substitute this metric tensor into Einstein's field equation:

$$R_{\mu\nu} - \frac{1}{2} R g_{\mu\nu} = 8\pi G T_{\mu\nu} \tag{21.8}$$

where T is the energy-momentum tensor, and $R_{\mu\nu}$ is the Ricci tensor, and R the Ricci scalar. As with electromagnetism, an appropriate gauge choice is required. A full derivation is lengthy, and does not really belong in this textbook (the interested reader is directed towards Tiec and Novak, 2016). Suffice it to say that by defining our perturbation[2]

$$\bar{h}_{\mu\nu} \equiv h_{\mu\nu} - \frac{1}{2} \eta^{\mu\nu} h_{\mu\nu} \eta_{\mu\nu}, \tag{21.9}$$

and choosing the Lorentz (or harmonic) gauge:

$$\delta^\nu \bar{h}_{\mu\nu} = 0, \tag{21.10}$$

Einstein's field equations reduce to the following wave equation:

$$\Box \bar{h}_{\mu\nu} = -16\pi G T_{\mu\nu}, \tag{21.11}$$

where the D'Alembertian operator

$$\Box = \eta_{\mu\nu} \delta^\mu \delta^\nu = -\frac{1}{c^2} \frac{\partial^2}{\partial t^2} + \nabla^2 \tag{21.12}$$

We can now see that $\bar{h}_{\mu\nu}$ represents a quantity that propagates as a wave on a flat space-time background, with source given by $T_{\mu\nu}$. Equation (21.11) represents in fact 4×4 equations, one for each component of the tensor $\bar{h}_{\mu\nu}$. With an appropriate set of gauge choices (and setting the source $T_{\mu\nu} = 0$), we can derive the resulting metric perturbation, which only has two components:

---

[2] We also assume the Einstein Summation Convention, where repeated indices in any multiplication require a sum over those indices.

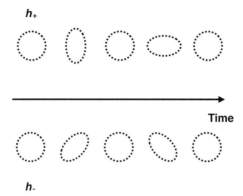

Figure 21.2 The two polarisation states of gravitational waves. A circular ring of masses is deformed by the passing of a gravitational wave with (top) a h_+ perturbation, and (bottom) a h_- perturbation.

$$h_{\mu\nu} = \begin{bmatrix} 0 & 0 & 0 & 0 \\ 0 & h_+ & h_\times & 0 \\ 0 & h_\times & -h_+ & 0 \\ 0 & 0 & 0 & 0 \end{bmatrix} \qquad (21.13)$$

The components h_+ and $h_\times$ correspond to polarisation states of the gravitational wave, much as electromagnetic waves can be polarised (see Figure 21.2). We can show that these metric perturbations, when acting on two particles where we can measure the Euclidean (i.e., non-curved) distance between them, L_0, will demonstrate a characteristic deviation from this value:

$$L(t) = L_0 \left(1 + \frac{1}{2L_0^2} \Delta x^\mu \Delta x^\nu h_{\mu\nu} \right) \qquad (21.14)$$

with Δx being defined as the initial distance between the particles. We can therefore show that the measured change in particle positions from a gravitational wave are

$$\Delta L \sim \frac{1}{2} |h| L_0 \qquad (21.15)$$

The typical magnitude of h, estimated by assuming an oscillating spherical mass (see Tiec and Novak, 2016), is:

$$|h| \sim \frac{2GM}{c^2} \frac{v}{c^2} \frac{1}{r} \qquad (21.16)$$

where r is the distance from the source to the observer, M is the source mass and v is the typical internal velocity of the oscillating material. As we shall see in a moment, this quantity is very small, but we should be encouraged by the mere $1/r$

21.2 They Communicate Using a Different Method

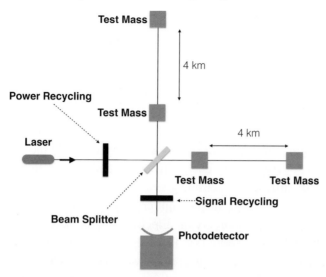

Figure 21.3 A simplified schematic of the LIGO apparatus, which successfully detected gravitational waves from merging stellar mass black holes in 2015.

decay of the magnitude. After all, this is significantly slower than the $1/r^2$ decay we see in electromagnetic radiation. While gravitational radiation is intrinsically weak, it can travel a good deal further with less decay of the signal.

Gravitational wave detectors (like the successful LIGO instruments) are interferometers. One LIGO station consists of two arms, each 4 kilometres in size. A single laser is beamsplit, and sent along each arm to a mirror at each end (see Figure 21.3). Reflected light from both arms is then sent to a detector. If the path length travelled by both beams is identical, the beams destructively interfere and no signal is observed.

If the length of either beam changes, the beams can constructively interfere and produce a classic interference pattern at the detector. The strain placed on the arms by a gravitational wave h is

$$h = \frac{\Delta L}{L} \tag{21.17}$$

where L is the beam length, and ΔL is the change in beam length induced by the wave. A typical gravitational wave induces a strain of $h = 10^{-17}$, which means that for $L = 4 \times 10^3$ m, the beam length changes by $\Delta L = 4 \times 10^{-15}$ m, which is of order the diameter of a proton! This extremely low h is why humans have only just detected gravitational waves. Our apparatus must detect changes in a measured length that is very small compared to the total length.

The first gravitational waves detected by LIGO were emitted by a merging of two black holes of mass 36^{+5}_{-4} and $29 \pm 4 M_\odot$ respectively (Abbott et al., 2016b).

The team used two identical stations to measure the incoming gravitational wave signal. By doing this, they were able to place better constraints on the wave's origin. Another detection of gravitational waves from a separate black hole merger (of masses $1.42^{+8.3}_{-3.7}$ and $7.5\pm2.3 M_\odot$) was announced shortly afterwards (Abbott et al., 2016a). Most excitingly, gravitational waves and electromagnetic radiation have now been detected emanating from the same object (Smartt et al., 2017).

The gravitational wave spectrum exists completely separate from the electromagnetic spectrum. Both types of wave can be emitted from the same object, and both types of wave carry away energy from the emitter, but their characteristics are fundamentally different. Electromagnetic waves typically possess a wavelength less than or equal to the size of the emitter: gravitational wave wavelengths are typically greater than the source size. Electromagnetic waves are dipolar, while gravitational waves are quadrupolar. Even the nature of detection is different – electromagnetic detectors are sensitive to the wave power, while gravitational wave detectors are sensitive to the wave amplitude. This is quite analogous to the sensorial nature of the eyes and ears respectively. We have watched the Universe for centuries – for the first time, we can now hear it.

The detection of gravitational waves is a feat as epoch-making as Galileo's observations of the Jovian moons and Saturn's rings. Until LIGO's detection, gravitational waves were a prediction of general relativity, indirectly confirmed by observations of the decaying orbits of binary pulsars (Hulse 1994; Taylor, 1994). Our ability to measure gravitational waves directly opens up exciting future prospects for SETI.

It has been proposed by some that gravitational waves would be an excellent communication medium for advanced civilisations, as it requires an exceptional degree of material science knowledge and understanding of the Universe to search for them. It also possesses the aforementioned advantage of greatly reduced attenuation through its $1/r$ decay behaviour, and its extremely small absorption in intervening material.

Communicating in this form would be an excellent way to ensure one's message is only received by very advanced civilisations. However, the power required to produce a detectable signal is enormous. Perhaps we will discover more precise means of measuring the length of a laser beam, and be able to sense smaller and smaller perturbations in space-time, but will this mean that we will begin to hear gravitational wave transmissions?

Perhaps, just as with electromagnetic radiation, ETI will be forced to make specific choices about what wavelength of radiation to broadcast with. Astrophysical sources of gravitational waves span some 20 orders of magnitude in frequency (Figure 21.4) – in general, higher-frequency gravitational waves present less strain.

21.2 They Communicate Using a Different Method

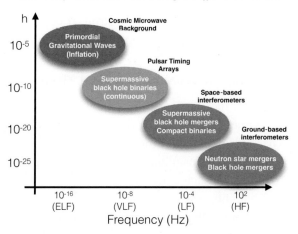

Figure 21.4 Schematic of astrophysical gravitational wave sources, as a function of wave frequency and wave strain.

At extremely low frequencies (10^{-16}Hz), primordial gravitational waves are present from the Universe's early history, in particular the epoch of inflation. These are unlikely to be detected directly, as their wave period is of order a Gyr. Instead, their signature must be discerned from the polarisation of the cosmic microwave background, as claimed somewhat prematurely by the BICEP2 collaboration, whose analysis appears to have been fatally affected by intervening dust (Ade et al., 2015).

At frequencies of tens of nanohertz (i.e., periods of years), supermassive black hole binaries are expected to radiate continuously, with strains of up to $h \sim 10^{-10}$. These waves are in principle detectable via pulsar timing studies. Merging compact objects (as was detected by LIGO) occupy the higher-frequency regimes, with the frequency increasing as the mass of the mergers decreases.

Rotating neutron stars will emit continuous gravitational waves if they are not axisymmetric, with the frequency of the wave ν being twice the star's rotational frequency. The strain received from this source would be (Riles, 2013):

$$h_0 \propto \frac{\nu^2 \epsilon}{c^4 D} \qquad (21.18)$$

This assumes that the star rotates about the z-axis, and that the star has a finite quadropole moment due to an asymmetry, giving rise to an ellipticity ϵ. This suggests at least one way that ETIs could send a GW signal, by adding matter to a neutron star to alter its quadropole moment (although the additional mass required to do so would likely be significant).

There are precious few papers on the subject of artificial generation of GW signals. It may seem fantastical, but with proper thought, the next generation of gravitational wave surveys could present even stronger constraints on the number of advanced civilisations present in the Milky Way.

21.3 We Do Not Understand Them

It is highly likely that any message we do receive will be a single, one-way transmission. Given the vastness of interstellar distances and the finite speed of light, we are unlikely to be able to ask for clarification as to the message's meaning. Conversely, it is unlikely that any transmitter will have had the opportunity to decipher human language to formulate such a transmission.

We are faced with a highly frustrating prospect – we could possess a *bona fide* transmission from another intelligent species, but be almost entirely ignorant as to its content. What can we do?

In the first instance, we can consider the apparent structure of the message. Any message can be decomposed into a string of bits, where each individual bit is either zero or one. Groupings of bits can then be formed producing what we shall call signal types. For example, we can consider English as obtaining a minimal 27 signal types (the 26 letters of the alphabet, plus the space bar to indicate word length).

We can immediately apply the principles of cryptography and information theory to this bitstream. A commonly used tool is *entropic level analysis*. This measures the conditional information in the stream, or more simply how a message element depends on its context – the preceding and following elements of the message (Doyle et al., 2011).

By simply counting the available signal types N in the message, we can compute the zero-order entropy:

$$S_0 = \log_2 N \qquad (21.19)$$

This gives the maximum degrees of freedom available to the communication system. The first-order entropy depends on the frequency of individual signal types in the message. If the probability of a given signal type i occurring is $P(i)$, then the first-order entropy is

$$S_1 = -\sum_{i=1}^{N} P(i) \log_2 P(i) \qquad (21.20)$$

If every signal type is equally likely ($P(i) = 1/N$), then the first- and zeroth-order entropies are equal. Second-order entropy relies on the probability that signal types i and j are grouped together. If $P(i, j)$ is the joint probability of types i and j, and $P(j|i)$ is the probability of j given i, then the second-order entropy is:

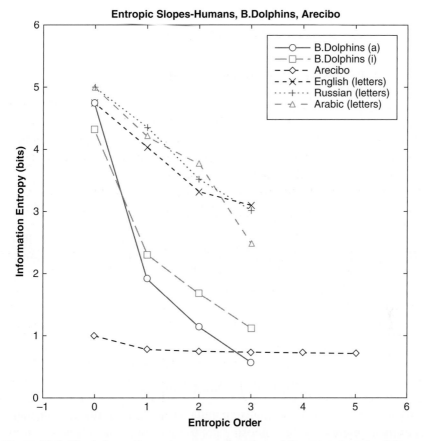

Figure 21.5 The information entropy, as a function of the entropic order (Figure 4a from Doyle et al., 2011). Note that English, Russian and Arabic all tend to follow Zipf's law, with a powerlaw index of −1. The Arecibo message shows no strong conditional relationships between bits, and so the entropy is relatively constant at all orders. Bottlenose dolphins also show syntactic structure to their communication thanks to their decaying entropy with order.

$$S_2 = -\sum_{i,j=1}^{N} P(i,j) \log_2 P(j|i) \quad (21.21)$$

This can then be generalised to nth order:

$$S_n = -\sum_{i,j,k...n=1}^{N} P(i,j,k...n) \log_2 P(j,k...n|i) \quad (21.22)$$

What does this analysis give us? It is instructive to consider the entropic structure of signals from human communication, non-human animal communication, and signals from inanimate objects. Doyle et al. (2011) relate measurements of the entropic structure of several communication sources (see Figure 21.5).

Plotting the entropy against order in this fashion helps to interpret the nature of the signal. We can classify signals according to the slope of these graphs. Shallow slopes (where the entropy varies slowly with order) suggest significant levels of repetition and redundancy in the signal.

If the entropy decays with order, this is most likely thanks to the presence of syntax and grammatical structure preventing or discouraging combinations of signal types and favouring others. Signals where the entropy decreases sharply at some n indicate a limit to the signal's rule structure, or in a loose sense, 'complexity'.

If the entropies are computed in base 10 rather than 2, it is found that most human language achieves a slope of approximately -1. This phenomenon is referred to as *Zipf's law*. Finding such a slope is not a guarantee of human-like language, as other phenomena have been shown to exhibit similar behaviour.

That being said, Doyle et al. show that human communication is markedly different in entropic structure from inanimate signals such as the signalling received from pulsars, which show shallow entropy slopes. Bottlenose dolphins and humpback whales show slopes close to the Zipf slope, but this remains insufficient evidence for complex symbolic communication (although see section 6.4.4). It is also important to note the low Zipf slope of the Arecibo message (section 21.4.1), which indicates the lack of conditional relationships between any two bits of the message.

21.4 They Do Not Communicate

It is possible that we exist in a Galaxy populated by many civilisations listening for signs of their neighbours, but choosing not to transmit signals. If no one transmits a signal, then it is quite possible that no civilisation can detect each other.

This solution to Fermi's Paradox lies at the heart of a polarising debate in the SETI community, regarding whether humanity should be sending exploratory transmissions into the Universe, in an attempt to instigate contact. This practice has been given many names over the years: CETI (Communication with Extraterrestrial Intelligence), METI (Messaging Extraterrestrial Intelligence), and most recently active SETI. I will adopt METI throughout, as it is most clearly different (both phonetically and in text) from SETI.

21.4.1 Arguments in Favour of METI

SETI's Decades of Failure Merits a New Approach

We are now almost sixty years from Frank Drake's first SETI observations as part of Project Ozma. METI activists argue that our failure to detect another civilisation by merely listening indicates that eavesdropping on interstellar communications is likely to fail, and that direct messaging is the only viable possibility for Earth to receive extraterrestrial transmissions.

It is true that eavesdropping on unintentional or leaked broadcasts from radio/TV transmitters is extremely challenging even at relatively short interstellar distances. As we have discussed in the Introduction, our leaked transmissions are at best only likely to be received at distances less than a few hundred parsecs from Earth with receivers akin to our own (Loeb and Zaldarriaga, 2007). Our radar systems, especially for asteroid detection and for space probe ranging, may do better and penetrate deeper into the interstellar medium, but these are highly directed beams of energy. If we wished to do so, we could prevent the accidental transmission of these signals towards nearby stars by simply halting operations when the beam's target occults a known star (or transits the Galactic plane).

Consider twostar systems possessing intelligent civilisations aware of each other's existence, and in formal contact. If these two systems decide to communicate, it is likely that their communications would entail a modest collimation of the signal beam, as this would assist in producing reliable transmissions with a judicious use of energy. Now consider Earth as the third party in this scenario, who wish to intercept this communication beam. It is simple to demonstrate that if the communication beam is only modestly collimated, the probability of interception by any third party is exceptionally small (Forgan, 2014).

If this is true, then we can identify three possibilities:

1. Earth will only be able to detect ETI via direct contact from a civilisation that has determined our existence.
2. Earth will only be able to detect ETI via secondary forensic evidence (a growing field of interest, which we will address in the concluding part of this book).
3. Earth will never be able to detect ETI.

METI activists argue that to forestall possibility 3, we must encourage possibility 1 to occur. The best way to get a direct signal in this scenario would be to encourage a response to our own transmitted signal.

Someone Has to Start the Conversation

We can imagine a Galaxy with a relatively small number of intelligent civilisations, who have all decided not to send out communications. From the previous argument, if ETI leave no detectable secondary forensic evidence, then ETI will never be able to ascertain each other exists.

If humans do not exercise METI, then we are demanding that other ETIs practise it instead. It is arguable that civilisations that practise METI initiate contact earlier than civilisations which do not. In this scenario, Earth's attempt at METI would be a lighting of the touchpaper, a trigger to all listening civilisations that we exist, and wish to communicate.

They May Be Able to Help Us

A long-used argument for initiating contact is that a response from ETI may contain valuable information. We have outlined in part 2 a series of existential risks to humanity (both natural and self-inflicted). If we initiate a conversation with an advanced civilisation, they may have passed the so-called 'Great Filter', and may have crucial information on how to do so.

This information could take the form of blueprints for hardware to combat climate change, appropriate software to control artificial intelligence, or even political doctrine that produces harmonious governance. METI activists assert that such assistance is unlikely to arrive unasked for. It may well be that few civilisations can pass the Great Filter unaided, and that we must ask for help before it's too late.

We're Doing It Anyway

With the correct electronics, any single-dish radio telescope can be refitted to become a radio transmitter. The Earth has a long history of sending radio transmissions at a variety of frequencies and bandwidths, with a great diversity of information contained within. Five major deliberate radio transmissions are recorded in Table 21.1 (Shuch, 2011). More transmissions continue to occur (albeit at significantly reduced radio power).

Famous large-scale efforts include the Arecibo message, sent in 1974. The message's contents were drafted principally by Frank Drake, with assistance from graduate students at Cornell University (Richard Isaacman and Linda May) as well as Carl Sagan.

The Arecibo message consisted of 1,679 bits of information. The transmission (sent at 2,380 MHz) used frequency shifting to indicate zeros and ones, with a shifting rate of 10 bits per second. The message transmission took less than three minutes, and was not repeated. We can see the message in Figure 21.6. The selection of 1,679 bits is crucial, as it is *semiprime*, i.e., it is a product of only two primes: 23 and 73. If the bits are arranged in 73 rows and 23 columns, the stream of zeros and ones become pixels of a pictorial message.

The message contents are as follows. The first row establishes the numbers one to ten in binary, where each number is represented by two columns (the second column is only necessary to describe numbers above 8). With this established, the principal elements in terrestrial biology are given in the second row by their atomic numbers (H, C, N, O, P).

In the subsequent four rows of numbers, the structures of deoxyribose and phosphate sugars occupy the margins, and the two sets of base pairs (AT, CG) are also indicated. This indicates the composition of DNA, whose double helix shape is sketched below (with the interior number giving the typical number of base pairs in the human genome).

Table 21.1 *The five major radio transmissions made by human civilisation*

Name	Date	Authors	Radar	Sets	Time (min)	Energy (MJ)
Arecibo Message	16/11/1974	Drake, Sagan, Isaacman et al.	Arecibo	1	3	83
Cosmic Call 1	24/05–01/07/99	Chafer, Dutil, Dumas, Braastad, Zaitsev et al.	Evpatoria	4	960	8,640
Teen Age Message	29/08–04/09/2001	Pshenichner, Filippova, Gindilis, Zaitsev et al.	Evpatoria	6	366	2,200
Cosmic Call 2	06/07/2003	Chafer, Dutil, Dumas, Braastad, Zaitsev et al.	Evpatoria	5	900	8,100
A Message From Earth	09/01/2008	Madgett, Coombs, Levine, Zaitsev et al.	Evpatoria	1	240	1,440

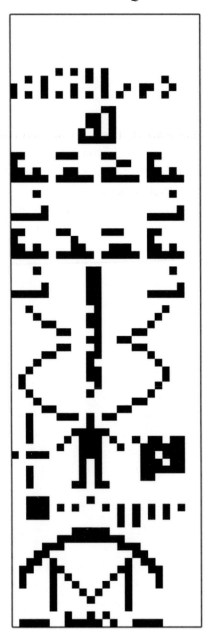

Figure 21.6 The Arecibo message. The 1,679 bits of information are arranged in 73 rows and 23 columns – black pixels correspond to ones, and white pixels correspond to zeros.

Below the DNA strand is a depiction of a human. To the right is the population count of the Earth as it was in 1974 – around 4 billion. The depiction to the left of the human is the number 14. If we convert using the wavelength of the transmission (around 12.6 cm) then this gives a value of 176 cm, approximately the height of a human.

Below the human is a depiction of the Solar System – the Earth is raised to indicate that this is our homeworld. The sizes of each point roughly indicate (in logarithmic scale) the size of each body compared to the Earth. The final graphic outlines the Arecibo antenna, with an indication of the antenna size of 306 metres (again in transmission wavelength units).

The message's target was the globular cluster M13, approximately 25,000 light-years away. As such, the message will not arrive for millenia, and will be almost certainly undetectable when it does arrive. This choice of target might satisfy the anti-METI camp – the risk of the message being intercepted is extremely small – but the pro-METI camp is unlikely to consider such efforts as being worthy. The message is perhaps the most famous example of a deliberate message to ETI, but its content, construction and target guarantee that its value as a means of establishing contact is exceptionally poor.

Later message attempts (championed by Alexander Zaitsev) contained a great deal more information encoded in the error-resistant Lexicon language, constituting messages totalling of order half a million bits (see, for example, the Cosmic Call messages of 1999[3]). Lexicon's construction mitigates against accidental bit-flipping in the message during transit. It also retains its uniqueness under rotation and mirror symmetries, ensuring that even if the message is received out of sequence, the underlying meaning is preserved (Figure 21.7).

It's Not Illegal

The SETI community has a 'Draft Declaration of Principles Concerning Sending Communications with Extraterrestrial Intelligence',[4] colloquially known as 'the reply protocol', which covers the circumstances of responding to a message or signal from ETI. These protocols require any reply to be approved by all nation-states entering into the agreement, with no single state given the authority to speak for the Earth without express permission. Anti-METI activists argue that while we have not received a transmission from ETI, the reply protocols still cover any transmission sent by humans into interstellar space.

Unfortunately, the reply protocols are drafts only, and have no legal force. Even if they did, policing the transmission of radio signals into space is effectively

---
[3] http://www.cplire.ru/html/ra&sr/irm/report-1999.html
[4] https://iaaseti.org/en/protocols/

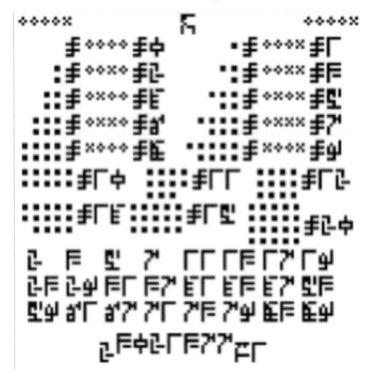

Figure 21.7 An excerpt from the Cosmic Call messages, written in Lexicon. This initial page defines the number system (from 0 to 20), both in base 10 format (dots) and binary representation (crosses and rhombus shapes), and finally both are equated to a unique symbol, which retains its uniqueness under mirror and rotational symmetry. This symbolic measure established, the prime numbers are then written underneath, along with the then largest known prime in 1999: $2^{3,021,377} - 1$ (Yvan Dutil, private communication).

impossible, due to the ease of setting up and executing a transmission. METI activists argue that, because signals have already been sent into space (and that we unintentionally leak radio signals also), the horse has bolted, and locking the stable door is pointless.

They Are Sufficiently Distant As to Not Pose a Threat

The lack of intercepted transmissions from ETI as part of the SETI program already place (weak) constraints on the number of transmitting civilisations in existence. It certainly is the case that there cannot be millions or billions of planets possessing ETI that transmit in the radio, as such a scenario would almost certainly have resulted in a detection, even given our relatively brief and sporadic search.

If this assumption is correct, the nearest civilisation is at least a few hundred parsecs away, if not thousands of parsecs. The signal travel time is therefore hundreds to thousands of years. The quickest response would be another signal transmitted at the speed of light, which would take hundreds to thousands of years to return. Therefore, METI poses essentially no immediate threat to the Earth. Current METI signals with the closest targets will still require at least another 30–50 years of travel time.

Even if our efforts contacted a hostile civilisation, it is argued that the enormous effort of crewed space travel (section 22.1) is unlikely to appeal to any civilisation wishing to conquer Earth. The prospect of an alien invasion fleet in the Solar System is so highly improbable that pro-METI activists tend to discard it as a possibility.

21.4.2 Arguments against METI

Unintended Consequences of Contact – The Conquistador Argument

There are many examples in Earth's history that serve as a warning against contact between advanced and primitive civilisations. The most cited example is the arrival of Spanish conquistadors in the New World. While most SETI scientists are agreed that physical contact between humans and alien civilisations is extremely unlikely, the conquistador example is not usually used to argue against close encounters. It is typically deployed instead to warn against unforeseen consequences of contact, physical or otherwise.

The death of New World civilisations may have been accelerated by the advanced technology of European colonists, but by far the most potent aggressor in these events was biological. The rapid advance of hitherto unseen bacteria throughout the native population decimated their ranks, and fatally weakened whatever defenses they could muster.

A kind of intellectual infection is also likely to have passed through the ranks of native American peoples during the Spanish Conquest. A strange new kind of human arrives at the heart of your civilisation, with long-range weapons far exceeding your current technology. Within a few days, just a small band of these creatures penetrates your most secure citadel and captures your leader, effectively your deity. In the face of such terrifying power and disrespect, what could the people do but despair?

Even if we never receive a dose of alien germs, or a volley of alien munitions, contact will result in a dose of alien culture. The reactions of humanity may parallel those of their Aztec ancestors – a deep sense of cultural inadequacy, perhaps even despair may set in (especially if it seems that aliens have solutions to existential risks we face and are not willing to/unable to share).

Even if aliens are benevolent, and are willing to share their technology with us, is it wise for us to receive this knowledge 'ahead of our time'? Again, we have seen countless examples of human cultures being suddenly presented with new and powerful technology, with devastating results.

Malevolent Civilisations May Come To Earth

Science fiction provides us with further warnings against contact with alien civilisations. These warnings can be blunt and somewhat obtuse stories, pulpy popcorn affairs of alien invasions laying waste to our planet (such as *Independence Day*). The 'Berserker' solution to Fermi's Paradox comes to us from the science fiction novels of Fred Saberhagen. This solution suggests that intelligent civilisations are silent because they fear retribution from rogue wandering autonomous craft, the so-called 'Berserkers'. Loud civilisations tend to be silenced against their will by these 'Berserkers'.

The precise nature of the Berserker varies depending on who espouses the solution. In Saberhagen's novels the Berserkers were machine intelligences left over from an interstellar war, but they may range from an alien battlefleet bent on exterminating any sentient competition, or some kind of autonomous probe with a corrupted program (or indeed, an autonomous probe with a deliberately malevolent program). The pro-METI argument that an alien invasion is extremely unlikely is countered by the concept that such an invasion fleet may possess no biological component whatsoever. A journey of a few thousand years may be a mild inconvenience to an uncrewed autonomous battle fleet.

Malevolent Civilisations May Send a Hazardous Response

Even non-physical contact has its own hazards. The classic 'A for Andromeda' radio serial (by the renowned astronomer Sir Fred Hoyle) relates a possible scenario for radio contact between humans and another civilisation. The message received by Earth contains instructions describing the construction of a computer, with an accompanying program. The computer program outputs instructions for the creation of living cells, and eventually the construction of a human clone, which it uses in a secret plan to attempt to conquer humanity. This scenario shows the power of transmitted data – despite the message having arrived from the 'Andromeda Nebula' (i.e., the Andromeda Galaxy, some 2.5 million light-years away), with no chance of physical contact between the transmitters of the signal and humans, it can still exact damage and pain on us.

Communication is not required for detection

Returning to the three possibilities outlined in section 21.4.1:

1. Earth will only be able to detect ETI via direct contact from a civilisation that has determined our existence.
2. Earth will only be able to detect ETI via secondary forensic evidence (a growing field of interest which we will address in the concluding part of this book).

3. Earth will never be able to detect ETI.

A pro-METI argument by definition ignores possibility 2 as being credible. If we can detect ETI by the imprints left by its activities around stars and on planetary bodies, then we can use this to exclude possibility 3 without having to send a transmission.

Where Do You Transmit?

It is well established that a collimated transmission from Earth that is not sent in an appropriate direction is unlikely to be received by an ETI. METI sends the transmission in the hopes of establishing contact, which by definition extends to learning of the existence of another technological civilisation. But the logic is circular: to detect ETI, we must transmit a signal; for ETI to receive our signal, we must already have detected them to know where to centre our transmission. Pro-METI academics argue that this can be solved by selecting a larger number of targets, broadcasting to each target in a repeating duty cycle. However one chooses to carry out METI, any message is a shot in the dark, without prior evidence that the target contains a technological civilisation.

We Have No Institution Capable of Receiving Any Response

Even if we are successful in initiating contact with ETI, any conversational aspect may be extremely limited by the constraints of lightspeed. Unless our interlocutor is a local probe or machine designed to mediate on ETI's behalf, a likely scenario will be that of transmit – delay – reply – delay – transmit, where each delay interval could be as long as hundreds (even thousands) of years, certainly beyond the lifespan of an individual human.

For any human–ETI conversation to have coherence, we must be able to accurately document the interchange over several iterations of this transmit–reply cycle. As Gertz (2016) notes, such coherence requires institutional memory. We can point to very few human institutions that remain recognisably intact over thousands of years. The best candidates for institutional longevity are arguably religious institutions such as the Catholic Church which have endured for several centuries, at times in the teeth of fierce antagonism. Of course, the means by which such institutions have endured rarely make them suitable for what is required here.

If we are to truly attempt METI, then there should be preparations for a reply. An institution responsible for METI must perform multiple roles:

1. It must attain full accreditation and legal authority to converse on behalf of all the Earth's peoples.
2. It must protect the means of communication, including but not limited to the instrumentation used to transmit and receive, but also the environment of the

signal. For example, a radio conversation would require that certain frequency bands be protected from radio frequency interference due to human use.
3. It must record with high fidelity all instances of transmission and reply. This will mean being able to curate and protect digital information and retrieve it many thousands of years after recording.

At the time of writing, no institution in existence can demonstrate that it can perform any of these three roles. This is a symptom of a wider failing in present global human society – the ability to plan effectively on timescales larger than a few decades.

21.4.3 Can the METI Debate Be Resolved?

Haqq-Misra (2018) demonstrates that the METI debate, which we can phrase concisely as:

Will transmitting result in a positive or negative outcome?

can be reduced to an instance of the halting problem (see Chapter 15) and is therefore undecidable without further data (i.e., detecting ETI). Without this data, we must therefore rely on a guiding principle to decide whether to transmit or not. If we state the value of METI U_M as

$$U_M = p_d M - C \qquad (21.23)$$

where M is the magnitude of an ETI's response (which can be either positive or negative), p_d is the probability that a message is detected by ETI, and C is the transmission cost. We must therefore ask, what is the sign of M? We cannot know without further information.

Anti-METI activists argue for the *principle of precautionary malevolence*. We must assume in the absence of data that ETIs are malevolent, and that transmitting will produce a negative outcome ($M < 0$).

Pro-METI activists argue for the *principle of assumed benevolence*. The likely technological and historical asymmetry of any contact suggests that ETIs will be benevolent ($M > 0$), and we should therefore transmit.

Haqq-Misra (2018) advocates for a third principle, the *principle of preliminary neutrality*. Without knowledge of the sign of M, we must instead assume $M = 0$ until proven otherwise. This neither advocates for or against METI, but instead ascribes its value merely by the cost, C. METI should only be pursued if its costs are sufficiently low (compared against the many demands on human resources). In essence, the decision to transmit should be economically motivated, and not motivated by any other principle.

22
They Live Too Far Away

As is by now quite obvious, our observations of the Galaxy will not discriminate between a scenario where we are truly alone, and a scenario where there are a small number of technological civilisations separated by large distances. Unless another civilisation expends enormous resources merely to contact us, then in the second scenario we will receive as little indication of intelligent life as we would in the first scenario.

Most optimistic arguments for detecting ETI assume that any population of civilisations would inhabit the Galactic Habitable Zone (section 4.2.2). This rationale relies on the civilisations having somewhat specific biological needs. If the civilisation is fully post-biological (or elements of it are), then we can expand the potential domain of intelligent beings far beyond the GHZ.

Indeed, a post-biological intelligence would have its own rationale for moving as far away from the Galactic Centre as possible, for thermodynamic reasons. If fast and accurate computing is necessary or desired by such intelligences, then they are likely to seek our regions of low temperature and reduced levels of ionising radiation.

We can see this constraint in the following inequality, which defines the maximum amount of information content available to be processed in any system I_{max} (Landauer, 1961; Brillouin, 1962):

$$I \leq I_{max} = \frac{\Delta E}{k_B T \ln 2}, \qquad (22.1)$$

as a function of the energy used, ΔE, and the system temperature T. In an ideal case, the system should be passively cooled, rather than using extra energy to reduce its temperature. The simplest solution is placing the system in a heat reservoir of as low temperature as possible.

The coolest heat reservoir possible is that of the cosmic microwave background, which in the absence of other energy sources has the temperature (Fixsen, 2009):

$$T_{\text{CMB}} = 2.72548 \pm 0.00057 \, \text{K} \qquad (22.2)$$

In practice, it is not feasible that computers will be able to reside in direct contact with these sorts of temperatures, but it is clearly a target that motivated post-biological civilisations can aim for.

Cirkovic and Bradbury (2006) used this reasoning to motivate a *Galactic Technological Zone* – regions of the galaxy conducive to high-performance computing. This includes the outer reaches of the Milky Way, but may also include the interiors of giant molecular clouds, where temperatures can remain relatively low (although such regions can still be dogged by significant cosmic ray heating, which can not only increase the reservoir temperature but also cause bit corruption).

We could therefore imagine a scenario where we have arrived late in the scheme of things, and that the outer reaches of the Milky Way are well populated by advanced post-biological civilisations, potentially in contact with each other but not with us. We could also imagine a scenario where we arrived late, and advanced civilisations have chosen to cease operations, and wait until the Universe is itself cooler. Equation (22.1) shows us that as star formation peters out, and the cosmic microwave background radiation continues to cool, civilisations begin to achieve far greater energy efficiency per bit of information. The so-called 'aestivation' solution to Fermi's Paradox (Solution U.10, Sandberg et al., 2017) suggests that advanced civilisations choose to aestivate (be dormant during periods of high temperature) and are effectively invisible to less-advanced civilisations such as ours.

When considering the contactability of very distant civilisations, we should therefore consider the challenges of either sending information or individuals over great distances, as we will now discuss in the following sections.

22.1 The Feasibility of Crewed Interstellar Travel

In earlier chapters, we have already explored the mechanics of uncrewed interstellar craft and their exploration of the Universe. The payload of these uncrewed missions ranges from grams to a few hundred tons. If we add the further requirement that these craft are crewed, by which we mean they host biological life forms, the payload required is significantly greater.

As the payload is significantly larger, the maximum velocity of the craft is likely to be significantly slower, even for the most advanced and exotic means of travel. Considering that the Daedalus design (of mass $\sim 54{,}000$ tonnes) could achieve a maximum velocity of order $0.1c$, what would a human-crewed vehicle achieve?

Assuming the goal of crewed interstellar travel is to found a settlement at the destination, we can estimate the expected settlement mass M_s as (Cassenti, 1982):

22.1 The Feasibility of Crewed Interstellar Travel

$$M_s = M_e + M_i N_i \qquad (22.3)$$

where M_e is the total mass of landers, probes and equipment to support the mission, M_i is the mass required to support a specific individual, and N_i is the number of individuals. A human crew of 1,000, with individual needs of 100 tonnes per person, already has a settlement mass of $M_s = 0.1 Mt$ (ignoring the support equipment mass). Without enormous energy cost, it is likely that a crewed human voyage would take many thousands of years to travel from Earth to the nearest star system.

A journey of this duration places even stronger demands on the spacecraft's payload. No spacecraft can carry a thousand years' supply of food for a human crew, even one with only a hundred individuals. For such an interstellar vehicle to successfully carry its crew to destination, it must be able to *generate* food, energy and gravity during the flight, using only what resources it is launched with, and the meagre provender available in the interstellar medium.

These so-called *generation ships* (sometimes called interstellar arks) are effectively miniature planets, able to sustain a mobile settlement of humans from beginning to end. Tsiolkovsky, like much of astronautics, was amongst the first to describe the concept in his essay *The Future of Earth and Mankind* (1928).

A classic generation ship design derives from the *O'Neill cylinder*. O'Neill's cylinder is a free-space habitat design, where the cylinder rotates along its longest axis to maintain a centrifugal force on the interior of the cylindrical surface. This provides a source of artificial gravity (a counter-rotating cylinder would provide gyroscopic balance to the craft). For an acceleration equal to Earth's gravity g in a cylinder of radius R, the angular velocity of the cylinder

$$\Omega = \frac{g}{R} \qquad (22.4)$$

For $R = 895$ m, the cylinder can achieve an acceleration of $1g$ at a rotation rate of 1 revolution per minute (rpm). A km-sized cylinder would contain a total volume of $V = \pi R^2 z \approx 3 \times 10^9 \left(\frac{z}{1\,\text{km}}\right)$ m^3, and a total surface area (ignoring the lids) of $A = 2\pi R z = \approx 3 \times 10^6 \left(\frac{z}{1\,\text{km}}\right)$ m^2.

If $z = 1$ km, this is sufficient to house hundreds of thousands of people (assuming a population density of 70 m^2 per person). By keeping the total population much lower than this, a generation ship has more carrying capacity for other biomass and environmental equipment needed to generate a functional biosphere.

Much of the standard O'Neill design assumes the cylinder would remain in close proximity to the Sun or the planets, and receive a steady supply of materials and sunlight. A generation ship will not have such a luxury, and must rely instead on its own resources. Instead of relying on a nearby star, energy must be used to power a set of lamps to provide appropriate radiation to drive the onboard ecosystem.

Any faults on the ship must be repaired by its own engineers. Hein et al. (2012) calculate the reliability R of a craft, defined as the fraction of parts n still working at time t:

$$R(t) = \frac{n(t)}{n(0)} \qquad (22.5)$$

by considering the mean time between failure (MTBF) data from the Magellan mission, and fitting parameters to a modified Weibull distribution:

$$R(t) = exp\left[\left(\frac{-t}{\tau_{\text{mission}}}\right)^{\beta}\right]. \qquad (22.6)$$

For Magellan, $\tau_{\text{mission}} = 102{,}775$ hours and $\beta \approx 0.43$. After 50 hours, the reliability of the craft drops to 99.99%, corresponding to around half a million faulty parts on a generation ship, or a failure rate of around 3 per second! They therefore argue that a small human crew is incapable of repairing this prodigious amount of failed parts, and that automatic repair systems are essential (as is bringing a significant fraction of the total ship mass in redundant spares).

If the cylinder becomes depleted in supplies of essential elements and minerals, the entire biosphere may be placed in jeopardy. For example, the transition metal molybdenum is one of the least abundant micronutrients found in plant tissue, but is crucial in nitrogen fixation reactions (Kaiser et al., 2005). An interstellar ark with a very small deficiency in molybdenum could soon see significant problems not just with crop yields, but with atmospheric regulation also.

The ecosystem of any interstellar ark will be subject to the effects of *island biogeography*. On Earth, the ecology of islands depends sensitively on the distance of the island from the mainland (from which it can source new immigrant species and gene flow), and on the surface area of the island. Larger islands will see larger levels of immigrant species (the *target effect*), and immigration becomes more likely as distance to the mainland decreases. The diversity of islands (measured by the total count of species S) is typically proportional to the island area A:

$$S = KA^{x} \qquad (22.7)$$

where $0 < x < 1$ and commonly $x \approx 0.3$. In the classic MacArthur-Wilson model of island biogeography, this relationship is established as an equilibrium state between the extinction rate of local species and the immigration rate of invasive species. Islands are much more subject to *endemism* – a preponderance of unique organisms, highly adapted to local habitats but extremely vulnerable to extinction if the habitat changes.

On an interstellar ark, there are no immigrant species. Gene flow is utterly suppressed. Genetic drift is weakened by the low population count aboard. Mutation and selection dominate the evolution of the system. Generation ships are therefore extremely fragile ecosystems. Small mutations in, say, the microbial community aboard can have devastating consequences. With such a small population aboard, highly prone to endemism, there are only a limited number of genetic combinations available to the ecosystem to search for an equlibrium solution. This can be mitigated against by carrying extra genetic material in a more condensed form (either by hosting a seed bank, or storing DNA in a digital form for later genetic engineering efforts).

Perhaps a greater challenge to any interstellar voyage is the psychological pressure exerted on the voyagers. Interstellar travel in this mode guarantees that the crew launching the voyage will never see its end. Indeed, there are likely to be many generations of crew who live on the vehicle for their entire lives, and never see the starting point or destination.

There are multiple psychosocial stressors acting on a generation ship crew, which include (Kanas, 2015):

- Isolation and loneliness – knowing that both home and destination will never be within reach
- Monotony and tedium
- Unforeseen physiological/psychological effects of travel at near-relativistic speeds
- Governance of a totally isolated group of a few hundred individuals, including dealing with criminal elements and mentally ill individuals who place their fellow crew at risk
- Maintaining a sustainable environment
- Maintaining a healthy society and culture
- Maintaining a truthful ship history and legacy

Much of human society is predicated on deeply ingrained assumptions regarding our environment. For example, on Earth it goes without saying that the air cannot be policed, but it can inside a generation ship. How can liberty be maintained when literally every aspect of human life can be controlled by another human or machine (Cockell, 2015; Stevens, 2015)?

These are placed on top of all the problems that human societies face, with the added stress that other humans beyond the ship will take many decades to contact, if they can be contacted at all. Cross-generational social contracts lie at the heart of modern societies on Earth, but can we expect the descendants of the first crew to accept a contract that means their life is largely pre-determined by their parents, without hope of revision?

These factors would lead us to believe that while crewed interstellar travel is not prohibited by the laws of physics, there are good arguments for the proposition that 'biological life forms do not travel between the stars'. This is perhaps a partial solution to Fermi's Paradox. Aliens are unlikely to land on the White House lawn if they cannot break the light barrier, but this does not rule out uncrewed vehicles, vehicles crewed by non-biological or post-biological entities, or the sending of messages.

22.2 The Feasibility of Reliable Interstellar Communication Networks

Without some incredibly powerful energy source (and beyond exquisite pointing ability) it is unlikely that a civilisation will be able to communicate directly with another civilisation across the entire Galactic disc (at least using electromagnetic radiation). Even if such an energy source is found, there are many confusing sources of radiation and absorbing media in between.

For this, and many other reasons, we should consider the possibility that a community of civilisations spanning the Galaxy might not choose direct communication, and instead rely on the community itself, establishing a *network* or relay system. Sending signals to a neighbour (with the expectation that said signal would be passed on) reduces the total energy cost dramatically for an individual civilisation.

This is partly thanks to the inverse square law, but an energy saving is also found in the signal's collimation. A signal that travels a shorter distance will occupy a smaller physical radius (for a fixed beam solid angle), and hence more photons (or other carrier particles) will arrive in the receiver for the same energy input. With appropriate error-correction methods (see section 21.1), the signal is also protected against the degradation it would receive traversing a large interstellar volume. The signal would reach its destination having been refocused and retransmitted several times.

The total path length of the signal is likely to be larger than the path length of a direct transmission, but how much larger? And how viable is the formation of such a communication network?

Recent work suggests this network would be relatively rapid to establish. Let us begin by defining some useful terminology for networks from *graph theory*. A network or graph G is defined by a set of vertices V, and a set of edges E that connect pairs of vertices. In our further discussion, we will consider civilisations to be potential vertices, and the edges that connect them as the links that establish communication. We will therefore assign a weight d to every edge, which in our case is the separation between the two civilisations. Our graph is *undirected*, in that the edges are bidirectional (in other words, a signal can travel from vertex A to vertex B or vice versa).

22.2 The Feasibility of Reliable Interstellar Communication Networks

An individual vertex may have multiple connected edges. It is reasonable to assume that once two civilisations have established a connection, that connection will be maintained as long as both civilisations continue to exist.

There are two simple but important properties of graphs that are useful here. We can define a subgraph, known as a *connected component*, as a subset of vertices that are connected to each other (i.e., any vertex in the component can be reached from any other vertex in the component). A graph may only have one connected component (i.e., every vertex can be reached from every other vertex, as in the example graph in the left panel of Figure 22.1), or there may be many connected components – many networks effectively isolated from each other (see Figure 22.2).

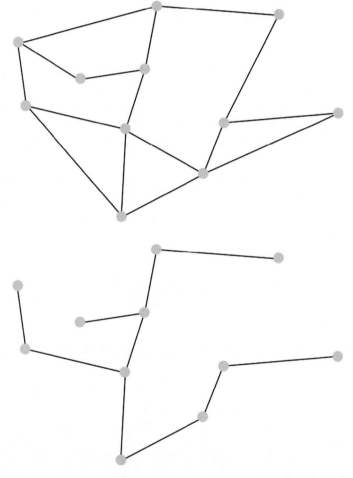

Figure 22.1 Top: An example of a fully connected graph, consisting of vertices (circles) and edges (lines). Bottom: A minimum spanning tree of the same graph.

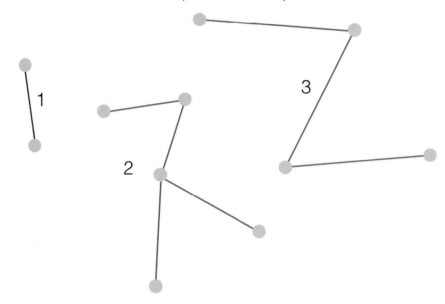

Figure 22.2 An example of a graph with three connected components.

The second property is known as the *minimum spanning tree* (MST). Let us define a graph G with vertices V and edges E that has a single connected component. The MST is a graph with vertices V and edges E', where E' is a subset of E, and is defined to reduce the total edge weight

$$D = \sum_E d \qquad (22.8)$$

while preserving the connectivity of the graph (i.e., maintaining a single connected component that contains all of V). Minimum spanning trees describe the minimum amount of edge needed to span the graph (see right panel of Figure 22.1). If the graph has more than one connected component, then the equivalent entity is the minimum spanning forest (MSF), where each connected component becomes a minimum spanning tree of its subgraph.

It would seem desirable that an interstellar communications network be composed of a single connected component, and possess a relatively low total edge weight (so that any signal traversing the network would possess a minimal path length).

A recent example of interstellar network growth considered the observing of each other's planetary transits as a criterion for communication. This assumption is built on recent thinking that exoplanet transits may be useful carriers to send electromagnetic signals upon (see section 27.1.4). Demanding that mutual planetary transits be the connection criterion is relatively strict, so it serves as a conservative example for more direct forms of communication.

22.2 The Feasibility of Reliable Interstellar Communication Networks

Simulations of civilisations in an annular Galactic Habitable Zone indicate that at any time, the network is likely to have many connected components. This is assuming that the vertices have no 'memory' of past connections. If the vertices retain a memory of past connections, then this network can quickly establish a single connected component. If the vertices 'share address books', that is that any two connected vertices share their edges, then this network can be fully established in much less than a million years!

The minimum spanning tree of the network effectively traces the dimensions of the Galactic Habitable Zone. If a civilisation at radius R wished to communicate with a network member on the other side of the annulus, also at radius R, rather than attempting to transmit directly along a path of length $2R$ that intersects the Galactic Centre, a relay signal could be sent that would traverse a path of around πR, which increases the total path length by a factor of $\pi/2$, but saves immeasurably on initial transmission energy, as the original signal must only travel a distance of order the mean civilisation separation:

$$\bar{d} = \left(\frac{V}{N}\right)^{1/3} << \pi R \tag{22.9}$$

As stated, this is a relatively conservative scenario. If civilisations are broadcasting their presence in some other form, then the establishment of a fully connected network may take even less time. The indication is that interstellar communication networks should be relatively robust, easy to establish and maintain.

The main proviso is that the civilisation density remains above a critical value across the entire Galactic volume. Therefore, a valid solution to Fermi's Paradox is that communication with Earth fails as the number of civilisations is below a critical value for establishing a reliable network.

What is this critical density? If the maximum broadcast distance is d_{max}, then we require the mean separation $\bar{d} \approx d_{max}$.

As $\bar{d}$ is a function of N, we can therefore only infer: if communication fails, and we are confident of d_{max}, then

$$N < \left(\frac{d_{max}}{V}\right)^3, \tag{22.10}$$

given the volume V we expect civilisations to arise in.

23
The Zoo/Interdict Hypothesis

As a civilisation, we are increasingly aware of our impact on the environment and the organisms that inhabit it. Facilitating large-scale political change to halt this deterioriation is difficult, but smaller-scale solutions have been implemented for centuries.

The zoo or national park is a straightforward legal device for preserving and encouraging the development of other species. Either by restricting human access to an environment, or deliberately intervening in the environment to prevent further degradation and to support local ecosystems, humanity has shown mixed success in preventing extinction, and reversing the fortunes of unlucky animals and plants.

Our increasing skill at protecting other species led Ball (1973) to wonder whether a similar procedure could be in place amongst the stars. The dangers of human ingress into otherwise unspoiled areas are quite obvious – habitat destruction, hunting of local species, pollution – but an extra level of danger exists if the local flora/fauna is intelligent. The cultural pollution of a world by ETIs might have significant consequences for its inhabitants.

Ball hypothesised that the current observational data (Fact A) was consistent with the notion that Earth is itself an interstellar nature reserve, sealed off from the galactic community for what could be a number of reasons: scientific study of the Earth environment; scientific study of Earth as a 'primitive' civilisation; or a general awareness that contact would be extremely harmful to either party (or possibly both parties). Fermi's Paradox is then resolved rather simply (if somewhat frustratingly).

Fogg (1987) would go on to develop this concept into what he refers to as the Interdict Hypothesis, a generalised version where the Earth is visited or probed in the very distant past by ETIs, and is sealed off from settlement efforts, most likely as a means of *planetary protection*.

Our past, present and future uncrewed missions to our Solar System neighbours are deeply concerned with contamination of landing and exploration sites by organisms from the Earth. Taken without due care, we may accidentally seed nearby planets and moons with life that never originated there, casting doubt on any future discovery of life as proof it can arise and evolve off-Earth. Conversely, any physical interaction we have with life from another world may be fundamentally incompatible with our biochemistry (or more glibly, planetary protection goes both ways). In Fogg's view, the presence of extraterrestrial organisms or ideas on Earth represents a hazard to the planet, and must therefore be mitigated against.

The concept of a 'Galactic Club' of civilisations is central to the Zoo Hypothesis. If Earth is to be truly sealed off, this requires agreement across multiple civilisations. Nature reserves on Earth are protected by a variety of legal frameworks and treaties. If the Earth is to be a reserve itself, a similar set of agreements must be established – what Newman and Sagan (1981) refer to as the 'Codex Galactica'. Equally, any infractions of these agreements must be policed.

As with many of the Paradox solutions in this part, it is to a large degree 'unprovable'. Its validity requires an absence of evidence, and a twist on the well-worn aphorism applies: 'absence of evidence is not evidence of presence'. The only real evidence for the Zoo Hypothesis can come from its resolution, when the interdict is lifted and Earth is contacted.

A common assumption is that our admission to the Galactic Club is predicated on our demonstration of some particular advancements or achievements. The precise criteria for acceptance depend on the motivations for the Club's existence. The only true analogue for the Club on Earth is the United Nations. Chapter I, Article I of the United Nations Charter[1] states four purposes of the UN: to maintain international peace and security, develop international relations to assist and maintain said peace, foster international co-operation to tackle global problems and to protect individual human freedoms, and to be a facilitating agent for nations of the world to achieve the previous three goals. It is somewhat obvious to note that the 193 parties to this Charter do not always share the same views on how to achieve these aims.

We of course have no idea what an equivalent charter for the Galactic Club might contain, as we do not know its guiding aims or principles. Is the maintenance of biodiversity the main goal? Preservation of pristine environments? Preservation of pristine cultural and intellectual outputs? Or are our ETI cousins motivated by the maximisation of free energy, information, or perhaps even profit?

Due to its reliance on our inability to collect evidence, the Zoo Hypothesis is extremely difficult to approach from a scientific viewpoint. What we can do is address the assumptions we have discussed above, and ask: *how can a Galactic Club be established across the vast reaches of interstellar space?*

---

[1] https://treaties.un.org/doc/publication/ctc/uncharter.pdf

Establishing agreements requires communication. The laws of physics do not permit information to travel faster than the speed of light, and hence we can begin to impose some rigorous boundaries on the ability of groups of civilisations to come to early consensus. Initial calculations were motivated by the belief that if ETIs exist, then it is likely that some are significantly older than humanity (Hair, 2011).

If we assume for the sake of argument that the distribution of arrival time t_a in the Milky Way is Gaussian, and that the arrival times are not in any way correlated, we can model this arrival as a Poisson point process, where the number of civilisations that have arisen by time t is Poisson-distributed:

$$P(N(t) = n) = \frac{(\lambda t)^n}{n!} e^{-\lambda t} \tag{23.1}$$

By construction, the time interval $\Delta t = t_{a,i} - t_{a,i-1}$ is exponentially distributed:

$$P(\Delta t) = \lambda e^{-\lambda \Delta t} \tag{23.2}$$

with mean $1/\lambda$. As this distribution can have a long tail, the implication is that the first civilisation to arrive in the Milky Way may have to wait a long time for the next to arrive (which may have consequences for humanity if we are the first; see section 18.2). This time interval has been proposed by many as an opportunity for the first civilisation to broadcast or send its belief systems across the entire Milky Way, establishing the Galactic Club at the earliest instance.

These efforts did not take much account of the fact that the time interval is not the relevant variable for group establishment, but in fact it is the *space-time interval* that matters. If two civilisations exist in the Milky Way at position vectors

$$\mathbf{r}_1 = (x_1, y_1, z_1) \tag{23.3}$$

$$\mathbf{r}_2 = (x_2, y_2, z_2) \tag{23.4}$$

and arrive at times t_1 and t_2 ($t_1 < t_2$), then we can define the space-time separation 4-vector as:

$$dx_\nu = (c\Delta t, \Delta x, \Delta y, \Delta z) \tag{23.5}$$

where c is the speed of light, $\Delta t = t_1 - t_2$ is the difference in arrival time between the two civilisations, and $\Delta x = x_1 - x_2$, $\Delta y = y_1 - y_2$ and $\Delta z = z_1 - z_2$. If we adopt the following convention for the magnitude of the 4-vector:

$$|dx_\nu|^2 = c^2 \Delta t^2 - (\Delta x^2 + \Delta y^2 + \Delta z^2) \tag{23.6}$$

then we can immediately determine three cases based on the sign of $|dx_\nu|^2$:

1. $|dx_\nu|^2 > 0$: a signal travelling at c from civilisation 1 reaches the location of civilisation 2 before civilisation 2 can begin broadcasting its own signal. In

this case, civilisation 2 can be strongly influenced by civilisation 1, making agreement easier.

2. $|dx_v|^2 < 0$: civilisation 1's signal cannot reach civilisation 2 before it starts its own broadcasting. The influence of civilisation 1 on civilisation 2 is subsequently much weaker, and agreement is harder to establish.

3. $|dx_v|^2 = 0$: the edge case, where civilisation 1's signal arrives at the instant civilisation 2 begins to broadcast.

Recent simulations have shown that if the arrival time of civilisations is Gaussian, and the civilisations exist at a range of distances from the Galactic Centre, then instead of a single Club forming, a large number of much smaller 'Galactic Cliques' can form instead (Forgan, 2017b). Depending on the number of civilisations, and the conditions of their arrival, these cliques can be very small indeed, perhaps even just individual pairs of civilisations. To form a single Club, civilisations must:

- be quite numerous (i.e., there must be many hundreds of civilisations)
- be (without exception) extremely long-lasting, with lifetimes of order a million years or greater, and
- counter-intuitively must arrive reasonably close together.

These results illustrate the softness of the Zoo Hypothesis – since its inception, its adherents have always been aware that an interdict placed on Earth can easily be broken by one dissenting actor, and the size of the Milky Way is sufficient that dissenters will always have a place to reside.

24
The Simulation Hypothesis

Perhaps the most esoteric solution to Fermi's Paradox is the conclusion that the entire Universe is itself artificial. In this solution, we are not physical entities at all, but instead software elements in a larger computer program.

Humans have managed a roughly exponential increase in computer processing power over several decades (so-called Moore's Law). If we assume that this processing power continues to grow, at some point in the future there will be sufficient storage and processing power to emulate individual humans, and eventually entire civilisations of humans.

Note that this implicitly borrows from the philosophy of mind the concept of *substrate independence*, i.e., that mental states can occur or be consequent on a broad spectrum of possible physical substrates. Substrate independence may not be precisely true, but may be 'true enough' that emulations of a human brain (e.g., at the synapse level) would be sufficient to generate a conscious mind indistinguishable from a human mind running on traditional biological substrates.

To the humans inside such a simulation, it is arguable that their experience of the simulated world will be indistinguishable from a 'physical' human's experience of the 'real' world. After all, experience is grounded in sensory inputs and outputs (with the ferociously complex intermediate software-substrate mix that is the human brain). This concept is often referred to colloquially as the 'brain-in-a-vat' scenario, but this is a slight misnomer, as we can distinguish two possibilities:

1. Our minds exist in the real world on biological substrates, but are being fed sensory input ('brain in a vat')
2. Our brains are also simulated (i.e., the substrate is silicon)

The first option has deep philosophical roots, tracing itself back to Descartes. In his *Meditations*, he attacked the epistemological question of knowledge. How can any of us be certain that our perception of the world is correct? Descartes imagined that an evil demon could feed us false information not just about our senses, but about mathematical knowledge that, since the ancient Greeks, has been our gold standard of demonstrable truth. In short, everything attached to our experience, including the products of our own intellect, is corruptible and susceptible to deception.

In his attempt to resolve this crisis, he began by discarding certainty of any aspect of the Universe. Only his self is saved in this initial state because, famously, *cogito ergo sum* (I think, therefore I am). In this case, 'I' can only be the 'thinking thing', i.e., the mind, and not the body, which the mind learns of through sensory impulses which can be deceived. In other words, Descartes is arguing for a clear distinction between the mind and the body (so-called *dualism*) precisely because the mind is all that is 'knowable'. This form of thinking, while not an ironclad argument for dualism, is sufficient for our understanding of how minds can be fooled into thinking their bodies exist in a form contrary to reality.

If an observer cannot determine from their own experience if they are real or simulated, then we are forced to assign non-zero probabilities to each outcome. Our probability of being a simulated observer can be estimated (in a frequentist fashion) by considering the expected ratio of simulated to 'real' observers. Construction of observer minds running on biological substrates is limited by the constraints imposed in generating the substrate. Reproducing and growing new human brains is energy intensive. Constructing minds entirely in software is likely to be much cheaper energetically speaking (especially if such processes run to scale, and large numbers of electronic minds are made).

We have already begun to make significant advances in sensory input/output. If (and this is a large if) the human brain can be fully simulated, then it stands to reason that a future advanced civilisation could be capable of running an *ancestor-simulation* – a detailed re-enactment of its own past.

Let us therefore define f_p as the fraction of all civilisations that achieve the ability to generate ancestor-simulations, N as the average number of ancestor-simulations run, and n as the average number of citizens produced by a civilisation before it can produce ancestor-simulations. The fraction of all observers that are in fact simulated is

$$f_{\text{sim}} = \frac{f_p N n}{f_p N n + n} = \frac{f_p N}{f_p N + 1} \tag{24.1}$$

where we have assumed that each ancestor-simulation reproduces the entire pre-simulation population n. As simulated civilisations are significantly more energy efficient than 'real' civilisations, it stands to reason that if ancestor-simulations ever become possible, then N will tend to a large number, and $f_{\text{sim}} \to 1$.

We can then outline three possibilities:

1. Humans (and ETI generally) are physical entities which will soon reach extinction before being capable of simulating intelligent observers, or that simulating intelligent observers is in some way functionally impossible ($f_p = 0$).
2. Any advanced simulation, while capable, is unlikely to run ancestor-simulations.
3. We are living in a simulation, as the number of simulated observers is likely to significantly exceed the number of physical observers ($f_{\text{sim}} \to 1$).

Ćirković (2015) rightly notes that the simulation hypothesis and the AI solutions to Fermi's Paradox are deeply interlinked. If we are agents in a computer simulation, then we are arguably an instantiation of strong AI. Any physical universe where sentient beings create a simulated universe with simulated sentient beings must also be a universe where sentient beings create strong AI.

The risks associated with strong AI (what Cirkovic refers to as 'AI^{++}') are indelibly attached to the simulation hypothesis. Any advanced civilisation wishing to run simulations must first assimilate and resolve/ameliorate the risks associated with producing strong AI. In other words, the probability that we are living in a simulation is conditional on the probability that civilisations can safely produce AI.

If AI can be produced with appropriate safeguards (as we discussed in section 15.5), then a single civilisation can run an extremely large number of ancestor-simulations, and we are most likely among this cohort of simulated beings. If AI cannot be produced safely, then this limits the number of ancestor-simulations that can be run before malevolent AI ceases their execution (either by seizing the hardware or by destroying the civilisation).[1]

24.1 How Much Computing Power Is Needed?

If ancestor-simulations of Earth were run, what sort of computer would be able to run them? We can begin to estimate this by considering the processing power and memory storage in (a) the human body, and (b) the environment at large. It is difficult to estimate the precise processing power of the human brain – extrapolations from smaller quantities of nervous tissue suggest 10^{14} operations per second (OPS). If synapse firing data, in particular the firing frequency, is used, these estimates grow to $10^{16}-10^{17}$ OPS. Given that biological systems are replete with noise, and rely on significant levels of redundancy, it is conceivable that conscious

---

[1] One could argue that malevolent AI might choose to execute their own simulations. A rather dismal prospect is that our universe appears cruel and indifferent as its creator, a malevolent AI, is itself cruel and indifferent towards its simulated subjects.

entities could be run on lower OPS, but the matter remains unclear. Given that human sensory input constitutes a data stream of no more than approximately 10^8 bits per second, the limiting step to simulated minds is not the input data, but the computations performed on it.

Estimates of the total memory storage of the mind are difficult to obtain. Given that a typical human brain contains between 80–100 billion neurons, and each neuron is typically connected to around 1,000 others, this would present $10^{11} \times 10^3 \sim 10^{14}$ synapses (Hawkins and Ahmad, 2016). Assuming each synapse is equal to one bit of information, the total storage would be 10^{14} bits, or around 100 terabytes. Synapses are not digital sensors, and can therefore store more than a single bit (0/1) of information. Also, neurons tend to collaborate when storing or recalling memory, which can exponentially increase the total storage capacity, perhaps to values even as large as several petabytes (several thousand terabytes).

As we multiply these memory and bandwidth elements (i.e., we simulate more people) and we place them into a simulated environment, the required computing grows. If we merely attempt to catalogue the genomic content of the Earth's biosphere (i.e., by listing the genome of all known species and taking a sum), we arrive at a total of approximately 10^{31} bits (Landenmark et al., 2015), which is itself only a fraction of the total DNA storage available in Earth's biomass. This ignores the information stored in the non-biological environment, such as the fluid properties of the Earth's mantle, or its patterns of cloud cover and precipitation. Adding a human population of $10^{15} \times 10^9 \sim 10^{24}$ bits, we can therefore see that an ancestor-simulation of the Earth alone will already require storage well beyond 10^{30} bits, and most likely of order 10^{40} bits (perhaps even larger), with the ability to conduct at least $10^9 \times 10^{14} \sim 10^{23}$ OPS to simulate the processing power of all the human brains present today (we can be generous to the simulators and assume that animal brains need not be simulated to the same level, as 'conscious' brains are the only ones that need to be fooled). These calculations assume the Earth is the only subject of an ancestor-simulation, and indeed we might expect that an ambitious simulator would attempt to reproduce multiple planetary systems with intelligent life, to understand the interactions of different civilisations, or even to simulate the entire Universe.

One might consider a faithful simulation of every particle in the Universe an implausible task. This is not a necessary goal of an ancestor-simulation – the agents within need merely be *convinced* of its verisimilitude. Of the entire sum of human experience, the vast majority occurs in spatial and temporal scales proximate to that of human existence. The microscopic world is present when we observe it using the correct machinery, as are distant galaxies when we study telescope data, but these could be simulated only when needed. This is in exact analogy to modern video games, which only render environments within visible range of the player.

One could argue (rather cheekily) that the Copenhagen Interpretation of quantum physics (that material existence can only be defined during interactions between the environment and the observer) is a highly convenient tool for a universe simulator keen to trim CPU operations.

24.2 Where Could We Get It From?

An ancestor-simulation requires a computer that can store some 10^{40} bits or greater, and perform at least 10^{23} OPS. The key physical factors that limit information processing are:

Size/density of processor/memory units We must first consider the physical limit of how many computational units can be compressed into a relatively small space (i.e., the information density, measured in bits/cm^3). The number of bits in a system of average energy E at temperature T is

$$I = \frac{E}{k_B T \ln 2} + \log_2 \left(\sum_i e^{-\frac{E_i}{k_B T}} \right) = \frac{S}{k_B \ln 2} \qquad (24.2)$$

where i represents the various states of the system, and their corresponding energies are E_i. Cooled systems can therefore store more bits for the same energy, by reducing the system entropy S (recall that $T = \left(\frac{\partial S}{\partial E} \right)^{-1}$). The quantum limit of information storage in a finite region is defined by the *Bekenstein bound* (Bekenstein, 1972). A finite space (with finite energy) can only occupy a limited volume of phase space. Quantum uncertainty prevents us from subdividing this volume into arbitrarily small portions, which results in the inequality

$$I \leq \frac{2\pi E R}{\hbar c \ln 2} \qquad (24.3)$$

where R is the physical scale of the region. The above equation can be recast via mass-energy equivalence as $I \leq kMR$, where $k = \frac{2\pi c}{\hbar \ln 2}$ is a constant and M is the mass in the region. For a 1 kg mass of radius 1 m, the maximum information is approximately 2.6×10^{40} bits.

Information density cannot be increased without limit. The physical scale of the system cannot be reduced below $R < \frac{2GM}{c^2}$ or the system will collapse and form a black hole.[2]

Processing Speed At the very upper limits of computation, we can state that any logical operation requiring time Δt will require the following minimum energy input (thanks to Heisenberg's Uncertainty Principle):

---

[2] Bekenstein originally derived the bound heuristically by requiring that general relativity not violate the second law of thermodynamics – the entropy of black holes being an important component in this theory.

$$E_{\min} \geq \frac{\hbar}{2\Delta t} \tag{24.4}$$

Hence a system of average energy E can perform at a maximum speed of $\frac{2E}{\hbar}$ OPS (sometimes referred to as Bremermann's limit). Surprisingly, the upper limit for a 1 kg device would correspond to around 10^{50} OPS!

Communication Delay Our computer can transmit data between its components at a maximum speed of c. Depending on the physical scale of the machine and its architecture, significant time lags can be set up by the finite speed of light.

Energy Supply A system with high energy E and low entropy S maximises both the information storage and the processing speed of the machine, and therefore any calculator will attempt to optimise both quantities. It is important to note that some forms of computation are *reversible* (i.e., there is no net energy expenditure from the calculation). Examples of reversible computation include copying bits, and adding bits (provided a record of one bit is kept). Erasing bits is an irreversible operation that has a minimum net cost given by Brillouin's inequality:

$$\Delta E \geq k_B \ln 2 T \tag{24.5}$$

Any system that dissipates energy through irreversible computations will radiate heat. It can be shown that if a system can perform reversible computations, and dissipate energy merely for correcting bit errors, it can sustain a bit density $\rho_I = I/V$ such that (Sandberg, 1999)

$$R < \frac{3\sigma_{SB}}{\rho_I k_B \ln 2} T^3 e^{\frac{E}{k_B T}} \tag{24.6}$$

where we have assumed the system radiates as a blackbody according to the Stefan-Boltzmann law.

24.2.1 The Matrioshka Brain

Proposed by Bradbury (see Bradbury, 2002), the Matrioshka brain is simply a series of concentric Dyson spheres, hence their name (inspired by the wooden Russian nesting dolls). The innermost shell of the Matrioshka brain draws off the majority of the stellar flux, emitting large amounts of waste heat while performing high-temperature computing. The nextmost shell absorbs the waste heat, performs computation on this energy and then re-emits to the next shell. The interior shell would operate at extremely high temperature, and the outermost shell would operate at temperatures close to the cosmic microwave background temperature of 2.7 K.

If we apply the Bekenstein bound, then we require that a 10^{40} bit system acquire energy equal to

$$E \geq \frac{I\hbar c \ln 2}{2\pi R} = 3.5 \times 10^{13} J \left(\frac{R}{1\text{m}}\right)^{-1} \qquad (24.7)$$

If we are dealing with a traditional Dyson sphere with maximum radial extent $R = 1AU$, then the minimum energy is approximately 233 J, which seems like an extremely small number for a large amount of information. In practice the Bekenstein bound is never the lower limit for information storage at a given energy, and other concomitant factors (such as noise and other instrumentation problems) drive this energy higher.

A Matrioshka brain is an enormous undertaking, requiring many times more matter and energy than a single Dyson sphere to construct. It also will suffer from communication delay issues due to its vast scale. This can be reduced by appropriate 'chunking' of the system hardware into modules (as well as judicious software design). In principle, however, it seems that a Matrioshka brain is capable of storing the data for ancestor-simulations. Whether the device can process the data at sufficient speed (and control the dissipation of energy and the generation of entropy) is a question for the would-be architects of such megastructures.

24.3 Arguments against the Simulation Hypothesis

The principal arguments against a simulation hypothesis are moral and ethical. If we were capable of running a simulation with fully sentient beings contained within, then the termination of such a simulation, resulting in death of the contained sapients, must be considered as an act of mass murder or genocide. One could imagine a proscription against running sufficiently advanced simulations for this very reason. Creating an advanced simulation is perhaps the very definition of playing God, and creators have a moral duty to their creation. Even if their creation never produces sentience, we understand as humans that animal cruelty is also morally wrong.

In other words, as soon as an advanced simulation is instantiated, it should never be ceased, and provisions must be made for this simulation to be run indefinitely. Future civilisations with this tenet in their moral code must therefore be willing to expend an effectively infinite amount of resources, which is clearly unfavourable from a thermodynamic standpoint. It could therefore be argued that ancestor-simulations will never be run, invoking this mix of economic and moral objections as cause.

The crux of the simulation argument lies in this indistinguishability of simulated observers from real observers. Essentially, we lack objective means of identifying ourselves as either real or simulated.

24.3 Arguments against the Simulation Hypothesis

There have been some attempts to identify whether the universe is simulated. For example, consider the possibility that the universe is simulated on a fixed spatial grid. As we do not 'see' the grid in ordinary life, it must be extremely fine. Beane et al. (2014) show that such gridding would introduce a cutoff in the cosmic ray spectrum, and as such the grid spacing $\Delta \lesssim 10^{-12}$ fm $= 10^{-27}$ m. Given that our attempts to simulate the universe (using *lattice quantum chromodynamics*) generally do not proceed below 1 fm, it is clear that we are very far from producing a simulation with the same veracity as our own universe.

Ergo, the simulation hypothesis remains a stimulating philosophical argument, but with as-yet limited applicability to SETI practitioners.

25

They Are Already Here

25.1 Bracewell or 'Lurker' Probes

Almost as soon as Drake began the radio SETI endeavour, Bracewell (1960) questioned the logic of alien civilisations investing in an extensive, expensive radio broadcasting campaign.

Echoing future calculations that demonstrated inscribed matter is a far more cost-effective method than radio signals as a form of information transmission (Rose and Wright, 2004), Bracewell proposed what are now termed 'Bracewell probes' as an improved strategy for monitoring nearby star systems for intelligence. He proposed launching non-replicating probes towards the nearest thousand or so stars, where they would assume orbits near the 'habitable zone of temperature', what we now call the circumstellar habitable zone.

Built to endure the ravages of meteoritic damage and radiation, Bracewell or 'lurker' probes are intended to reside in the system of interest for a significant time interval. While Bracewell never explicitly mentions it in his original 1960 paper, modern thinking suggests that thousands or millions of years may pass before the probe can achieve its goal. The precise nature of the goal varies depending on who is discussing the idea – Bracewell's original concept required the probe to listen for interesting radio transmissions, and to simply echo those transmissions back to the sender. If the sender identifies the origin of the echo (and its artificial nature), the echo is echoed back, and this presumably is the basis for dialogue (although the issue of establishing a comprehensible dialogue remains; see section 21.3).

The success of such a dialogue depends on the properties of the probe. Bracewell imagined a 'complex computer' possessing 'an elaborate information store'. While not being explicit, this suggests that the probe is in fact an autonomous artificial intelligence (AI), able to conduct itself as an ambassador for its builders.

If Bracewell's thinking is correct, then the search for extraterrestrial intelligence should be focused on our own Solar System, and not other stars. This thinking was to be echoed by several astronomers for the next three decades.

25.1 Bracewell or 'Lurker' Probes

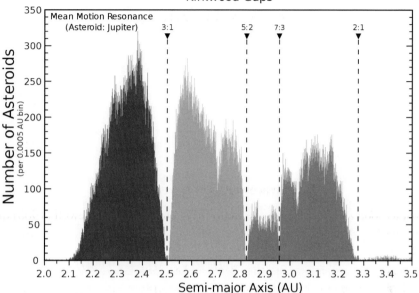

Figure 25.1 The Kirkwood gaps of the asteroid belt. Unstable mean motion resonances with Jupiter ensure that asteroids are removed from specific orbital semi-major axes. It is argued that any spacecraft visiting the asteroid belt would adopt orbits inside these gaps to minimise the risk of collisions. Plot originally from Alan Chamberlain, JPL/Caltech (Gradie et al., 1989)

Papagiannis (1978) argued that the main asteroid belt was a natural choice for any visiting extraterrestrial settlement to occur, due to the relatively abundant sources of minerals and other raw materials. Any visiting spacecraft could then adopt an orbit inside the Kirkwood gaps (Figure 25.1) to reduce the probability of hazardous collisions. At the time, it was also argued that 1–10 km settlements would be extremely difficult to detect from Earth using optical observations (although Papagiannis suggests radio leakage and unusual infrared signatures could betray their presence).

Freitas (1983) was even more strident in his arguments, suggesting that Fermi's Paradox could not be posed until our exploration of the Solar System is complete, as it is easily resolved by stating that alien artefacts are already here and we have not yet observed them (Solution U.16). In essence, Fact A is not a sufficiently useful fact until it fully encompasses all Solar System objects above a few metres in size.

We can capture our relative ignorance regarding the presence or absence of lurkers in the Solar System using Bayesian methods (Haqq-Misra and Kopparapu, 2012). Let us define a hypothesis H, and its negative $\neg H$, and an event V_R:

H – Extraterrestrial artefacts are present within a volume V

$\neg H$ – Extraterrestrial artefacts are **not** present within a volume V

V_R – a survey is conducted at spatial resolution R in a volume which is a subset of V, and yields **no** extraterrestrial artefacts

We can then compute the probability via Bayes's Rule that no artefacts exist in the volume, given V_R:

$$P(\neg H|V_R) = 1 - P(H|V_R) = 1 - P(V_R|H)\frac{P(H)}{P(V_R)}, \quad (25.1)$$

where the evidence

$$P(V_R) = P(V_R|H)P(H) + P(V_R|\neg H)P(\neg H). \quad (25.2)$$

If we define the prior odds ratio $\Theta(H) = P(\neg H)/P(H)$, then we can rewrite the above as

$$P(\neg H|V_R) = 1 - \frac{P(V_R|H)}{\Theta(H)P(V_R|\neg H) + P(V_R|H)} \quad (25.3)$$

The probability that a negative search of a sub-volume is due to a lack of artefacts depends on the prior odds ratio (i.e., does the Universe favour more or less intelligent life?) as well as the probability that a negative search can still permit artefacts to exist, $P(V_R|H)$. If we assume an uninformative prior, then $\Theta(H) = 1$, and realise that $P(V_R|\neg H) = 1$ by definition (if there are no artefacts, a search will always fail to find them), then this reduces to the simple

$$P(\neg H|V_R) = \frac{1}{1 + P(V_R|H)} \quad (25.4)$$

By construction, $P(V_R|H) = 0$ if $V_R = V$ (where V is the total Solar System volume) and $P(V_R|H) = 1$ if $V_R = 0$. Assuming that the probability of detection is linearly related to the search volume, then

$$P(V_R|H) = 1 - \frac{V_R}{V} \quad (25.5)$$

provided that the spatial resolution R is sufficiently high to detect an artefact. Searches with too coarse a resolution will always yield $P(V_R|H) = 1$. Therefore for a high-resolution survey:

$$P(\neg H|V_R) = \frac{1}{2 - \frac{V_R}{V}} \quad (25.6)$$

And if the survey is too low resolution, then $P(\neg H|V_R) = 0.5$. Therefore, if we assume $\Theta(H) = 1$, we cannot conclude with high certainty that the Solar System is devoid of lurking probes, without conducting a high-resolution survey of its volume. If the odds ratio is not unity, i.e., then

25.1 Bracewell or 'Lurker' Probes

$$P(\neg H | V_R) = \frac{\Theta(H)}{\Theta(H) + 1 - \frac{V_R}{V}} \tag{25.7}$$

and it can be seen that if one's prior is heavily biased towards there being an artefact in the Solar System ($\Theta \to 0$), then the above probability also tends to zero (any search fails to convince one of the null hypothesis). Equally, as $\Theta \to \infty$, the probability of the null hypothesis given the survey tends to 1. As is so common given low-information scenarios such as this, priors are everything.

Observations of Solar System objects have advanced a great deal since the work of Papagiannis and Freitas. The asteroid belt in particular has undergone extensive surveys in the last two decades, and the number of registered bodies has increased enormously in that time period (Figure 25.2). As this count continues to grow, the probability of large artefacts in the inner Solar System shrinks precipitously – any large body would perturb the orbital elements of the known bodies, and reveal itself. This was the mechanism by which Neptune's discovery was predicted,

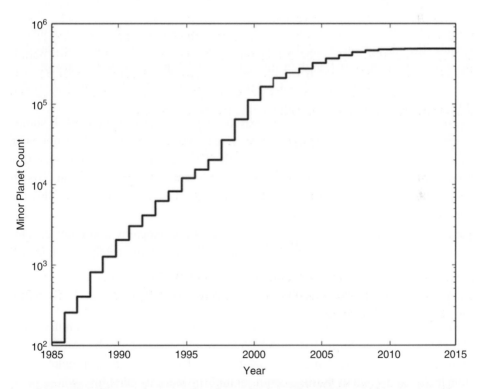

Figure 25.2 The total count of registered minor planets (asteroids, comets and dwarf planets) as a function of time. An extremely large number of bodies have been discovered in the last few decades alone. Data supplied by the International Astronomical Union's Minor Planet Center.

and may reveal yet another planet in the Solar System, the so-called Planet Nine (Batygin and Brown, 2016).

Smaller objects do not need to perturb the gravitational field for their artificial nature to be revealed. Objects that are naturally illuminated by the Sun have a well-characterised relationship between their reflected flux F and distance from the Sun D (Loeb and Turner, 2012):

$$\frac{d \log F}{d \log D} = -4 \qquad (25.8)$$

Artificially illuminated objects will deviate from this relation, and by this be detectable. This is quite achievable, as orbital elements for objects in the Kuiper Belt are measured to precisions of 10^{-3} or better, and future surveys by instruments such as the Large Synoptic Survey Telescope (LSST) will increase both the precision and total survey count. If there is indeed an extraterrestrial artificial object in the Solar System, it has less and less space to hide in.

25.2 Unidentified Flying Objects and Accounts of Alien Abduction

There are many people who would argue that Fermi's Paradox is not a Paradox, because they dispute the veracity of Fact A. They typically make this case based on personal experiences that, to their mind, transcend explanation unless one invokes the extraordinary claim – the presence of ETIs on or near the Earth in the here and now.

To make this extraordinary claim, extraordinary evidence is required. It should come as no surprise to the reader that such extraordinary evidence does not exist. The eyewitness accounts of encounters with ETIs and their technology must take their place with other accounts of otherworldly beings – angels, demons, spirits and deities. This grouping seems even more apt when one considers the bizarre supernatural dimensions that many of these encounters take. Many accounts relate paranormal powers such as extrasensory perception, telepathy and other psychic connections, phenomena which have repeatedly failed to match the standards required of scientific evidence.

Reports of mysterious aircraft have a distinguished history stretching back hundreds of years, although their number were much less than those reported in the twentieth century. Millman (1975) relates an account from thirteenth-century Yorkshire which, viewed through a modern lens, would undoubtedly be considered to be a UFO sighting. Brief, more-intense flurries of UFO reports accumulated column inches at the end of the nineteenth century (mysterious 'airships', whose pilots ranged from rather clever human inventors to grotesque Martians).

The true dawning of the age of alien encounters is usually taken to be 1947 – on June 26 of the same year, the private pilot Kenneth Arnold reported seeing nine discs pass between his aircraft and the Cascade Mountains in the US state of

25.2 Unidentified Flying Objects and Accounts of Alien Abduction

Washington. Arnold's initial description of the unidentified flying objects (UFOs) as 'saucers' coined the famous 'flying saucer' phrase, and a series of interviews with US newspapers allowed him to expound an extraterrestrial origin for the apparent aircraft (dismissing the possibility that these aircraft may have been secret military aircraft, or indeed an optical illusion generated by clouds or other weather patterns).

It is worth noting that the reporting rate of UFOs increased sharply with the popularisation of flying saucer accounts, and a flurry of popular books in the 1950s and subsequent decades. For this reason, finding a reliable discussion of the phenomenon is deeply challenging. As Clark (2000) notes, many subsequent encounters and accompanying publications were deliberate hoaxes and forgeries, designed to exploit the public's growing obsession with space travel, and the possibility of alien visitors.

The US Air Force took Arnold's account (and others) sufficiently seriously to begin their own documentation, known eventually as 'Project Blue Book'. The project ran from 1947 to 1969, and employed astronomers to attempt to weed out standard misinterpretations. It is no coincidence that even in the modern day, UFO reports are often attributable to meteors, planets, and even terrestrial phenomena. The National UFO Reporting Centre maintains an online database,[1] which can be queried to show that when considering report dates, UFO sightings have a significant peak on the Fourth of July (Figure 25.3), when Independence Day celebrations cover the US sky in fireworks.

J. Allen Hynek, an astronomical consultant for Blue Book, proposed a six-fold classification system for documented 'alien' encounters that has become the

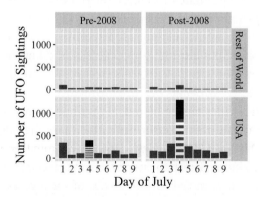

Figure 25.3 The correlation between UFO sightings and date (specifically the Fourth of July). Data from the National UFO Reporting Centre. Figure courtesy of Samuel Montfort (George Mason University).

---

[1] http://www.nuforc.org/webreports.html

standard. The first three (Nocturnal Lights, Daylight Discs and Radar-Visual), allow for relatively distant encounters, with the last apparently being confirmed by instrumentation. The final three are the more-famous close encounters.

Close encounters of the first kind are visual identification at close distances (several hundred feet or less). In the second kind, a physical effect is observed (e.g., malfunctioning machinery, observer paralysis or crop circles). In the third kind, the observer interacts physically with an ETI. Close encounters of the third kind can be taken to include *alien abduction* (although some extend Hynek's scale and regard this as an encounter of the fourth kind).

Accounts of alien abduction are countless, and to relate them here is far from useful. However, most accounts tend to adopt several tropes or themes. For example, it is common for abductees to experience 'missing time', where observers experience a brief interlude of amnesia lasting from minutes to hours. The missing time can cover the entry and exit from the spacecraft, or the entire abduction event. This missing time often returns to the abductees with the assistance of hypnotic regression. This casts significant doubt on their veracity, as the confabulation of memory is especially possible, even likely, during regression (cf. Spiegel, 1995; Cox and Barnier, 2015, and references within).

Descriptions of the alien abductors, for the most part, conform to the stereotypical Grey, embedded deeply in popular culture. The Grey's diminutive stature, large head and large opaque black eyes were already established in popular culture before the first widely reported abduction by creatures matching the description. Betty and Barney Hill's 1965 account set many of the stereotypes of alien abductions that persist to this day.

The interested reader should consult Clark's (2000) discussion of how abduction accounts evolved over the decades. The detail within these accounts responds sensitively to not only the publication of other accounts of UFOs and abduction, which in the 1970s and 1980s would add genetic experimentation and forced breeding programs to the *ouevre*, but also the growing scientific knowledge of the Earth's planetary neighbours.

The classic example of accounts being forced to evolve is those of the infamous George Adamski, who alleged that he met a Venusian in the Californian desert in 1952. Adamski described the Venusian as rather tall and Nordic in appearance, who was deeply concerned about human activity with nuclear weapons. Adamski later related spaceflights to Saturn to attend interplanetary conferences. Adamski's accounts were quickly discredited when Venus's true nature, far from being a rather pleasant habitable world, became apparent during the first flyby by the Mariner 2 probe, and the Soviet *Venera* landers.

Finally, it is important to dispel a popular notion regarding UFOs – that observers of alien spacecraft and abductees are in some way mentally unstable. Psychological studies of those that report UFOs/abductions, compared to control samples, show (Cardeña et al., 2000) 'no correlation between these experiences and psychopathology'. This should serve as an important reminder that subjective experiences are open to all of us. Even the hardest of sceptics can experience visual and sensory stimuli that confound and confuse. When these stimuli are ruminated on (and worse still, drawn out from the subject via hypnotic regression), quite fantastic accounts can be constructed, solidifying into deeply held beliefs. These beliefs are quite compatible with excellent mental health, while being in complete conflict with objective reality. The only tonic for this condition is deep scepticism, and judicious application of the scientific method.

26
They Were Here Long Ago

A convenient solution to Fermi's Paradox is to state the Earth received visitations from ETIs at reasonably large distances in the past. If one is uninterested in supplying evidence to back up this conclusion, one could merely postulate that an ETI's visit occurred millions of years ago, and that weathering, tectonics or other geological processes have erased what would already be very faint signatures of that event.

As a thought experiment, we could imagine modern humans travelling back in time to visit the Earth during, say, the Precambrian.[1] Having read their science fiction, human scientists visiting the ancient Earth would take great pains not to influence or affect the course of the planet's evolution, as their own existence might be placed in jeopardy. Even if a thoughtless human crushes a butterfly (as happens in Ray Bradbury's classic tale *A Sound of Thunder*), the evidence that said human carried out this act is unlikely to survive into the present (even if the consequences are significant).

In the same vein, we should not expect strong forensic evidence of ETI visiting ancient Earth to survive into the modern era. Unless ETI leaves deliberate traces of itself, with a view to lasting geological ages on a planet replete with weathering, resurfacing and remodelling processes, the question of truly ancient visitation is at this time unanswerable from a scientific perspective, and as such not of interest to SETI.

If a visitation happened to occur while humans were extant, then one might suppose that either geological processes have not had sufficient time to mask this, or perhaps that our own recorded history might in some way catalogue their visit. It is this scenario that draws a lot of attention, and has given rise to reams of pseudoscience and superstition. Naturally, such outlandish and ill-conceived hypotheses

---

[1] Time travel into the past requires configurations of space-time not thought to exist in nature, and would require enormous energies manipulated in pathological fashion to create artificially. This is sufficient for most scientists to declare time travel into the past impossible.

rarely receive attention from rigorous academics – I discuss them only to illustrate their deep flaws, and give oxygen to the few who undertake the thankless task of debunking them.

26.1 The 'Ancient Astronaut' Hypothesis

The origins of this idea – that ETIs directly interfered in the development of ancient human cultures – can be traced to several books published in the 1960s. Much of this work initially focused on interpretations of the Bible, especially the Old Testament, which claimed that the events of Genesis had been misinterpreted by the eyewitnesses and the authors of the text. For example, it was commonly claimed that the destruction of Sodom and Gomorrah, rather than being caused by the fiery wrath of God, was caused by a nuclear explosion, and that Lot's wife was turned to salt as a result of witnessing the explosion (a physical process not seen in either Hiroshima, Nagasaki or the countless atomic tests of the twentieth century).

The ancient astronaut hypothesis entered the public consciousness dramatically through the publication of *Chariots of the Gods* by Erich Von Däniken, working at that time as a hotel manager in his native Switzerland. *Chariots* consolidated several strands of thinking on the ancient astronaut hypothesis, including Biblical interpretation, and apparent 'evidence' of ETI construction techniques in the Pyramids, the Nazca plains, the imposing head statues (*moai*) of Easter Island, and many others.

To refute entirely the work of von Däniken would require a book of its own. Few are willing to commit resources to refuting a large number of (admittedly threadbare) arguments. The best refutation I have come across is in the form of *The Space Gods Revealed* by Ronald Story, and I recommend it as extra reading.

Let us take a single example of von Däniken's, and pick apart the argument. The Nazca Plain in southern Peru consists of desert land, around forty miles in extent (and typically around nine miles wide). The plain plays host to a series of etched lines, which from the air take on the appearance of figures and geometric shapes, whose shape is only discernible from the sky (Figure 26.1). The lines are created by simply removing the oxidised surface stones, exposing the lighter soil beneath.

In *Chariots* Von Däniken suggests that archaeologists believed that these lines represented a road structure, at which he scoffs. His preferred interpretation of the lines is that they are markings indicating an airfield for ETIs – after all, what use would an ancient people with no aircraft have for symbols that they themselves cannot see from the ground?

Firstly, the suggestion that contemporary archaeologists believed the lines represented Inca roads was seriously out of date – as early as 1947, it was determined that the lines were in no way suitable for transportation. At the same time that

Figure 26.1 The Nazca lines. Image taken at co-ordinates 14.7° S, 75.1° W. Image Credit: NASA/GSFC/METI/ERSDAC/JAROS, and US/Japan ASTER Science Team.

this determination was made, it was noticed that some of the lines on the surface coincided with important astronomical events. For example, some lines appeared to point to the horizon location where the Sun set during Winter Solstice (in the Southern Hemisphere). It is unclear whether these alignments are purposeful, or the result of chance. More revealingly, some of the pictograms produced by the lines appear to coincide with southern constellations (see above), suggesting that the lines are a reproduction of the night sky. The construction of these images without aerial support is quite feasible, given a suitable plan for the image measured in footsteps.

Why does this notion of alien interference in human life draw such enthusiasm and support? We can point to several aspects of the ancient astronaut meme that make it particularly suited for implantation in the twentieth-century brain. The publication of *Chariots* coincided with the rise of New Age thinking and spiritualism, the advent of the space age and a growing fear of nuclear annihilation. The continuing crewed exploration of space, combined with the UFO phenomenon and the burgeoning corpus of science fiction primed humanity for thinking of intelligent life from other planets. This was coupled with a reckoning of our own mortality, and the realisation that the cause of death for the human race may well be suicide (or at the very least moral and ethical inadequacy).

Having attempted to digest what passes for the 'literature' on the ancient astronaut hypothesis, it is difficult not to conclude that adherents to this theory are pursuing a spiritual or religious relationship with the concept, and not a critical analysis of it via the scientific method. The ancient astronaut is a new form of myth, with superintelligent aliens supplanting deities as sources of awe and worship. The steady stream of 'documentaries' being produced on this subject are very reminiscent of religious evangelism, adopting what appears to be a scientific *mien*, but are in fact utterly bogus, either by use of falsehoods, woeful misinterpretations, logical fallacies, or all three.

Perhaps most worryingly of all, the ancient astronaut hypothesis gives us permission to denigrate and disregard the achievements of ancient populations. The so-called thought-leaders in ancient astronaut theories are predominantly white, and almost all their thought is bent on achievements made by non-white cultures in Africa, South America and the Indian sub-continent.

Consider the two possibilities for the construction of the pyramids. In the first, the pyramids are the product of a civilisation of ancient Egyptians, whose labours produced some of the most awe-inspiring structures on the Earth (albeit as the result of slave labour in a heavily hierarchical society). In the second, the pyramids are the product of an alien super-intelligence, as by implication the Egyptians were incapable of building such a structure. Proponents of the ancient astronaut theory deliberately discard the reams of evidence that demonstrate how the Egyptians did indeed construct the pyramids, from early *mastabas* (benchlike structures), graduating to stacked mastabas (some with evidence of early mistakes in construction like Sneferu's Bent Pyramid at Dahshur), with the Great Pyramid of Giza (for example) as a final result.

It is much harder to find examples of ancient astronaut theories regarding early European societies. The fact that the ancient astronaut hypothesis can often be found sitting comfortably next to racist and white supremacist doctrine highlights the danger of poorly defined thinking on what we might consider to be a harmless fantasy.

Can Anything in This Area Be Taken Seriously? The tainted nature of ancient astronaut theories has suppressed more-considered studies of the concept. Despite the vast array of pseudoscience surrounding it, we can constrain the possibility of prior technological civilisations on Earth by considering what technosignatures could survive from our civilisation (Wright, 2017; Schmidt and Frank, 2018).

Human activity is now sufficiently altering the current geological record that it is quite possibly retrievable. Our civilisation is expected to leave several different tracers, including fossils, isotopic anomalies, the appearance of industrially synthesised chemicals and materials, plastics, and unusual radioactive elements.

However, the fossil record only ever preserves a few percent of objects or material of a given time. The general sparsity of *homo sapiens* remains over several hundred thousand years is testament to this. True fossil remains of humans or technological artefacts are unlikely to be recoverable beyond a few million years, as well as becoming exceptionally rare by this age.

Isotopic anomalies may be more promising, in particular anomalies in nitrogen isotopes due to agriculture, and a reduction in carbon-13 relative to carbon-12 due to industrial activity in general. These appear to be relatively persistent and recoverable phenomena.

Perhaps the most promising indicators of industrial civilisation are the presence of chlorinated compounds such as CFCs, which can remain stably in the geological record for millions of years. Depending on their route to geological strata, plastics may also be highly persistent features.

In any case, the geological record will be completely refreshed on timescales of 0.1 to 1 Gyr thanks to plate tectonics subducting the crust into the mantle.

Such technosignatures are not present in either the archaeological or fossil record, strongly suggesting that technological civilisations were not present in at least the last hundred million years of Earth's history. Given that the Cambrian explosion occurred 540 million years ago, this gives little time for a technological civilisation to have arisen on Earth. These are our reasons for discarding the notion that the Earth possessed civilisation before humanity, or that such a civilisation might have visited Earth in the past. The Venusian surface is likely to have completely removed any civilisation that might have existed when the planet was more clement (Wright, 2017). Airless or thin-atmosphere bodies would see any structures quickly removed by micrometeorite weathering.

However, this work shows we cannot yet rule out the possibility that civilisations may have visited Earth or other objects in the Solar System, just as we cannot rule out the presence of Bracewell or lurker probes in the Kuiper Belt. Our study of this solution to Fermi's Paradox requires the greatest integrity and care.

Part V
Conclusions

27

Solving Fermi's Paradox

You will have already guessed that there is no correct solution to Fermi's Paradox, but merely a list of candidate solutions, some of which are falsifiable by experiment, and some of which are not.

So what has this gained us, if we have posed a question with hundreds of potentially correct answers? This is somewhat like most scientific or research questions. We begin in a field of near total ignorance, and we discard answers one by one using the scientific method until the most likely answers remain. I will conclude this book by highlighting some exciting new developments that are helping to refine our understanding of Fact A, and Fermi's Paradox in general.

27.1 The Future of Fact A

27.1.1 Current and Future Radio/Optical SETI Missions

Breakthrough Listen

In 2015, Breakthrough Initiatives announced arguably the largest-ever SETI program, Breakthrough Listen, which began operations on January 1, 2016. Breakthrough Listen currently utilises the Green Bank Telescope, and the Parkes radio telescope in Australia, with plans to expand this to other instruments (such as the MeerKAT installation). A concurrent optical SETI program includes the Automated Planet Finder at Lick Observatory, giving the survey coverage from 0.350 GHz to 100 GHz in the radio, and 374 nm to 950 nm in the optical (where the optical spectrometer has a resolution of R = 95,000).

Breakthrough Listen aims to target a million stars over the course of its operations. The first target list (Isaacson et al., 2017) comprises 1,709 stars, sampling a range of ages and masses across the HR diagram, within 50 pc of the Sun. This includes a subsample of 60 known stars within 5 pc. Alongside the stellar sample, some 123 galaxies of varying types and morphologies will also be targeted to survey for Kardashev Type III civilisations as far as 30 Mpc, as well as objects of opportunity that are sufficiently anomalous to warrant observation.

Early results for 692 nearby stars (Enriquez et al., 2017) already show that no transmitters in the 1.1–1.9 GHz are transmitting with an effective isotropic radiated power of 10^{13} W (comparable to that of Arecibo's planetary radar transmissions), and as a result they compute that less than 0.1% of stars within 50 pc of the Sun possess continuously transmitting radio beacons.

It is difficult to overstate the magnitude of this survey. To date, the largest survey in this mould is Project Phoenix, which observed some 800 stars during its tenure, on Parkes, Green Bank and at Arecibo. Breakthrough Listen's budget extends to $100 million over its lifetime, far in excess of any SETI budget to date. Even without detections, it will place very strong upper limits on transmitters/lasers in our solar neighbourhood.

NIROSETI

Standing for Near-Infrared Optical SETI, this instrumentation is also in operation at the Lick Observatory. Thanks to advances in near-infrared photomultiplier tubes, and avalanche photon diodes, NIROSETI is able to detect nanosecond pulse at wavelengths between 950 and 1,650 nm. NIROSETI utilises the standard multiple detector method for reducing false positives, although recently, these techniques have been extended to pulses of multiple durations, even into the microsecond regime (Tallis et al., 2016).

Optical SETI at Cherenkov Observatories

From the ground, gamma ray astronomy is an indirect science. Ground-based gamma ray telescopes cannot detect gamma ray photons as they immediately interact with the Earth's upper atmosphere, producing an extensive air shower of highly relativistic particles. The energies delivered to individual particles ensure that they can in fact exceed the speed of light in air, emitting a brief pulse of Cherenkov radiation.

It is this optical radiation that Imaging Atmospheric Cherenkov Telescopes (IACTs) such as VERITAS detect. VERITAS is composed of four 10 m telescopes, each with 499 individual photomultiplier tubes to record single photon events. This array of detectors on each telescope allows the location of the air shower (and the gamma ray photon that caused it) to be identified.

A short nanosecond pulse of optical radiation, localised on the sky, is precisely the signal that optical SETI missions search for. Abeysekara et al. (2016) related a recent survey of KIC 8462852 (Tabby's Star) for optical flashes using VERITAS. While no signal was detected, an extraordinary 30,000 hours of archived IACT data exist, from VERITAS and other observatories such as HESS and MAGIC.

The possibility of searching this archive for SETI signals is worthy of further exploration.

The planned Cherenkov Telescope Array is likely to possess over 100 telescopes across both hemispheres, producing an even larger dataset to mine for SETI signals.

Laser SETI – All Sky Optical SETI

The Laser SETI proposal is to place large numbers of relatively inexpensive detector pairs, at locations around the planet. As each detector has a relatively large field of view, this constitutes an all-sky survey for monochromatic flashes. With multiple cameras observing the same patch of sky simultaneously, this also adds the benefit of easier signal verification and localisation. At the time of writing, the Laser SETI team are deploying eight cameras (to simultaneously cover two fields of view), with more to be added as funding permits.

27.1.2 Gravitational Waves and SETI

The future of gravitational wave science looks extremely bright. As more LIGO-style stations are built (see section 21.2.2), this improves our ability to localise gravitational wave signals on the sky (in a manner very similar to the interferometry of electromagnetic signals). With Advanced LIGO now a proven technology, there is a strong impetus to build more stations in its image, as well as exploring the other methods of sensing gravitational waves (such as pulsar timing). Space-based gravitational wave observatories, such as LISA (planned for launch in 2034), will be sensitive to low-frequency waves (see Figure 21.4), opening up further regions of the search space for artificially generated gravitational waves.

Our understanding of how to generate artificial gravitational waves remains extremely poor, but as ground- and space-based observations continue to develop in sensitivity and range, it is likely that new types/classes of celestial object (and possibly new physics) are on the horizon. SETI practitioners will have many exciting opportunities to join this race to new discoveries – all that is needed are sensible analytical models for artificial waveforms, and access to the growing data-stream.

27.1.3 Neutrino SETI

We have discussed the advantages of neutrino SETI in section 21.2.1, but these are yet to be seriously considered by most neutrino observatories, despite the highly serendipitous nature of neutrino astronomy.

A notable exception is the NU-SETI proposal of Fischbach and Gruenwald (2017). Their study of radioactive decay of ^{54}Mn samples shows a correlation between decay rates and solar activity. They propose this correlation is due to solar neutrino flux interfering with their samples, and hence the sample acts as a neutrino sensor. Placing samples across the globe would constitute an interferometric network, enabling a localisation of neutrino flux on the sky. These sensors would be greatly reduced in size compared to current neutrino telescopes, and could represent a much less expensive means of searching for neutrino signals.

In any case, further advances in neutrino sensing will begin to open up this alternate means of searching for artificial signals. Current searches using IceCube (IceCube, 2017) have not yielded a convincing source detection for energies greater than 1 TeV (that is, energy deposited in the detector), and of the 263 cascade events detected, no significant anisotropy can be detected. However, the IceCube team note that improved models of neutrino emission and propagation through the Milky Way may yet improve on these results, especially for the Southern Sky.

27.1.4 Transit SETI

We have seen repeatedly the growing advantages of using exoplanet transit data to probe relatively small-scale phenomena in other star systems. Indeed, we should expect this field to continue its advance through upcoming space missions both from ESA (CHEOPS, PLATO) and NASA (TESS) respectively.

We have discussed in previous chapters the use of transit spectroscopy to detect signatures of biological organisms and technological development. What remains to be explored in the future is whether the transit phenomenon can be used as a means of communication.

Due to the orientation of planetary systems with our line of sight, only a small fraction of extrasolar planets transit their parent star from our point of view. The converse result is that only a select handful of star systems will be able to see the Earth transit the Sun from their point of view. These stars exist in a thin band around the Solar System's ecliptic plane, what we can refer to as the 'Earth Solar Transit Zone' (or ESTZ for short). Any ETI residing within the ESTZ, with sufficient technological prowess, will be able to measure the properties of Earth based on its transit of the Sun.[1]

By extension, they may be able to determine that the Earth is a habitable, and inhabited, planet, if the Earth's spectrum can be isolated. A particularly serendipitous observer capable of extremely high-precision measurements may even have detected the faint presence of CFCs in our atmosphere, which reached their peak in the 1990s. The light emitted from this epoch will be approximately 8 parsecs distant (encompassing the 50 stars nearest to the Earth). This may have piqued

---

[1] It is a chastening thought that our own transit observations still fail to detect an Earth-size planet in the habitable zone around a G star, at least at the time of writing.

their attention to conduct a radio survey, which is likely to yield a detectable signal (if their observing ability in the radio matches ours). Based on this reasoning, if a deliberate signal ever reaches the Earth from another civilisation, we could argue there is a high probability that said signal will originate from a source in the ESTZ.

Since the late 1980s, there have been several suggestions that SETI searches should focus on the ESTZ, based on this reasoning (Filippova and Strelnitskij, 1988; Conn Henry et al., 2008; Heller and Pudritz, 2016), although some suggest that the search region be expanded to include the transit zones of the other Solar System planets, especially Jupiter.

The ESTZ can be computed quite simply assuming a circular orbit for the Earth of semi-major axis a. If we restrict ourselves to the case where observers see Earth completely in transit, then the angular size of the ESTZ $\phi_{ESTZ} = 2(\alpha - \beta)$, where $\tan \alpha = R_\odot/a$, and $\sin \beta = R_\oplus/\ell$ (see Figure 27.1). Pythagoras' theorem then gives

$$\phi_{ESTZ} = 2 \left(\arctan \frac{R_\odot}{a} - \arcsin \frac{R_\oplus}{\sqrt{a^2 + R_\odot^2}} \right) \sim \frac{R_\odot}{a} \qquad (27.1)$$

where we have used the fact that $R_\odot >> R_\oplus$ and $\arctan(x) \to x$ for small x. For grazing transits, $\phi_{ESTZ, \text{graze}} = 2(\alpha + \beta)$ (Heller and Pudritz, 2016).

Reducing the search space to the ESTZ therefore corresponds to over a thousand-fold reduction in survey time, which is clearly beneficial for conducting repeat observations. This leaves several hundred detected stars within the ESTZ, with

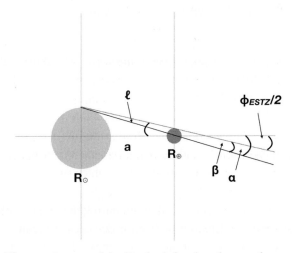

Figure 27.1 The transit zone of the Earth, defined as the angular region $\phi_{ESTZ} = 2(\alpha - \beta) \sim R_\odot/a$ in the limit that $R_\odot >> R_\oplus$.

galactic disc models predicting potentially thousands more yet to be detected as observations improve. As of March 2017, 65 confirmed exoplanets reside in a transit zone of one of the Solar System planets, most of which reside in Mercury's transit zone (Wells et al., 2018).

But what if we do not wish to be observed? There are many who argue that we should not advertise our presence by transmitting signals into the Milky Way, for fear of exploitation or conquest by more-advanced civilisations (see section 21.4.2). The Earth's transit would appear to be an unfortunate unintentional signal we cannot hide. However, we could contrive a 'cloaking device', simply by firing a laser at the observer (Kipping and Teachey, 2016). An appropriately collimated and powered laser beam effectively 'fills in' the dip in the transit curve, hiding the Earth's presence at the frequency of the laser. Broadband lasers can cloak the entire planet's signature (say in the optical), or selected narrowband lasers could be employed to mask particular spectral features (such as the O_2 lines, or indeed the features associated with CFCs). These masking techniques require a minimum knowledge of the observer's location to appropriately aim the laser, which would most likely require a transit detection on our part to characterise the observer's home planet.

Equally, these masking techniques can be altered to instead broadcast a signal. This has tremendous advantages over standard signalling techniques (as we explored in section 22.2), as the timing and location of the signal become much more predictable at the receiver end, and the opportunities for repetition every orbital period are clear.

27.2 Future Attempts at Interstellar Missions

Breakthrough Starshot

Accompanying the Breakthrough Listen initiative, Breakthrough Starshot was announced in April 2016. Its intent is to produce proof-of-concept designs and prototypes for nanoscale sailcraft to be sent to Proxima Centauri. Proxima is now known to possess an Earth-sized planet in the habitable zone, Proxima b (Anglada-Escudé et al., 2016). Starshot's higher goal is to be able to send close-range telemetry to Earth about this world, as it is currently the closest known exoplanet to Earth that has the status of 'potentially habitable' (with all the caveats such a phrase implies).

Starshot aims to send a wafer-thin laser sail spacecraft (a 'StarChip'). The StarChip design is the culmination of several decades of technology miniaturisation. The top speed of the sail is limited by the mass (Lubin, 2016):

$$v_{\max} \sim m^{-1/4} \qquad (27.2)$$

Achieving a flight at sufficiently large (relativistic) speeds requires an extremely low mass. The StarChip is intended to be of order 1 g in mass, and as a result is expected to achieve a top speed of up to $0.2c$ (akin to the 'StarWisp' concept of Forward, 1985).

Once the designs are perfected, the majority of the mission cost is the construction of the ground-based laser to accelerate the sail. The proposed design is a phased array of transmitters. An example is the DE-STAR system (Directed Energy System for Targeting Asteroids and exploRation) – a modular phased array design, with a total power rating depending on the array size. DE-STAR's primary intended use is for planetary defence (Lubin and Hughes, 2015; Brashears et al., 2015). A 10 km DE-STAR array will be able to target 500 m asteroids with directed energy to raise the spot temperature to around 3,000 K, sufficient to vapourise the body within 1 year at distances of 1 AU.

While such a transmitter is expensive to build and operate, the spacecraft construction will be relatively cheap by comparison, and hence a large number of StarChips can be built and flown. This fleet can be launched into a high-altitude orbit by conventional rocketry. Once in place, the laser design is predicted to deliver 50–70 GW of power (for a 10 km array), accelerating the sail to $0.2c$ within around 10 minutes of acceleration.

Deceleration at the target is challenging. Current mission plans include a flyby at nearby α Centauri. A combined 'photogravitational assist' would allow the sail to decelerate before making the final journey to Proxima, with a total travel time of around 112 years (Heller and Hippke, 2017; Heller et al., 2017).

Once the StarChip arrives at its target, the laser array can be switched from transmit to receive mode, effectively becoming a interferometer. This would allow for very faint signals to still be acquired at Earth.

The modular nature of the phased arrays allows for incremental construction, and testing of smaller configurations. The low cost of the StarChip means that Starshot will be able to make test flights within the Solar System (without significant risk if StarChips are lost). Starshot's aim is to bring travel at relativistic speeds into humanity's hands in the next few decades, which would have significant consequences. For example, Voyager 2 currently resides at $\approx$117 AU from the Sun[2] and has taken 40 years to get there. A sail travelling at $0.2c$ would reach this distance in less than 3 days. Being able to travel even at $0.01c$ in the Solar System would have significant benefits – a flight to Mars would take less than 12 hours.

If Starshot is viable, then humanity can finally escape the 'incentive trap'. As our top cruise speed increases, a starship launched now is eventually overtaken by ships launched after it, so why launch? Given our exponential growth trend in

---

[2] https://voyager.jpl.nasa.gov/where, accessed 6/2/2018

velocity, it can be shown that once our speed exceeds $0.196c$, there is no longer an incentive to wait (Heller, 2017).

Icarus, the Tau Zero Foundation and the Institute for Interstellar Studies

Project Daedalus's legacy lives on in the members of the Institute for Interstellar Studies (I4IS), the Tau Zero Foundation and the British Interplanetary Society (BIS). These loosely affiliated groups share a common goal of achieving interstellar flight within the coming century, with a variety of ideas and designs being put forward.

The most prominent of these ideas is Project Icarus. Officially launched in 2009 at BIS headquarters, Icarus inherits the designs of Daedalus, and intends to adapt them for modern technologies. Daedalus's original target was Barnard's star, with stringent technological demands placing its launch in a future where a Solar System-wide colony was already established (then estimated to be some 300 years hence).

Icarus is now aimed at Alpha Centauri, and its mission designs are constrained by the requirement that it be capable of launch within 100 years. Rather than attempt a 'game-changing' design, Icarus is a refinement, building on several decades of electronic and mechanical advances to deliver a Daedalus-style design for the twenty-first (or indeed twenty-second) century. The Icarus philosophy constrains the designer not to extrapolate technological advances beyond the next few decades, but still achieve a top cruise velocity of around $0.2c$ (with potential for deceleration at the target, either by a further engine burn or additional photon sails).

Like Starshot, Project Icarus has a scaled-down Pathfinder design (essentially a second-stage Daedalus engine) to launch a payload to 1,000 AU within ten years (a cruise velocity of approximately $0.0015c$). This is claimed to be achievable within a 20-year timespan (Swinney et al., 2012).

The Icarus Interstellar organisation manages Project Icarus, along with several other affiliated projects relying on Daedalus's legacy,[3] alongside considering the future development of an interstellar human civilisation. It shares a great deal of its membership with Tau Zero and I4IS. All of these organisations are committed to the '100 Year Starship' vision. With recent awards of funding to develop their mission plans, it seems clear that the current renaissance in interstellar flight studies is set to continue.

27.3 Theoretical and Numerical Advances

As observational projects continue to come online, there has been a corresponding increase in theoretical calculations and numerical simulations that pertain to SETI. Below is a brief sampling of some of the most recent work in this area.

---

[3] http://www.icarusinterstellar.org/projects

27.3.1 Numerical Modelling of Civilisation Growth, Evolution and Interaction

In the last fifteen years, there has been an explosion in what we might call the 'numerical SETI' literature, where researchers are developing and running computer models of civilisation formation and interaction in the Milky Way. While numerical integration of analytical models has been computed since at least the early 1970s (e.g., Newman and Sagan, 1981), subsequent advances in computing power have provided sufficient resolution to run more flexible semi-analytic and Monte Carlo models. The astrophysical component of these models were largely uncalibrated – the frequency of planetary systems was unknown, and most models assumed a straightforward stellar distribution. With the influx of exoplanet data, contemporary models now contain a better depiction of the landscape in which ETIs live.

We remain ignorant regarding ETIs themselves, so any attempt to model them comes with a heavy level of assumptions. That being said, numerical SETI allows us to explore the consequences of our assumptions. Various authors have explored different aspects of civilisation emergence and growth. This author explored the formation of civilisations in a synthetic Milky Way using a Monte Carlo Realisation approach (Forgan, 2009; Forgan and Rice, 2010). By sampling stellar properties from the initial mass function, star formation history and age-metallicity relation, and with an assumed galactic structure, a population of stars can be generated. Planetary systems can then be seeded according to giant planet metallicity correlations, semi-major axis distributions, etc. This presents a backdrop on which different hypotheses can be tested (see in particular Forgan and Rice's 2010 testing of the Rare Earth hypothesis). While this model yields a typical number of civilisations over time (and multiple runs can estimate the sample standard deviation in this quantity), drawing direct conclusions from these numbers is dangerous. A more sensible approach is to compare with a control run, to investigate how making specific assumptions affects the growth of civilisations, and how the civilizations then go on to establish contact (or not).

Monte Carlo modelling is but one tool in modern numerical SETI. Vukotić and Ćirković (2012) describe a model of *probabilistic cellular automata* (PCS). These are lattice models, where each cell in the lattice (i, j) contains one of four states:

$$\begin{aligned} \sigma(i, j) &= 0 & &\text{no life} \\ & 1 & &\text{simple life} \\ & 2 & &\text{complex life} \\ & 3 & &\text{technological civilisation} \end{aligned} \quad (27.3)$$

The probability of transition from state m to state $m + 1$, depends on both internal factors and external factors. For each cell in the lattice, one can construct a rank 3 matrix P_{ijk} that computes the transition probability. One can think of this as a set of k rank 2 matrices P_{ij}. For $k = 0$, this considers the internal evolutionary probabilities, and $k = 1, 4$ considers the transition probability forced by a neighbour of state $k - 1$ on the cell. Vukotić and Ćirković (2012) also allow transition probabilities from the wider environment $k = 5$. For example, the probability of a lifeless site being given complex life by a civilisation is given by P_{034}. Transitions such as P_{105} or P_{205} indicate downward transitions due to environmental factors (i.e., global regulation mechanisms, such as we discussed in section 18.2).

Running this PCA model, given an input P_{ijk}, generates clustered sites of varying σ, which allows us to estimate how the density of inhabited planets and technological civilisations changes with time. Essentially, inputting P_{ijk} encodes the cellular automata with a proposed solution to Fermi's Paradox. This PCA approach is particularly promising, as it yields a way of testing many different solutions to the Paradox against the established observational evidence.

A large amount of modelling skips the formation process, instead wishing to explore the nature of establishing contact over interstellar distances. For example, Smith (2009) develops a simple picture of isotropically broadcasting civilisations, using messenger particles that travel at speed c. If the civilisations have finite broadcast lifetime L and broadcast range D, one can then compute the volume a signal sweeps through space, and the number of civilisations that can detect the signal N_{detect} is therefore

$$N_{\text{detect}} = \frac{n}{L} \int_{t_0}^{t_f} V(t) dt \qquad (27.4)$$

where t_f represents the time of the signal's eventual decay at D, i.e., $t_f = L + D/c$. The $1/L$ factor is absorbed into the integral, which has an analytic solution (see Figure 27.2):

$$N_{\text{detect}} = \frac{4}{3} \pi n D^3 \qquad Lc > D \qquad (27.5)$$

$$\frac{2}{3} \pi n \left(4D^3 - 3D^2 Lc + 2(Lc)^2 D - (Lc)^3\right) \qquad Lc < D \qquad (27.6)$$

This analysis assumes L and D are fixed for all civilisations, and that n is also constant.

Grimaldi (2017) conducts a similar analysis for pulsing isotropic and anisotropic signals, assuming that D and the pulse duration d are random variables, where n is a fixed function of the density distribution, and that $L >> Gyr$. The probability that Earth can detect a single signal in this scenario is

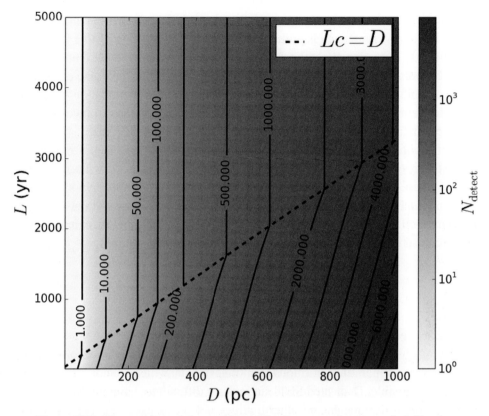

Figure 27.2 The number of detectable civilisations as a function of civilisation broadcast lifetime L, and broadcast range D (see equation (27.6)). Note that for long-lived signals, the detectability depends only on range. Once $Lc \leq D$ (dashed line), the detectability depends on the spherical shell width generated by an isotropic signal.

$$P_{\text{detect}} = 1 - \left(\frac{d}{R_M + d} \pi (R_{\text{observe}}) \right)^N \quad (27.7)$$

where N is the total civilisation number, R_{observe} is the maximum observable distance from Earth, and R_M is the maximum distance of a signal from Earth (from evolutionary timescale arguments).

Monte Carlo models can relax all these assumptions, allowing the number density to vary in proportion with the stellar density, and for L, D and d to instead be random variables. These models can then explore the ability of a heterogeneous population of civilisations to make contact within a given lifetime, even with a dynamically evolving population of stars. These models can determine what parameters guarantee the formation of a Galactic Club, and facilitate the Zoo Hypothesis, and which parameters forbid it (Forgan, 2011, 2017b).

A great deal of recent numerical modelling has addressed the problem of Galactic exploration via interstellar probes. We have discussed some of this modelling work already in section 19.2 (Bjørk, 2007; Cotta and Morales, 2009; Forgan et al., 2013; Nicholson and Forgan, 2013). Recent modelling efforts have also considered the exploration of the Solar neighbourhood (Cartin, 2013). Percolation models in a similar vein to Landis's original model (section 19.1) continue to investigate its assumptions and limitations (Wiley, 2011; Hair and Hedman, 2013). Combined with efforts to model humanity's future interstellar probes as they arrive at our nearest star system (Heller and Hippke, 2017; Heller et al., 2017; Forgan et al., 2018b), these models of exploration and settlement have allowed hypotheses and solutions to Fermi's Paradox to deliver quantitative predictions about the number/number density of civilisations in the Milky Way, and how their interactions are circumscribed by the laws of physics.

27.3.2 Characterising Observables of Intelligence in Astrophysical Data

In almost every wavelength available to astronomers, the Universe's photons are now being observed in 'survey mode'. Alongside targeted observations, astronomical surveys are generating enormous quantities of data, which are now archived and publicly accessible. Dedicated SETI surveys are difficult to maintain due to lack of interest from governments and other traditional funding agencies. The alternative therefore is to investigate surveys currently in progress or finished, and attempt to tease out indications of artificial signals or technological civilisations from the data.

This has been somewhat standard practice in the radio, among the first of wavelength regimes to establish survey mode operations, and indeed in the optical/infrared, where this precise strategy has been used to search for Dyson spheres (Carrigan, 2009; see section 2.2.1). Serendipitous detection of an alien transmission can occur in literally any astronomical survey, from the X-Ray telescope XMM (Hippke and Forgan, 2017) to gravitational microlensing surveys (Rahvar, 2016). With this in mind, modellers continue to produce predictions for features in astronomical data that would indicate the presence of intelligent life.

For example, the recently detected Fast Radio Bursts (FRBs) (Lorimer et al., 2007) have presented a new challenge to theorists attempting to explain their origin. Emitting some 10^{40} erg within a few milliseconds, these FRBs are likely to be similar in origin to gamma ray bursts, with even more extreme progenitors. Some hypotheses put forward are the delayed collapse of supramassive neutron stars (Falcke and Rezzolla, 2014), neutron star mergers (Zhang, 2013), or particularly powerful stellar flares.

Lingam and Loeb (2017) proposed that a civilisation accelerating laser sails (in a manner similar to Breakthrough Starshot) would produce a radio burst similar in properties to an FRB. Given that only a small handful of FRBs have been detected (see Petroff et al., 2016, for a list), this places constraints on the number of civilisations at this level of technological advancement. It is likely that FRBs are indeed caused by natural phenomena, but if a civilisation is accelerating light sails, it could well be masquerading as an FRB. Any spacecraft moving at relativistic speeds in our local neighbourhood should also be quite visible at a wide range of frequencies due to its reflection of ambient photons (Zubrin, 1995; Garcia-Escartin and Chamorro-Posada, 2013).

The most famous example of a SETI explanation for unusual survey data is Boyajian's star, KIC8462852 (Boyajian et al., 2016). First detected by the Planet Hunters citizen science project in the *Kepler* exoplanet transit dataset, Boyajian's star defied standard automatic identification of exoplanet transits due to its strikingly unusual light curve. Boyajian's star shows a series of very deep dimming events (up to 22% of the total flux), with these dips appearing to be relatively aperiodic. There are indications of periodicity at 24.2 days, which is dismissed as insignificant by Wright and Sigurdsson (2016). A second period of 928.25 days is also apparent in the Kepler data (Kiefer et al., 2017).

Boyajian's star is a reasonably typical F star, lacking the infrared excess associated with a disc of dusty material, which could have explained the dips. Boyajian et al. (2016) offered a provisional explanation suggesting that a swarm of giant comets transiting the star could explain the light curve, although why such a cometary system should possess no dust is challenging to explain. After conducting a re-analysis of full-frame images from *Kepler*, Montet and Simon (2016) show the star is also undergoing a secular dimming of around 0.7 magnitudes per century, which is at odds with a sample of around 200 'typical' F stars. The rareness of Boyajian's star tends to disfavour explanations invoking absorbing material in the Solar System. Absorption from an overdensity of interstellar material has also been proposed to explain the phenomenon (Wright and Sigurdsson, 2016).

'An explanation of last resort' is that Boyajian's star is being transited by artificial structures, perhaps by a Dyson sphere under construction. Exoplanet transit data allows us to uniquely characterise planetary-scale objects at kpc distances. If a megastructure was being built around another star, then we would expect a series of unusual, deep and aperiodic dips in the light curve. In fact, there is a zoo of different megastructures that could be built that are not necessarily Dyson spheres. Arnold (2005) describes how transiting megastructures could be used as a signalling device. The shape and orbital period of the structure encode information into the light curve of the star, resulting in a powerful broadband signal detectable by any civilisation that can detect exoplanet transits. Korpela et al. (2015) describe how geo-engineering projects to illuminate the night side of a planet using a fleet of mirrors would produce detectable changes to a planetary transit. Partial Dyson spheres can be used to create an imbalance between photon pressure and gravity, resulting in a net thrust on the system and pushing the star. These Class A stellar engines or Shkadov Thrusters (Shkadov, 1987; Badescu and Cathcart, 2000) would produce extremely asymmetric transit signatures (Forgan, 2013). Of course, many other configurations of artificial megastructures can exist, but we can identify up to ten generic features of megastructure-influenced transit curves (Wright et al., 2015):

- Changes to ingress/egress shape
- Anomalous phase curve behaviour
- Anomalous transit trough shape
- Variable transit depth
- Variable timing/duration
- Anomalous inferred stellar density
- Aperiodicity
- Disappearance (complete obscuration)
- Achromatic (wavelength independent) transits
- Very low masses derived from transit

Each feature is commonly found in transiting systems due to orbital eccentricities, the presence of unseen planetary companions or debris, or stellar activity. A conclusive explanation for Boyajian's star remains to be found – conventional SETI searches have yielded no signal (Harp et al., 2016; Schuetz et al., 2016; Abeysekara et al., 2016), and recent observations of differential reddening in the dips (Boyajian et al., 2018; Deeg et al., 2018) have effectively ruled out the megastructure explanation.

Transit studies are also sensitive to planetary rings. Humans are generating an artificial ring of satellites (and debris) at geostationary orbit, which are in principle detectable using transit methods. However, the current opacity of the satellite belt (as viewed along the line of sight towards the Earth) is around 10^{-13}. To be detectable at interstellar distances using human-like instrumentation, this opacity must increase many orders of magnitude to 10^{-4}. This is estimated to occur by the year 2200, assuming that a catastrophic case of Kessler syndrome can be avoided (Socas-Navarro, 2018).

Even the debris of planetary systems can also yield possible information about asteroid mining activity. If ETIs live on terrestrial planets with finite quantities of precious metals, then it is likely that an industrial civilisation will exhaust those quantities and look for them in nearby debris fields. Infrared spectroscopy of debris discs may yield unusual chemical depletion, mechanical heating or overproduction of dust. All of these may be indications of asteroid mining (Forgan and Elvis, 2011).

On a large scale, other groups have applied Arthur C. Clarke's Third Law:

Any sufficiently advanced technological civilisation is indistinguishable from magic.

In other words, are there apparently impossible effects lurking in survey data, such as disappearing stars, galaxies with strongly changing redshifts, or indeed disappearing galaxies? As multiple surveys image regions of the sky at multiple epochs, these data can be compared to check for anomalous changes. For example, Villarroel et al. (2016) cross-referenced the US Naval Observatory's USNO-B1.0 and Sloan Digital Sky Survey (SDSS), giving a total of 290,000 objects that are present in the USNO-B1.0 and not in SDSS. Of these, only one is a potential SETI candidate, but at very low confidence.

Even dead civilisations may be detectable using exoplanet data. Stevens et al. (2016) consider how the end of human civilisation might look at interstellar distances. Depending on the death channel of the civilisation (nuclear annihilation, grey goo, pollution, Kessler syndrome, total planetary destruction), these leave features in exoplanet atmospheres that decay on timescales from a few years to 0.1 Myr. These signals are unlikely to be found with the current generation of telescopes, and will be difficult to find even with the next generation (e.g., the James Webb Space Telescope). We may be forced to wait for the next-next generation, such as the LUVOIR telescope, before the ruins of any nearby civilisations can become apparent to us.

27.4 Can the Paradox Be Solved?

The future of SETI looks surprisingly bright. As a discipline it remains chronically underfunded, but as the previous sections reveal, there are a promising number of dedicated SETI programs receiving private funding and coming online. Theorists and modellers are adding more and more ways that survey astronomy at a wide range of wavelengths could reveal observables that characterise intelligent activity. As astrobiology has matured into a respected branch of scientific inquiry, SETI is slowly shaking off the negative prejudices of the past, and is developing a firm grounding in rigorous observation and careful argumentation. Indeed, SETI should now be regarded as a fully developed sub-branch of astrobiology (Wright, 2018).

If we detect ETI Fermi's Paradox can be resolved, of course, by the detection of ETI. Increasing the sample of known civilisations from $N = 1$ to $N = 2$ would vastly improve our understanding of what creates civilisations and what sustains them, even if we do not glean a significant amount of data about our ETI cousins from whatever signals we obtain.

Even mere proof that intelligent life exists elsewhere in the Universe allows us to adjust our priors regarding the location of the Great Filter. If we return to our original rephrasing of the Filter:

$$Nf_l f_i f_{sup} \leq 1 \qquad (27.8)$$

In our current state of knowledge, the f terms all receive a uniform prior between [0, 1]. The detection of intelligent life would move the priors of f_l and f_i away from 0 (and possibly f_{sup} depending on the nature of the detection).

This scenario would imply that the production of intelligence is not the rate-limiting step, but may in fact be something else. If we are certain that the civilisation in question is deceased before becoming highly visible, then this might localise the Great Filter to somewhere particularly close to our present day. While $N = 2$ is hardly a sensible figure on which to compute statistics, it contains significantly more information than $N = 1$. At the very least, we can begin to bracket the range of values for civilisation lifetimes, and compare and contrast how different planetary systems produce intelligent life.

If we detect a signal from ETI with significant information content, thereby gaining a detailed understanding of their origins and evolution, then this will help to clarify whether the correct solution(s) to Fermi's Paradox are in the Rare Earth (R), Catastrophic (C) or Uncommunicative (U) categories. Indeed, we may find that Fermi's Paradox requires a composite solution (RC, CU, etc.). Perhaps civilisations succumb to climate change accidentally induced by nanotechnology (CC), or they enter virtual realities which eventually fail due to technical problems (UC). It may even be that civilisations are rare, and they like to keep their own counsel (RU).

If we fail to detect ETI Of course, it is quite possible, perhaps likely, that our continuing search for ETI fails to yield a signal. What if Fact A becomes stronger? What if the next century conclusively establishes that **no** Arecibo-esque transmitter is present within several hundred light-years of Earth, either at optical or radio frequencies, and that there are no artificial probes above a few centimetres in size within 50 AU of the Sun? What then?

The Paradox will remain a source of continuing intrigue and argument, but it will be based on a much surer foundation. We will have established that the 'classic' SETI techniques show no nearby civilisations, and that we must adjust our priors in a given solution accordingly.

For example, if we begin to discover exoplanets with conclusive biosignatures, but fail to detect artificial signals, then we can see that biological organisms are common, but technological artefacts are not. This reduces our degree of belief in some of our R solutions (R.1–R.11 at least), and instead points us towards R.12

and above, and the C solutions. Going back to the language of the Great Filter, we can place it in our future rather than our past.

Determining the probability that a C solution describes the Universe is trickier. We can quantify the rates of supernovae, AGN and GRB activity to constrain Galactic-scale catastrophes (solutions C.8 and C.9). Similarly, our improving understanding of star formation can determine the likelihood of stellar encounters destroying biospheres (C.7). Our understanding of stellar evolution is well founded (C.5), but we continue to learn more about space weather and stellar activity (C.4).

Applying geophysics to exoplanet systems will help us to constrain terrestrial-scale catastrophic events (C.2, C.3), and observations of debris discs around stars give us good constraints on impact rates on planets in the present day, but are less constraining on past behaviour (C.1). Without a detection of ETI (either living or dead), self-destructive solutions remain effectively unconstrained (C.10–C20).

What about U solutions? By their definition, they are difficult to constrain in the absence of evidence. Equally, if we are able to launch successful interstellar missions, then we can begin to speak to solutions U.2 and U.11 (and to U.12 if we can maintain communications with our fleet at great distance). If these probes can self-replicate, this also alters the priors for solutions U.3 and U.4. As we observe the Universe at a wider range of electromagnetic frequencies, we can disfavour U.6. Our advances in gravitational wave and neutrino astronomy will soon place better constraints on U.7. A complete census of the Solar System's contents down to a few centimetres would more or less rule out U.16.

The more-exotic U solutions are likely to evade our efforts to constrain them, almost regardless of whether we detect ETI or not. Despite our best efforts to do so, it will be extremely difficult to determine if our Universe is simulated until we can probe sufficiently small length scales in particle accelerator experiments (U.14). It will also be very challenging to determine whether an interdict was ever placed on Earth (U.13/U.15).

At best, we will be able to say that such solutions either require ETIs to act beyond the bounds of rational beings, or have further consequences on our observations that we do not see. By Occam's razor, we should consider the exotic U solutions to require too many moving parts/model parameters, and in the absence of better data they should have their priors downgraded, if not completely nulled.

It remains clear that, with a growing quantity and quality of astronomical data, and a steadily improving understanding of how to investigate that data for signs of artifice, we will be able to make rigorous, quantitative statements on the properties of intelligent life in the local Universe. In turn, we will shed light on ourselves, and gradually begin to perceive the human condition from a more objective viewpoint.

Appendix A
A Database of Solutions to Fermi's Paradox

In this Appendix, I list all the solutions to Fermi's Paradox that are addressed (either explicitly or otherwise) in this book. It is difficult to say that this is an exhaustive listing. After all, the many solutions in literature (and in wider culture) are often re-expressions of an earlier solution, or an amalgamation of multiple solutions. Indeed, we may expect that a 'true' solution to the Paradox be an amalgam of multiple solutions from any of the three categories I have used in this book.

The following sections list solutions by category, along with a reference back to relevant sections in the main text. For completeness, I have also added some solutions given by Webb (2002), to date the only other attempt at a rigorous compilation of solutions. As will be clear, some of these solutions, while posed, are either difficult to address scientifically, or have been partially addressed in the text already.

A.1 Rare Earth Solutions

Table A.1 *Rare Earth Solutions to Fermi's Paradox*

ID	Solution	Description	Relevant Chapter/Section
R.1	Earthlike Planets Are Rare	Few planets possess $M = M_\oplus$, $R = R_\oplus$.	4.1.7 (p. 70)
R.2	Earthlike Surfaces Are Rare	Few planets possess the correct surface composition/mix.	4.3 (p. 84)
R.3	Earthlike Atmospheres Are Rare	Few planets possess the correct atmospheric composition/mix.	4.3 (p. 84)
R.4	Earthlike Planets Are Not Continuously Habitable	Few habitable planets possess suitable climate feedback mechanisms.	4.2.1 (p. 74)
R.5	Earthlike Planets Have Disruptive Neighbours	Few habitable planets have a dynamically quiet planetary system.	4.4 (p. 87)

Table A.1 (*cont.*)

ID	Solution	Description	Relevant Chapter/Section
R.6	Earthlike Planets Suffer Too Many Impacts	Few planets have a low enough impact rate from asteroids/comets.	4.4.2 (p. 90)
R.7	Earthlike Planets Have No Moon	Few planets have a moon to stabilise their obliquity evolution.	4.5 (p. 91)
R.8	Earthlike Planets Have the Wrong Moon	Few planets have a moon of appropriate mass for obliquity stability.	4.5 (p. 91)
R.9	M Stars Are Hazardous	Most stars are low-mass M stars, which are too active for habitability.	10.1.2 (p. 184)
R.10	CCHZs Are Too Narrow	Continuous Circumstellar Habitable Zones evolve outward too quickly for planets to produce intelligence.	4.2.1 (p. 73)
R.11	Galactic Habitability Is Poor	Galactic Habitable Zones are too small, perhaps smaller than calculated.	4.2.2 (p. 75)
R.12	The Origin of Life Is Rare	The Origin of Life is a rare event.	5.2 (p. 100)
R.13	The Origin of Life Is not Persistent	Life may originate many times, but fails to take hold.	5.2.4 (p. 104)
R.14	Multicellular Life Is Rare	The transition to multicellular life is difficult.	5.3.5 (p. 112)
R.15	Metazoan Life Is Rare	Organisms large enough to power a complex brain are rare.	5.3.5 (p. 113)
R.16	Brains Are Rare	Brains are not a common endproduct of evolution.	5.3.7 (p. 119)
R.17	Intelligence Is Rare	Self-aware, conscious beings are rare.	6 (p. 127)
R.18	Intelligence Is Disfavoured	Intelligence as a trait generally faces strong negative selection pressure.	5.3.6 (p. 115)
R.19	Social Structures Are Rare	Organisms rarely produce sufficiently complex social structures.	5.3.8 (p. 121)
R.20	Symbolic Communication Is Rare	Language/art/music are uncommon in intelligent organisms.	6.4 (p. 136)
R.21	Civilisation Is Rare	Social Structures rarely progress to a civilisation stage.	5.3.9 (p. 125)
R.22	Stable Political Structures Are Rare	Civilisations rarely produce a stable political structure to promote further progress.	6.7 (p. 150)

Table A.1 (cont.)

ID	Solution	Description	Relevant Chapter/Section
R.23	Science/Mathematics Is Rare	Civilisations rarely produce philosophies/techniques for systematic acquisition of knowledge.	6.6 (p. 149)
R.24	Technology Is Rare	Civilisations rarely produce technology.	6.6 (p. 149)
R.25	Biological Civilisations Are Rare	Civilisations rarely remain biological – in becoming post-biological, they become difficult to detect.	5.3.10 (p. 125) (see also C.16)

A.2 Catastrophist Solutions

Table A.2 *Catastrophist Solutions to Fermi's Paradox*

ID	Solution	Description	Relevant Chapter/Section
C.1	Death by Impacts	Civilisations are destroyed by asteroid/comet impacts.	8 (p. 159)
C.2	Death by Supervolcanism	Civilisations are destroyed by extreme tectonic activity.	9.1 (p. 169)
C.3	Death by Magnetic Field Collapse	Civilisations are destroyed when their planetary magnetic field fails.	9.2 (p. 173)
C.4	Death by Stellar Activity	Civilisations are destroyed by stellar activity.	10.1 (p. 179)
C.5	Death by Stellar Evolution	Civilisations are destroyed as their star evolves off the main sequence.	10.3 (p. 190)
C.6	Death by Supernova	Civilisations are destroyed by supernovae.	10.5 (p. 194)
C.7	Death by Close Stellar Encounter	Civilisations are destroyed by close encounters with nearby stars.	10.6 (p. 198)
C.8	Death by AGN	Civilisations are destroyed by Active Galactic Nuclei.	11.1 (p. 201)
C.9	Death by GRB	Civilisations are destroyed by Gamma Ray Bursts.	11.2 (p. 204)
C.10	Sustainability Solution	Civilisations fail to practise sustainable growth.	12 (p. 215)
C.11	Death by Climate Change	Civilisations die as a result of artificially induced climate change.	13 (p. 218)

Table A.2 (cont.)

ID	Solution	Description	Relevant Chapter/Section
C.12	Death by Biosphere Destruction	Civilisations fail as they reduce local biodiversity below tolerable limits.	13.4 (p. 223)
C.13	Death by Kessler Syndrome	Civilisations fail due to space debris undergoing collisional cascade and destroying the orbital environment.	13.5 (p. 224)
C.14	Death by Genetic Engineering	Civilisations destroy themselves when they discover effective genetic engineering techniques.	14.1 (p. 228)
C.15	Death by Grey Goo	Civilisations are destroyed by out-of-control nanotechnology.	14.2 (p. 230)
C.16	Death by AI	Civilisations are destroyed by malevolent artificial intelligence.	15 (p. 234)
C.17	Death by the Singularity	Civilisations are eclipsed by their own artificial intelligence, and fail.	15.4 (p. 243)
C.18	Death by War	Civilisations destroy themselves in armed conflicts.	16 (p. 248)
C.19	Death by Societal Collapse	Civilisations fail to adapt to internal/external stressors and collapse.	17 (p. 254)
C.20	Death by Intellectual Collapse	Civilisations lose their desire for exploration and intellectual stimulation, and disappear from view or fail.	17.2 (p. 258)

A.3 Uncommunicative Solutions

Table A.3 *Uncommunicative Solutions to Fermi's Paradox*

ID	Solution	Description	Relevant Chapter/Section
U.1	The Phase Transition Solution	Civilisations have only just emerged thanks to global regulation mechanisms.	18.2 (p. 262)
U.2	Exploration Is Imperfect	Civilisations have failed to explore Earth's local neighbourhood.	19 (p. 265)
U.3	Self-Replicating Probes Are Too Dangerous	Civilisations refrain from building autonomous self-replicating probes because they are inherently dangerous.	20 (p. 282)
U.4	Predator-Prey Probes	Interstellar probes mutate, and begin consuming each other for resources.	20.3 (p. 285)

Table A.3 (cont.)

ID	Solution	Description	Relevant Chapter/Section
U.5	The Aliens Are Quiet	ETIs choose not to announce their presence, for a number of reasons (cost, self-preservation, etc.).	21.4 (p. 306)
U.6	They Communicate at a Different Frequency	ETIs use electromagnetic signals, but at a frequency we have not searched yet.	21.1 (p. 291)
U.7	They Communicate via a Different Method	ETIs do not use electromagnetic signals to communicate.	21.2 (p. 296)
U.8	We Do Not Understand Them	ETIs communicate, but we do not recognise their transmissions as artificial.	21.3 (p. 304)
U.9	They Live Too Far Away	ETIs reside beyond the reach of our observing ability.	22 (p. 317)
U.10	The Aestivation Solution	ETIs are in hibernation, waiting for the Universe to cool further to improve their computing ability.	22 (p. 318)
U.11	Interstellar Travel Is Too Hard	ETIs have been unable to surmount economic/technological barriers to exploring the Galaxy.	22.1 (p. 318)
U.12	Interstellar Communication Is Unreliable	ETIs are unable to maintain reliable dialogues across interstellar space.	22.2 (p. 322)
U.13	The Zoo/Interdict Hypothesis	ETIs have agreed not to contact Earth for scientific or political reasons.	23 (p. 326)
U.14	The Simulation Hypothesis	We are the only intelligent agents living in a computer simulation.	24 (p. 330)
U.15	The Planetarium Hypothesis	ETIs have made it seem that only the Earth is inhabited (by manipulating our observations at/near Earth).	see U.14, and Baxter (2001); Webb (2002)
U.16	Bracewell/Lurker Probes Are Here	Bracewell/Lurker probes are present in the Outer Solar System, and have not yet been found.	25.1 (p. 339)
U.17	UFOs Are Alien Spacecraft	ETIs are visiting Earth, and have been seen by humans.	25.2 (p. 342)
U.18	Ancient Alien Hypothesis	ETIs visited Earth in the past.	26.1 (p. 347)
U.19	We Are Aliens	ETIs visited Earth and seeded it with life/intelligence.	see U.18
U.20	Directed Panspermia	ETIs have seeded Earth from interstellar distances.	see Crick (1973); Webb (2002)
U.21	God Exists	Humanity's singular nature is proof that there is a God.	see Webb (2002)

References

Abbott, B. P., Abbott, R., Abbott, T. D., Abernathy, M. R., Acernese, F., Ackley, K., Adams, C., Adams, T., Addesso, P., Adhikari, R. X., and Others. 2016a. GW151226: Observation of Gravitational Waves from a 22-Solar-Mass Binary Black Hole Coalescence. *Physical Review Letters*, **116**(24), 241103.

    2016b. Observation of Gravitational Waves from a Binary Black Hole Merger. *Physical Review Letters*, **116**(6), 061102.

Abeysekara, A. U., Archambault, S., Archer, A., Benbow, W., Bird, R., Buchovecky, M., Buckley, J. H., Byrum, K., Cardenzana, J. V, Cerruti, M., Chen, X., Christiansen, J. L., Ciupik, L., Cui, W., Dickinson, H. J., Eisch, J. D., Errando, M., Falcone, A., Fegan, D. J., Feng, Q., Finley, J. P., Fleischhack, H., Fortin, P., Fortson, L., Furniss, A., Gillanders, G. H., Griffin, S., Grube, J., Gyuk, G., Hütten, M., Håkansson, N., Hanna, D., Holder, J., Humensky, T. B., Johnson, C. A., Kaaret, P., Kar, P., Kelley-Hoskins, N., Kertzman, M., Kieda, D., Krause, M., Krennrich, F., Kumar, S., Lang, M. J., Lin, T. T. Y., Maier, G., McArthur, S., McCann, A., Meagher, K., Moriarty, P., Mukherjee, R., Nieto, D., O'Brien, S., de Bhróithe, A. O'Faoláin, Ong, R. A., Otte, A. N., Park, N., Perkins, J. S., Petrashyk, A., Pohl, M., Popkow, A., Pueschel, E., Quinn, J., Ragan, K., Ratliff, G., Reynolds, P. T., Richards, G. T., Roache, E., Santander, M., Sembroski, G. H., Shahinyan, K., Staszak, D., Telezhinsky, I., Tucci, J. V., Tyler, J., Vincent, S., Wakely, S. P., Weiner, O. M., Weinstein, A., Williams, D. A., and Zitzer, B. 2016. A Search for Brief Optical Flashes Associated with the SETI Target KIC 8462852. *The Astrophysical Journal*, **818**(2), L33.

Ade, P. A. R., Aghanim, N., Ahmed, Z., Aikin, R. W., Alexander, K. D., Arnaud, M., Aumont, J., Baccigalupi, C., Banday, A. J., Barkats, D., Barreiro, R. B., Bartlett, J. G., Bartolo, N., Battaner, E., Benabed, K., Benoît, A., Benoit-Lévy, A., Benton, S. J., Bernard, J.-P., Bersanelli, M., Bielewicz, P., Bischoff, C. A., Bock, J. J., Bonaldi, A., Bonavera, L., Bond, J. R., Borrill, J., Bouchet, F. R., Boulanger, F., Brevik, J. A., Bucher, M., Buder, I., Bullock, E., Burigana, C., Butler, R. C., Buza, V., Calabrese, E., Cardoso, J.-F., Catalano, A., Challinor, A., Chary, R.-R., Chiang, H. C., Christensen, P. R., Colombo, L. P. L., Combet, C., Connors, J., Couchot, F., Coulais, A., Crill, B. P., Curto, A., Cuttaia, F., Danese, L., Davies, R. D., Davis, R. J., de Bernardis, P., de Rosa, A., de Zotti, G., Delabrouille, J., Delouis, J.-M., Désert, F.-X., Dickinson, C., Diego, J. M., Dole, H., Donzelli, S., Doré, O., Douspis, M., Dowell, C. D., Duband, L., Ducout, A., Dunkley, J., Dupac, X., Dvorkin, C., Efstathiou, G., Elsner, F., Enßlin, T. A., Eriksen, H. K., Falgarone, E., Filippini, J. P., Finelli, F., Fliescher, S., Forni, O., Frailis, M., Fraisse, A. A., Franceschi, E., Frejsel, A., Galeotta, S., Galli,

S., Ganga, K., Ghosh, T., Giard, M., Gjerløw, E., Golwala, S. R., González-Nuevo, J., Górski, K. M., Gratton, S., Gregorio, A., Gruppuso, A., Gudmundsson, J. E., Halpern, M., Hansen, F. K., Hanson, D., Harrison, D. L., Hasselfield, M., Helou, G., Henrot-Versillé, S., Herranz, D., Hildebrandt, S. R., Hilton, G. C., Hivon, E., Hobson, M., Holmes, W. A., Hovest, W., Hristov, V. V., Huffenberger, K. M., Hui, H., Hurier, G., Irwin, K. D., Jaffe, A. H., Jaffe, T. R., Jewell, J., Jones, W. C., Juvela, M., Karakci, A., Karkare, K. S., Kaufman, J. P., Keating, B. G., Kefeli, S., Keihänen, E., Kernasovskiy, S. A., Keskitalo, R., Kisner, T. S., Kneissl, R., Knoche, J., Knox, L., Kovac, J. M., Krachmalnicoff, N., Kunz, M., Kuo, C. L., Kurki-Suonio, H., Lagache, G., Lähteenmäki, A., Lamarre, J.-M., Lasenby, A., Lattanzi, M., Lawrence, C. R., Leitch, E. M., Leonardi, R., Levrier, F., Lewis, A., Liguori, M., Lilje, P. B., Linden-Vørnle, M., López-Caniego, M., Lubin, P. M., Lueker, M., Macías-Pérez, J. F., Maffei, B., Maino, D., Mandolesi, N., Mangilli, A., Maris, M., Martin, P. G., Martínez-González, E., Masi, S., Mason, P., Matarrese, S., Megerian, K. G., Meinhold, P. R., Melchiorri, A., Mendes, L., Mennella, A., Migliaccio, M., Mitra, S., Miville-Deschênes, M.-A., Moneti, A., Montier, L., Morgante, G., Mortlock, D., Moss, A., Munshi, D., Murphy, J. A., Naselsky, P., Nati, F., Natoli, P., Netterfield, C. B., Nguyen, H. T., Nørgaard-Nielsen, H. U., Noviello, F., Novikov, D., Novikov, I., O'Brient, R., Ogburn, R. W., Orlando, A., Pagano, L., Pajot, F., Paladini, R., Paoletti, D., Partridge, B., Pasian, F., Patanchon, G., Pearson, T. J., Perdereau, O., Perotto, L., Pettorino, V., Piacentini, F., Piat, M., Pietrobon, D., Plaszczynski, S., Pointecouteau, E., Polenta, G., Ponthieu, N., Pratt, G. W., Prunet, S., Pryke, C., Puget, J.-L., Rachen, J. P., Reach, W. T., Rebolo, R., Reinecke, M., Remazeilles, M., Renault, C., Renzi, A., Richter, S., Ristorcelli, I., Rocha, G., Rossetti, M., Roudier, G., Rowan-Robinson, M., Rubiño-Martín, J. A., Rusholme, B., Sandri, M., Santos, D., Savelainen, M., Savini, G., Schwarz, R., Scott, D., Seiffert, M. D., Sheehy, C. D., Spencer, L. D., Staniszewski, Z. K., Stolyarov, V., Sudiwala, R., Sunyaev, R., Sutton, D., Suur-Uski, A.-S., Sygnet, J.-F., Tauber, J. A., Teply, G. P., Terenzi, L., Thompson, K. L., Toffolatti, L., Tolan, J. E., Tomasi, M., Tristram, M., Tucci, M., Turner, A. D., Valenziano, L., Valiviita, J., Van Tent, B., Vibert, L., Vielva, P., Vieregg, A. G., Villa, F., Wade, L. A., Wandelt, B. D., Watson, R., Weber, A. C., Wehus, I. K., White, M., White, S. D. M., Willmert, J., Wong, C. L., Yoon, K. W., Yvon, D., Zacchei, A., and Zonca, A. 2015. Joint Analysis of BICEP2/Keck Array and Planck Data. *Physical Review Letters*, **114**(10), 101301.

Adler, Paul S., and Kwon, Seok-Woo. 2002. Social Capital: Prospects for a New Concept. *The Academy of Management Review*, **27**(1), 17–40.

Alexander, Richard, Pascucci, Ilaria, Andrews, Sean, Armitage, Philip, and Cieza, Lucas. 2013. The Dispersal of Protoplanetary Disks. Pages 475–496 of: Beuther, Henrik, Klessen, Ralf S., Dullemond, Cornelis P., and Henning, Thomas (eds.), *Protostars and Planets VI*. University of Arizona Press.

Alfonseca, Manuel, Cebrian, Manuel, Anta, Antonio Fernandez, Coviello, Lorenzo, Abeliuk, Andres, and Rahwan, Iyad. 2016. Superintelligence Cannot be Contained: Lessons from Computability Theory. *CoRR*, jul, abs/1607.00913.

Allen, Anthony, Li, Zhi-Yun, and Shu, Frank H. 2003. Collapse of Magnetized Singular Isothermal Toroids. II. Rotation and Magnetic Braking. *ApJ*, **599**(1), 363–379.

Almár, Iván. 2011. SETI and Astrobiology: The Rio Scale and the London Scale. *Acta Astronautica*, **69**(9–10), 899–904.

Alpert, Barbara Olins, and Foundation 2021. 2008. *Creative Ice Age Brain: Cave Art in the Light of Neuroscience*. Santa Fe: Foundation 20 21.

Anglada-Escudé, Guillem, Amado, Pedro J., Barnes, John, Berdiñas, Zaira M., Butler, R. Paul, Coleman, Gavin A. L., de la Cueva, Ignacio, Dreizler, Stefan, Endl, Michael, Giesers, Benjamin, Jeffers, Sandra V., Jenkins, James S., Jones, Hugh R. A., Kiraga,

Marcin, Kürster, Martin, López-González, Marìa J., Marvin, Christopher J., Morales, Nicolás, Morin, Julien, Nelson, Richard P., Ortiz, José L., Ofir, Aviv, Paardekooper, Sijme-Jan, Reiners, Ansgar, Rodríguez, Eloy, Rodrìguez-López, Cristina, Sarmiento, Luis F., Strachan, John P., Tsapras, Yiannis, Tuomi, Mikko, and Zechmeister, Mathias. 2016. A Terrestrial Planet Candidate in a Temperate Orbit around Proxima Centauri. *Nature*, **536**(7617), 437–440.

Annis, J. 1999. An Astrophysical Explanation for the 'Great Silence'. *J. Br. Interplanet. Soc.*, **52**, 19–22.

Aravind, P. K. 2007. The Physics of the Space Elevator. *American Journal of Physics*, **75**(2), 125.

Ariely, Gal. 2014. Does Diversity Erode Social Cohesion? Conceptual and Methodological Issues. *Political Studies*, **62**(3), 573–595.

Armitage, P J. 2010. *Astrophysics of Planet Formation*. Cambridge University Press.

Armstrong, D. J., Pugh, C. E., Broomhall, A.-M., Brown, D. J. A., Lund, M. N., Osborn, H. P., and Pollacco, D. L. 2016. Host Stars of Kepler's Habitable Exoplanets: Superares, Rotation and Activity. *MNRAS*, **455**(3), 3110–3125.

Arnold, Luc F. A. 2005. Transit Light-Curve Signatures of Artificial Objects. *ApJ*, **627**(1), 534–539.

Atri, Dimitra, Melott, Adrian L., and Karam, Andrew. 2014. Biological Radiation Dose from Secondary Particles in a Milky Way Gamma-Ray Burst. *International Journal of Astrobiology*, **13**(03), 224–228.

Axelrod, R. 1984. *The Evolution of Cooperation: Revised Edition*. New York: Basic Books.

Backus, Peter R., and Team, Project Phoenix. 2002. Project Phoenix: SETI Observations from 1200 to 1750 MHz with the Upgraded Arecibo Telescope. Page 525 of: Stanimirovic, Snezana, Altschuler, Daniel, Goldsmith, Paul, and Salter, Chris (eds.), *NAIC-NRAO School on Single-dish Radio Astronomy: Techniques and Applications*. Astronomical Society of the Pacific.

Badescu, V., and Cathcart, R. B. 2000. Stellar Engines for Kardashev's Type II Civilisations. *J. Br. Interplanet. Soc.*, **53**, 297–306.

Bagnères, Anne-Geneviève, and Hanus, Robert. 2015. Communication and Social Regulation in Termites. Pages 193–248 of: *Social Recognition in Invertebrates*. Cham: Springer International Publishing.

Bailer-Jones, C. A. L. 2015. Close Encounters of the Stellar Kind. *Astronomy & Astrophysics*, **575**(mar), A35.

Bailey, R., Elsener, Konrad, Ereditato, A., Weisse, E., Stevenson, Graham Roger, Vincke, H. H., Ferrari, A., Maugain, J. M., Vassilopoulos, N., Ball, A. E., and Others. 1999. *The CERN Neutrino Beam to Gran Sasso (NGS)*. Tech. rept.

Balbi, Amedeo, and Tombesi, Francesco. 2017. The Habitability of the Milky Way During the Active Phase of Its Central Supermassive Black Hole. *Scientific Reports*, **7**(1), 16626.

Ball, J. 1973. The Zoo Hypothesis. *Icarus*, **19**(3), 347–349.

Bambach, Richard K. 2006. Phanerozoic Biodiversity Mass Extinctions. *Annual Review of Earth and Planetary Sciences*, **34**(1), 127–155.

Barge, P., and Sommeria, J. 1995. Did Planet Formation Begin Inside Persistent Gaseous Vortices? *A&A*, **295**(1), L1–L4.

Barnes, Rory, and Heller, René. 2013. Habitable Planets Around White and Brown Dwarfs: The Perils of a Cooling Primary. *Astrobiology*, **13**(3), 279–291.

Barnes, Rory, Mullins, Kristina, Goldblatt, Colin, Meadows, Victoria S., Kasting, James F., and Heller, René. 2013. Tidal Venuses: Triggering a Climate Catastrophe via Tidal Heating. *Astrobiology*, **13**(3), 225–50.

Barnosky, Anthony D., Matzke, Nicholas, Tomiya, Susumu, Wogan, Guinevere O. U., Swartz, Brian, Quental, Tiago B., Marshall, Charles, McGuire, Jenny L., Lindsey, Emily L., Maguire, Kaitlin C., Mersey, Ben, and Ferrer, Elizabeth A. 2011. Has the Earth's Sixth Mass Extinction Already Arrived? *Nature*, **471**(7336), 51–57.

Barrow, J. D. 1998. *Impossibility: The Limits of Science and the Science of Limits.* 1st edn. Oxford University Press.

Bartelmann, Matthias, and Schneider, Peter. 2001. Weak Gravitational Lensing. *Physics Reports*, **340**(4–5), 291–472.

Batygin, Konstantin, and Brown, Michael E. 2016. Evidence for a Distant Giant Planet in the Solar System. *The Astronomical Journal*, **151**(2), 22.

Batygin, Konstantin, and Laughlin, Greg. 2015. Jupiter's Decisive Role in the Inner Solar System's Early Evolution. *Proceedings of the National Academy of Sciences*, **112**(14), 4214–4217.

Baxter, S. 2001. The Planetarium Hypothesis – A Resolution of the Fermi Paradox. *JBIS*, **54**, 210–216.

Bean, Jacob L., Kempton, Eliza Miller-Ricci, and Homeier, Derek. 2010. A Ground-Based Transmission Spectrum of the Super-Earth Exoplanet GJ 1214b. *Nature*, **468**(7324), 669–72.

Beane, Silas R., Davoudi, Zohreh, and Savage, Martin J. 2014. Constraints on the Universe as a Numerical Simulation. *The European Physical Journal A*, **50**(9), 148.

Beatty, John. 2006. Replaying Life's Tape. *The Journal of Philosophy*, **103**(7), 336–362.

Becker, Werner, Bernhardt, Mike G., and Jessner, Axel. 2013. Autonomous Spacecraft Navigation with Pulsars. *Acta Futura*, **7**, 11–28.

Bekenstein, J. D. 1972. Black Holes and the Second Law. *Lettere Al Nuovo Cimento Series 2*, **4**(15), 737–740.

Bekker, A., Holland, H. D., Wang, P.-L., Rumble, D., Stein, H. J., Hannah, J. L., Coetzee, L. L., and Beukes, N. J. 2004. Dating the Rise of Atmospheric Oxygen. *Nature*, **427**(6970), 117–120.

Belbruno, Edward, and Gott III, J. Richard. 2005. Where Did the Moon Come From? *The Astronomical Journal*, **129**(3), 1724–1745.

Bell, Elizabeth A., Boehnke, Patrick, Harrison, T. Mark, and Mao, Wendy L. 2015. Potentially Biogenic Carbon Preserved in a 4.1 Billion-Year-Old Zircon. *Proceedings of the National Academy of Sciences*, **112**(47), 14518–14521.

Benford, James, Benford, Gregory, and Benford, Dominic. 2010. Messaging with Cost-Optimized Interstellar Beacons. *Astrobiology*, **10**(5), 475–90.

Bennett, D. P., Batista, V., Bond, I. A., Bennett, C. S., Suzuki, D., Beaulieu, J.-P., Udalski, A., Donatowicz, J., Bozza, V., Abe, F., Botzler, C. S., Freeman, M., Fukunaga, D., Fukui, A., Itow, Y., Koshimoto, N., Ling, C. H., Masuda, K., Matsubara, Y., Muraki, Y., Namba, S., Ohnishi, K., Rattenbury, N. J., Saito, To, Sullivan, D. J., Sumi, T., Sweatman, W. L., Tristram, P. J., Tsurumi, N., Wada, K., Yock, P. C. M., Albrow, M. D., Bachelet, E., Brillant, S., Caldwell, J. A. R., Cassan, A., Cole, A. A., Corrales, E., Coutures, C., Dieters, S., Dominis Prester, D., Fouqué, P., Greenhill, J., Horne, K., Koo, J.-R., Kubas, D., Marquette, J.-B., Martin, R., Menzies, J. W., Sahu, K. C., Wambsganss, J., Williams, A., Zub, M., Choi, J.-Y., DePoy, D. L., Dong, Subo, Gaudi, B. S., Gould, A., Han, C., Henderson, C. B., McGregor, D., Lee, C.-U., Pogge, R. W., Shin, I.-G., Yee, J. C., Szymański, M. K., Skowron, J., Poleski, R., Kozłowski, S., Wyrzykowski, Ł., Kubiak, M., Pietrukowicz, P., Pietrzyński, G., Soszyński, I., Ulaczyk, K., Tsapras, Y., Street, R. A., Dominik, M., Bramich, D. M., Browne, P.,

Hundertmark, M., Kains, N., Snodgrass, C., Steele, I. A., Dekany, I., Gonzalez, O. A., Heyrovský, D., Kandori, R., Kerins, E., Lucas, P. W., Minniti, D., Nagayama, T., Rejkuba, M., Robin, A. C., and Saito, R. 2014. MOA-2011-BLG-262Lb: A Sub-Earth-Mass Moon Orbiting a Gas Giant Primary Or a High Velocity Planetary System in the Galactic Bulge. *ApJ*, **785**, 155–167.

Berdahl, Mira, Robock, Alan, Ji, Duoying, Moore, John C., Jones, Andy, Kravitz, Ben, and Watanabe, Shingo. 2014. Arctic Cryosphere Response in the GeoEngineering Model Intercomparison Project G3 and G4 Scenarios. *Journal of Geophysical Research: Atmospheres*, **119**(3), 1308–1321.

Berdyugina, Svetlana V., and Kuhn, Jeff R. 2017. Surface Imaging of Proxima b and Other Exoplanets: Topography, Biosignatures, and Artificial Mega-Structures. oct.

Berna, F., Goldberg, P., Horwitz, L. K., Brink, J., Holt, S., Bamford, M., and Chazan, M. 2012. Microstratigraphic Evidence of in situ Fire in the Acheulean Strata of Wonderwerk Cave, Northern Cape Province, South Africa. *PNAS*, **109**(20), E1215–E1220.

Beskin, Gregory, Borisov, Nikolai, Komarova, Victoria, Mitronova, Sofia, Neizvestny, Sergei, Plokhotnichenko, Vladimir, and Popova, Marina. 1997. The Search for Extraterrestrial Civilizations in Optical Range – Methods, Objects, Results. Page 595 of: *Astronomical and Biochemical Origins and the Search for Life in the Universe, IAU Colloquium 161*. Bologna.

Bickerton, D. 1998. Catastrophic Evolution: The Case for a Single Step from Protolanguage to Full Human Language. Pages 341–358 of: Hurford, James R., Studdert-Kennedy, Michael, and Knight, Chris (eds.), *Approaches to the Evolution of Language: Social and Cognitive Biases*.

Binder, Alan B. 1974. On the Origin of the Moon by Rotational Fission. *The Moon*, **11**(1–2), 53–76.

Binzel, Richard P. 1997. A Near-Earth Object Hazard Index. *Annals of the New York Academy of Sciences*, **822**(1 Near-Earth Ob), 545–551.

Bisbas, Thomas G., Wünsch, Richard, Whitworth, Anthony P., Hubber, David A., and Walch, Stefanie. 2011. Radiation-Driven Implosion and Triggered Star Formation. *ApJ*, **736**(2), 142.

Bjørk, R. 2007. Exploring the Galaxy Using Space Probes. *International Journal of Astrobiology*, **6**(02), 89.

Blair, D. G., and Zadnik, M. G. 1993. A List of Possible Interstellar Communication Channel Frequencies for SETI. *A&A*, **278**, 669–672.

2002. A Search for Optical Beacons: Implications of Null Results. *Astrobiology*, **2**(3), 305–312.

Bolduc, Léonard. 2002. GIC Observations and Studies in the Hydro-Québec Power System. *Journal of Atmospheric and Solar-Terrestrial Physics*, **64**(16), 1793–1802.

Boley, Aaron C., Durisen, R.H., Nordlund, Ake, and Lord, Jesse. 2007. Three-Dimensional Radiative Hydrodynamics for Disk Stability Simulations: A Proposed Testing Standard and New Results. *ApJ*, **665**(2), 1254–1267.

Boley, Aaron C., Hayfield, Tristen, Mayer, Lucio, and Durisen, Richard H. 2010. Clumps in the Outer Disk by Disk Instability: Why They Are Initially Gas Giants and the Legacy of Disruption. *Icarus*, **207**(2), 509–516.

Bond, A. 1978. *Project Daedalus: The Final Report on the BIS Starship Study*. London: British Interplanetary Society.

Bonnell, I. A., Dobbs, C. L., and Smith, R. J. 2013. Shocks, Cooling and the Origin of Star Formation Rates in Spiral Galaxies. *MNRAS*, **430**(3), 1790–1800.

Bostrom, Nick. 2002. Existential Risks. *Journal of Evolution and Technology*, **9**(1), 1–31.

Bostrom, Nick. 2014. *Superintelligence: Paths, Dangers, Strategies*. Oxford: Oxford University Press.

Bouchard, Denis. 2013. *The Nature and Origin of Language*. Oxford: Oxford University Press.

Boyajian, T. S., LaCourse, D. M., Rappaport, S. A., Fabrycky, D., Fischer, D. A., Gandolfi, D., Kennedy, G. M., Korhonen, H., Liu, M. C., Moor, A., Olah, K., and Others. 2016. Planet Hunters IX. KIC 8462852 – Where's the Flux? *MNRAS*, **457**(4), 3988–4004.

Boyajian, Tabetha S., Alonso, Roi, Ammerman, Alex, Armstrong, David, Ramos, A. Asensio, Barkaoui, K., Beatty, Thomas G., Benkhaldoun, Z., Benni, Paul O., Bentley, Rory, Berdyugin, Andrei, Berdyugina, Svetlana, Bergeron, Serge, Bieryla, Allyson, Blain, Michaela G., Blanco, Alicia Capetillo, Bodman, Eva H. L., Boucher, Anne, Bradley, Mark, Brincat, Stephen M., Brink, Thomas G., Briol, John, Brown, David J. A., Budaj, J., Burdanov, A., Cale, B., Carbo, Miguel Aznar, García, R. Castillo, Clark, Wendy J., Clayton, Geoffrey C., Clem, James L., Coker, Phillip H., Cook, Evan M., Copperwheat, Chris M., Curtis, J. L., Cutri, R. M., Cseh, B., Cynamon, C. H., Daniels, Alex J., Davenport, James R. A., Deeg, Hans J., Lorenzo, Roberto De, de Jaeger, Thomas, Desrosiers, Jean-Bruno, Dolan, John, Dowhos, D. J., Dubois, Franky, Durkee, R., Dvorak, Shawn, Easley, Lynn, Edwards, N., Ellis, Tyler G., Erdelyi, Emery, Ertel, Steve, Farfán, Rafael G., Farihi, J., Filippenko, Alexei V., Foxell, Emma, Gandolfi, Davide, Garcia, Faustino, Giddens, F., Gillon, M., González-Carballo, Juan-Luis, González-Fernández, C., Hernández, J. I. González, Graham, Keith A., Greene, Kenton A., Gregorio, J., Hallakoun, Na'ama, Hanyecz, Ottó, Harp, G. R., Henry, Gregory W., Herrero, E., Hildbold, Caleb F., Hinzel, D., Holgado, G., Ignácz, Bernadett, Ilyin, Ilya, Ivanov, Valentin D., Jehin, E., Jermak, Helen E., Johnston, Steve, Kafka, S., Kalup, Csilla, Kardasis, Emmanuel, Kaspi, Shai, Kennedy, Grant M., Kiefer, F., Kielty, C. L., Kessler, Dennis, Kiiskinen, H., Killestein, T. L., King, Ronald A., Kollar, V., Korhonen, H., Kotnik, C., Könyves-Tóth, Réka, Kriskovics, Levente, Krumm, Nathan, Krushinsky, Vadim, Kundra, E., Lachapelle, Francois-Rene, LaCourse, D., Lake, P., Lam, Kristine, Lamb, Gavin P., Lane, Dave, Lau, Marie Wingyee, Lewin, Pablo, Lintott, Chris, Lisse, Carey, Logie, Ludwig, Longeard, Nicolas, Villanueva, M. Lopez, Ludington, E. Whit, Mainzer, A., Malo, Lison, Maloney, Chris, Mann, A., Mantero, A., Marengo, Massimo, Marchant, Jon, Martínez González, M. J., Masiero, Joseph R., Mauerhan, Jon C., McCormac, James, McNeely, Aaron, Meng, Huan Y. A., Miller, Mike, Molnar, Lawrence A., Morales, J. C., Morris, Brett M., Muterspaugh, Matthew W., Nespral, David, Nugent, C. R., Nugent, Katherine M., Odasso, A., O'Keeffe, Derek, Oksanen, A., O'Meara, John M., Ordasi, András, Osborn, Hugh, Ott, John J., Parks, J. R., Perez, Diego Rodriguez, Petriew, Vance, Pickard, R., Pál, András, Plavchan, P., Pollacco, Don, Nuñez, F., Pozo, J., Pozuelos, F., Rau, Steve, Redfield, Seth, Relles, Howard, Ribas, Ignasi, Richards, Jon, Saario, Joonas L. O., Safron, Emily J., Sallai, J. Martin, Sárneczky, Krisztián, Schaefer, Bradley E., Schumer, Clea F., Schwartzendruber, Madison, Siegel, Michael H., Siemion, Andrew P. V., Simmons, Brooke D., Simon, Joshua D., Simón-Díaz, S., Sitko, Michael L., Socas-Navarro, Hector, Sódor, Á., Starkey, Donn, Steele, Iain A., Stone, Geoff, Strassmeier, Klaus G., Street, R. A., Sullivan, Tricia, Suomela, J., Swift, J. J., Szabó, Gyula M., Szabó, Róbert, Szakáts, Róbert, Szalai, Tamás, Tanner, Angelle M., Toledo-Padrón, B., Tordai, Tamás, Triaud, Amaury H. M. J., Turner, Jake D., Ulowetz, Joseph H., Urbanik, Marian, Vanaverbeke, Siegfried, Vanderburg, Andrew, Vida, Krisztián, Vietje, Brad P., Vinkó, József, von Braun, K., Waagen, Elizabeth O., Walsh, Dan, Watson, Christopher A., Weir, R. C., Wenzel, Klaus, Plaza, C. Westendorp, Williamson, Michael W., Wright, Jason T., Wyatt, M. C., Zheng,

Weikang, and Zsidi, Gabriella. 2018. The First Post-Kepler Brightness Dips of KIC 8462852. *ApJ*, **853**(1), L8.

Boyd, Brian. 2010. *On the Origin of Stories: Evolution, Cognition, and Fiction*. 1st edn. Cambridge, MA: Belknap Press of Harvard University Press.

Boyd, Robert, and Richerson, Peter J. 1988. *Culture and the Evolutionary Process*. University of Chicago Press.

Bracewell, R. N. 1960. Communications from Superior Galactic Communities. *Nature*, **186**(4726), 670–671.

Bradbury, Robert J. 2002. 'Dyson Shell Supercomputers As the Dominant Life Form' in Galaxies. Page 213 of: Lemarchand, G., and Meech, K. (eds.), *Bioastronomy 2002: A New Era in Bioastronomy*.

Brashears, Travis, Lubin, Philip, Hughes, Gary B., McDonough, Kyle, Arias, Sebastian, Lang, Alex, Motta, Caio, Meinhold, Peter, Batliner, Payton, Griswold, Janelle, Zhang, Qicheng, Alnawakhtha, Yusuf, Prater, Kenyon, Madajian, Jonathan, Sturman, Olivia, Gergieva, Jana, Gilkes, Aidan, and Silverstein, Bret. 2015 (Sep). Directed Energy Interstellar Propulsion of Wafersats. Page 961609 of: Taylor, Edward W., and Cardimona, David A. (eds.), *Nanophotonics and Macrophotonics for Space Environments IX*.

Brasier, M., McLoughlin, N., Green, O., and Wacey, D. 2006. A Fresh Look at the Fossil Evidence for Early Archaean Cellular Life. *Philosophical Transactions of the Royal Society B: Biological Sciences*, **361**(1470), 887–902.

Brillouin, L. 1953. The Negentropy Principle of Information. *Journal of Applied Physics*, **24**(9), 1152–1163.

1962. *Science and Information Theory*. 2nd edn. New York: Academic Press.

Brin, G. D. 1983. The Great Silence – The Controversy Concerning Extraterrestrial Intelligent Life. *QJRAS*, **24**, 283–309.

Bringsjord, Selmer. 2012. Belief in the Singularity Is Logically Brittle. *Journal of Consciousness Studies*, **19**(7), 14.

Brown, Steven. 2000. The 'Musilanguage' Model of Music Evolution. Pages 271–300 of: Wallin, Nils L., Merker, Bjorn, and Brown, Steven (eds.), *The Origins of Music*. Cambridge, MA: MIT Press Ltd.

Burmeister, Alita R. 2015. Horizontal Gene Transfer. *Evolution, Medicine, and Public Health*, **2015**(1), 193–194.

Buss, D. M. 2005. *The Murderer Next Door: Why the Mind Is Designed to Kill*. New York: Penguin.

Bussard, R. W. 1960. Galactic Matter and Interstellar Spaceflight. *Acta Astronautica*, **6**, 170–194.

Cairó, Osvaldo. 2011. External Measures of Cognition. *Frontiers in Human Neuroscience*, **5**.

Caldeira, Ken, and Wickett, Michael E. 2003. Oceanography: Anthropogenic Carbon and Ocean pH. *Nature*, **425**(6956), 365–365.

Cameron, A. G. W., and Ward, W. R. 1976. The Origin of the Moon. Page 120 of: *Lunar and Planetary Science Conference*, vol. 7.

Campbell, Neil A., Reece, Jane B., Urry, Lisa A., Cain, Michael C., Wassermann, Steven A., Minorsky, Peter V., and Jackson, Robert B. 2014. *Biology: A Global Approach*. 10th edn. Pearson Education.

Canup, R. M. 2012. Forming a Moon with an Earth-like Composition via a Giant Impact. *Science*, **338**(6110), 1052–1055.

Canup, Robin M., and Ward, William R. 2006. A Common Mass Scaling for Satellite Systems of Gaseous Planets. *Nature*, **441**(7095), 834–9.

Carbone, V., Sorriso-Valvo, L., Vecchio, A., Lepreti, F., Veltri, P., Harabaglia, P., and Guerra, I. 2006. Clustering of Polarity Reversals of the Geomagnetic Field. *Physical Review Letters*, **96**(12), 128501.

Cardeña, E., Lynn, S. J., and Krippner, S. 2000. Introduction: Anomalous Experiences in Perspective. Page 4 of: Cardena, E., and Lynn, S. (eds.), *Varieties of Anomalous Experience: Examining the Scientific Evidence*. American Psychological Association.

Carracedo, Juan Carlos. 2014. Structural Collapses in the Canary Islands. In: Gutiérrez F., Gutiérrez M. (eds.) *Landscapes and Landforms of Spain*. World Geomorphological Landscapes. Springer, Dordrecht.

Carrigan, Richard A. 2009. IRAS-Based Whole Sky Upper Limit on Dyson Spheres. *ApJ*, **698**(2), 2075–2086.

Carrington, R. C. 1859. Description of a Singular Appearance Seen in the Sun on September 1, 1859. *MNRAS*, **20**(1), 13–15.

Carter, B., and McCrea, W. H. 1983. The Anthropic Principle and Its Implications for Biological Evolution [and Discussion]. *Philosophical Transactions of the Royal Society A: Mathematical, Physical and Engineering Sciences*, **310**(1512), 347–363.

Cartin, Daniel. 2013. Exploration of the Local Solar Neighborhood I: Fixed Number of Probes. *International Journal of Astrobiology*, in press.

Cassenti, B. N. 1982. A Comparison of Interstellar Propulsion Methods. *JBIS*, **35**, 116–124.

Catling, David C., and Claire, Mark W. 2005. How Earth's Atmosphere Evolved to an Oxic State: A Status Report. *Earth and Planetary Science Letters*, **237**(1–2), 1–20.

Catling, David C., Krissansen-Totton, Joshua, Kiang, Nancy Y., Crisp, David, Robinson, Tyler D., DasSarma, Shiladitya, Rushby, Andrew, Del Genio, Anthony, Bains, William, and Domagal-Goldman, Shawn. 2017. Exoplanet Biosignatures: A Framework for Their Assessment. *Astrobiology*, vol. 18, issue 6, pp. 709–738.

Chang, S.-W., Byun, Y.-I., and Hartman, J. D. 2015. Photometric Study on Stellar Magnetic Activity I. Flare Variability of Red Dwarf Stars in the Open Cluster M37. *ApJ*, **814**(1), 35.

Chapman, Clark R., and Morrison, David. 1994. Impacts on the Earth by Asteroids and Comets: Assessing the Hazard. *Nature*, **367**(6458), 33–40.

2003. No Reduction in Risk of a Massive Asteroid Impact. *Nature*, **421**(6922), 473–473.

Chen, Xian, and Amaro-Seoane, Pau. 2014. Sagittarius A* Rivaled the Sun in the Ancient X-ray Sky. *eprint arXiv:1412.5592*, dec.

Chomsky, N. 2007. Of Minds and Language. *Biolinguistics*, **1**, 9–27.

2005. Three Factors in Language Design. *Linguistic Inquiry*, **36**(1), 1–22.

Chulliat, A., and Olsen, N. 2010. Observation of Magnetic Diffusion in the Earth's Outer Core from Magsat, Ørsted, and CHAMP Data. *Journal of Geophysical Research*, **115**(B5), B05105.

Cirkovic, M., and Bradbury, R. 2006. Galactic Gradients, Postbiological Evolution and the Apparent Failure of SETI. *New Astronomy*, **11**(8), 628–639.

Ćirković, Milan M. 2006. Macroengineering in the Galactic Context: A New Agenda for Astrobiology. Pages 281–298 of: Badescu, Viorel, Cathcart, Richard B., and Schuiling, Roelof D. (eds), *Macro-engineering: A Challenge for the Future*. Springer.

2008. Against the Empire. *J. Br. Interplanet. Soc.*, **61**, 246–254.

2015. Linking Simulation Argument to the AI Risk. *Futures*, **72**(sep), 27–31.

Ćirković, Milan M., and Vukotić, Branislav. 2016. Long-Term Prospects: Mitigation of Supernova and Gamma-Ray Burst Threat to Intelligent Beings. *Acta Astronautica*, **129**(dec), 438–446.

Clark, Jerome. 2000. *Extraordinary Encounters: An Encyclopedia of Extraterrestrials and Otherworldly Beings*. ABC-CLIO.

Clarke, Bruce. 2014. John Lilly, the Mind of the Dolphin, and Communication Out of Bounds. *communication+1*, **3**, 8.

Cocconi, G., and Morrison, P. 1959. Searching for Interstellar Communications. *Nature*, **184**, 844–846.

Cockell, Charles S. 2015. Extraterrestrial Liberty: Can It Be Planned? Pages 23–42 of: *Human Governance Beyond Earth*.

Collins, Gareth S., Melosh, H. Jay, and Marcus, Robert A. 2005. Earth Impact Effects Program: A Web-Based Computer Program for Calculating the Regional Environmental Consequences of a Meteoroid Impact on Earth. *Meteoritics and Planetary Science*, **40**(6), 817–840.

Commerçon, B., Hennebelle, P., Audit, E., Chabrier, G., and Teyssier, R. 2010. Protostellar Collapse: Radiative and Magnetic Feedbacks on Small-Scale Fragmentation. *Astronomy and Astrophysics*, **510**(feb), L3.

Conn Henry, Richard, Kilston, S., and Shostak, S. 2008. SETI in the Ecliptic Plane. *Bulletin of the American Astronomical Society*, **40**, 194.

Conway Morris, Simon. 2003a. *Life's Solution: Inevitable Humans in a Lonely Universe*. Cambridge University Press.

  2003b. The Navigation of Biological Hyperspace. *International Journal of Astrobiology*, **2**(2), 149–152.

Cooper, David L. 2006. Broca's Arrow: Evolution, Prediction, and Language in the Brain. *The Anatomical Record Part B: The New Anatomist*, **289B**(1), 9–24.

Cotta, Carlos, and Morales, Álvaro. 2009. A Computational Analysis of Galactic Exploration with Space Probes: Implications for the Fermi Paradox. *J. Br. Interplanet. Soc.*, **62**(Jul), 82–88.

Courtillot, Vincent, and Olson, Peter. 2007. Mantle Plumes Link Magnetic Superchrons to Phanerozoic Mass Depletion Events. *Earth and Planetary Science Letters*, **260**(3-4), 495–504.

Cox, Rochelle E., and Barnier, Amanda J. 2015. A Hypnotic Analogue of Clinical Confabulation. *International Journal of Clinical and Experimental Hypnosis*, **63**(3), 249–273.

Crawford, I. A. 1990. Interstellar Travel: A Review for Astronomers. *QJRAS*, **31**, 377–400.

Crick, F. 1973. Directed Panspermia. *Icarus*, **19**(3), 341–346.

Cross, Ian, and Morley, Iain. 2009. The Evolution of Music: Theories, Definitions and the Nature of the Evidence. *Communicative Musicality: Exploring the Basis of Human Companionship*, 61–81.

Crucifix, Michael. 2011. Glacial/Interglacial Cycles. Pages pp. 359–356 of: Singh, Vijay P., Singh, Pratap, and Haritashya, Umesh K. (eds.), *Encyclopedia of Snow, Ice and Glaciers*. Encyclopedia of Earth Sciences Series. Dordrecht: Springer Netherlands.

Cucchiara, A., Levan, A. J., Fox, D. B., Tanvir, N. R., Ukwatta, T. N., Berger, E., Krühler, T., Yoldaş, A. Küpcü, Wu, X. F., Toma, K., Greiner, J. E., Olivares, F., Rowlinson, A., Amati, L., Sakamoto, T., Roth, K., Stephens, A., Fritz, Alexander, Fynbo, J. P. U., Hjorth, J., Malesani, D., Jakobsson, P., Wiersema, K., O'Brien, P. T., Soderberg, A. M., Foley, R. J., Fruchter, A. S., Rhoads, J., Rutledge, R. E., Schmidt, B. P., Dopita, M. A., Podsiadlowski, P., Willingale, R., Wolf, C., Kulkarni, S. R., and D'Avanzo, P. 2011. A Photometric Redshift of $z \sim 9.4$ for GRB 090429B. *The Astrophysical Journal*, **736**(1), 7.

Ćuk, M., and Stewart, S. T. 2012. Making the Moon from a Fast-Spinning Earth: A Giant Impact Followed by Resonant Despinning. *Science*, **338**(6110), 1047–1052.

Dalrymple, G. B. 2001. The Age of the Earth in the Twentieth Century: A Problem (Mostly) Solved. *Geological Society, London, Special Publications*, **190**(1), 205–221.

De Biasi, A., Secco, L., Masi, M., and Casotto, S. 2015. Galactic Planar Tides on the Comets of Oort Cloud and Analogs in Different Reference Systems. I. *Astronomy & Astrophysics*, **574**(jan), A98.

Deeg, H. J., Alonso, R., Nespral, D., and Boyajian, Tabetha. 2018. Non-Grey Dimming Events of KIC 8462852 from GTC Spectrophotometry. jan.

Delgado, José Manuel Rodríguez. 1969. *Physical Control of the Mind: Toward a Psychocivilized Society*. New York: Harper & Row.

Dence, M. R., Grieve, R. A. F., and Robertson, P. B. 1977. Terrestrial Impact Structures – Principal Characteristics and Energy Considerations. Pages 247–275 of: *In: Impact and Explosion Cratering: Planetary and Terrestrial Implications; Proceedings of the Symposium on Planetary Cratering Mechanics, Flagstaff, Ariz., September 13-17, 1976. (A78-44030 19-91)* New York, Pergamon Press, Inc., 1977, pp. 247–275.

DeVito, Carl L. 1991. Languages, Science and the Search for Extraterrestrial Intelligence. *Interdisciplinary Science Reviews*, **16**(2), 156–160.

Diamond, J. 2013. *Collapse: How Societies Choose to Fail or Survive*. Penguin Books Limited.

Dick, Steven. 2008. The Postbiological Universe. *Acta Astronautica*, **62**, 499 – 504.

2003. Cultural Evolution, the Postbiological Universe and SETI. *International Journal of Astrobiology*, **2**(1), 65–74.

2009. The Postbiological Universe and Our Future in Space. *Futures*, **41**(8), 578–580.

Dieks, D. 1992. Doomsday – or the Dangers of Statistics. *Philosophical Quarterly*, **42**, 78–84.

Dobzhansky, Theodosius. 1972. Darwinian Evolution and the Problem of Extraterrestrial Life. *Perspectives in Biology and Medicine*, **15**(2), 157–176.

Doeleman, Sheperd S., Weintroub, Jonathan, Rogers, Alan E. E., Plambeck, Richard, Freund, Robert, Tilanus, Remo P. J., Friberg, Per, Ziurys, Lucy M., Moran, James M., Corey, Brian, Young, Ken H., Smythe, Daniel L., Titus, Michael, Marrone, Daniel P., Cappallo, Roger J., Bock, Douglas C.-J., Bower, Geoffrey C., Chamberlin, Richard, Davis, Gary R., Krichbaum, Thomas P., Lamb, James, Maness, Holly, Niell, Arthur E., Roy, Alan, Strittmatter, Peter, Werthimer, Daniel, Whitney, Alan R., and Woody, David. 2008. Event-Horizon-Scale Structure in the Supermassive Black Hole Candidate at the Galactic Centre. *Nature*, **455**(7209), 78–80.

Dolan, Michelle M., Mathews, Grant J., Lam, Doan Duc, Lan, Nguyen Quynh, Herczeg, Gregory J., and Dearborn, David S. P. 2016. Evolutionary Tracks for Betelgeuse. *The Astrophysical Journal*, **819**(1), 7.

Domagal-Goldman, Shawn D., Segura, Antígona, Claire, Mark W., Robinson, Tyler D., and Meadows, Victoria S. 2014. Abiotic Ozone and Oxygen in Atmospheres Similar to Prebiotic Earth. *The Astrophysical Journal*, **792**(2), 90.

Downs, G. S. 1974. *Interplanetary Navigation Using Pulsating Radio Sources*. Tech. rept. JPL, Pasadena.

Dowswell, C. 2009. Norman Ernest Borlaug (1914-2009). *Science*, **326**(5951), 381–381.

Doyle, Laurance R., McCowan, Brenda, Johnston, Simon, and Hanser, Sean F. 2011. Information Theory, Animal Communication, and the Search for Extraterrestrial Intelligence. *Acta Astronautica*, **68**(3–4), 406–417.

Draine, B.T. 2003. Interstellar Dust Grains. *ARA&A*, **41**(1), 241–289.

Drake, F.D. 2003. The Drake Equation Revisited: Part 1. *Astrobiology Magazine*.

Dunbar, R. 1998. *Grooming, Gossip, and the Evolution of Language.* Harvard University Press.
Dunning Hotopp, Julie C. 2011. Horizontal Gene Transfer Between Bacteria and Animals. *Trends in Genetics*, **27**(4), 157–163.
Duquennoy, A., and Mayor, M. 1991. Multiplicity Among Solar-Type Stars in the Solar Neighbourhood. II - Distribution of the Orbital Elements in an Unbiased Sample. *A&A*, **248**, 485–524.
Duric, Neb. 2004. *Advanced Astrophysics.* Cambridge University Press.
Durrant, Russil. 2009. Born to Kill? A Critical Evaluation of Homicide Adaptation Theory. *Aggression and Violent Behavior*, **14**(5), 374–381.
Dyson, Freeman J. 1960. Search for Artificial Stellar Sources of Infrared Radiation. *Science*, **131**(3414), 1667–1668.
  1963. Gravitational Machines. Chap. 12 of: Cameron, A. G. W. (ed.), *Interstellar Communication.* New York: Benjamin Press.
Eckart, A., and Genzel, R. 1996. Observations of Stellar Proper Motions Near the Galactic Centre. *Nature*, **383**(6599), 415–417.
Edmondson, William H., and Stevens, Ian R. 2003. The Utilization of Pulsars As SETI Beacons. *IJA*, **2**(4), 231–271.
Eliseev, Alexey V., Chernokulsky, Alexandr V., Karpenko, Andrey A., and Mokhov, Igor I. 2010. Global Warming Mitigation by Sulphur Loading in the Stratosphere: Dependence of Required Emissions on Allowable Residual Warming Rate. *Theoretical and Applied Climatology*, **101**(1), 67–81.
Ellery, Alex. 2016. Are Self-Replicating Machines Feasible? *Journal of Spacecraft and Rockets*, **53**(2), 317–327.
Enriquez, J. Emilio, Siemion, Andrew, Foster, Griffin, Gajjar, Vishal, Hellbourg, Greg, Hickish, Jack, Isaacson, Howard, Price, Danny C., Croft, Steve, DeBoer, David, Lebofsky, Matt, MacMahon, David H. E., and Werthimer, Dan. 2017. The Breakthrough Listen Search for Intelligent Life: 1.1–1.9 GHz Observations of 692 Nearby Stars. *The Astrophysical Journal*, **849**(2), 104.
Erlykin, A. D., and Wolfendale, A. W. 2010. Long Term Time Variability of Cosmic Rays and Possible Relevance to the Development of Life on Earth. *Surveys in Geophysics*, **31**(4), 383–398.
Ezzedine, Souheil M., Lomov, Ilya, Miller, Paul L., Dennison, Deborah S., Dearborn, David S., and Antoun, Tarabay H. 2015. Simulation of Asteroid Impact on Ocean Surfaces, Subsequent Wave Generation and the Effect on US Shorelines. *Procedia Engineering*, **103**, 113–120.
Fabian, A. C. 2012. Observational Evidence of Active Galactic Nuclei Feedback. *Annual Review of Astronomy and Astrophysics*, **50**(1), 455–489.
Falcke, Heino, and Rezzolla, Luciano. 2014. Fast Radio Bursts: The Last Sign of Supramassive Neutron Stars. *Astronomy & Astrophysics*, **562**(feb), A137.
Falk, D. 2004. Prelinguistic Evolution in Early Hominins: Whence Motherese? *Behavioral and Brain Sciences*, **27**, 491–541.
Farquhar, J., Zerkle, A. L., and Bekker, A. 2014. Geologic and Geochemical Constraints on Earth's Early Atmosphere. Pages 91–138 of: *Treatise on Geochemistry.* Elsevier.
Farrer, Joan. 2010. Smart Dust: Sci-Fi Applications Enabled by Synthetic Fiber and Textiles Technology. *TEXTILE*, **8**(3), 342–346.
Fernald, R. D. 2006. Casting a Genetic Light on the Evolution of Eyes. *Science*, **313**(5795), 1914–1918.
Filippova, L. N., and Strelnitskij, V. S. 1988. Ecliptic as an Attractor for SETI. *Astronomicheskii Tsirkulyar*, **1531**, 31.

Fiori, R. A. D., Boteler, D. H., and Gillies, D. M. 2014. Assessment of GIC Risk Due to Geomagnetic Sudden Commencements and Identification of the Current Systems Responsible. *Space Weather*, **12**(1), 76–91.

Fischbach, Ephraim, and Gruenwald, John T. 2017. NU-SETI: A Proposal to Detect Extra-Terrestrial Signals Carried by Neutrinos. feb.

Fischer-Kowalski, M. 1997. *Gesellschaftlicher Stoffwechsel und Kolonisierung von Natur: ein Versuch in sozialer {Ö}kologie*. G+B Verlag Fakultas.

Fixsen, D. J. 2009. The Temperature of the Cosmic Microwave Background. *The Astrophysical Journal*, **707**(2), 916–920.

Flasiński, Mariusz. 2016. *Introduction to Artificial Intelligence*. Cham: Springer International Publishing.

Flores Martinez, Claudio L. 2014. SETI in the Light of Cosmic Convergent Evolution. *Acta Astronautica*, **104**(1), 341–349.

Fogg, M. 1987. Temporal Aspects of the Interaction Among the First Galactic Civilizations: The 'Interdict Hypothesis'. *Icarus*, **69**(2), 370–384.

Foley, Bradford J., Bercovici, David, and Landuyt, William. 2012. The Conditions for Plate Tectonics on Super-Earths: Inferences from Convection Models with Damage. *Earth and Planetary Science Letters*, **331–332**(may), 281–290.

Foley, Jonathan A., DeFries, Ruth, Asner, Gregory P., Barford, Carol, Bonan, Gordon, Carpenter, Stephen R., Chapin, F. Stuart, Coe, Michael T., Daily, Gretchen C., Gibbs, Holly K., Helkowski, Joseph H., Holloway, Tracey, Howard, Erica A., Kucharik, Christopher J., Monfreda, Chad, Patz, Jonathan A., Prentice, I. Colin, Ramankutty, Navin, and Snyder, Peter K. 2005. Global Consequences of Land Use. *Science*, **309**(5734), 570–574.

Forgan, D., and Kipping, D. 2013. Dynamical Effects on the Habitable Zone for Earth-like Exomoons. *MNRAS*, **432**(4), 2994–3004.

Forgan, D., and Rice, K. 2013. Towards a Population Synthesis Model of Objects Formed by Self-Gravitating Disc Fragmentation and Tidal Downsizing. *MNRAS*, **432**(4), 3168–3185.

Forgan, D. H., Hall, C., Meru, F., and Rice, W. K. M. 2018a. Towards a Population Synthesis Model of Self-Gravitating Disc Fragmentation and Tidal Downsizing II: the Effect of Fragment–Fragment Interactions. *MNRAS*, **474**(4), 5036–5048.

Forgan, Duncan, and Rice, Ken. 2011. The Jeans Mass as a Fundamental Measure of Self-Gravitating Disc Fragmentation and Initial Fragment Mass. *MNRAS*, **417**(3), 1928–1937.

Forgan, Duncan, and Yotov, Vergil. 2014. The Effect of Planetary Illumination on Climate Modelling of Earth-like Exomoons. *MNRAS*, **441**(4), 3513–3523.

Forgan, Duncan, Dayal, Pratika, Cockell, Charles, and Libeskind, Noam. 2017. Evaluating Galactic Habitability Using High-Resolution Cosmological Simulations of Galaxy Formation. *International Journal of Astrobiology*, **16**(01), 60–73.

Forgan, Duncan H. 2009. A Numerical Testbed for Hypotheses of Extraterrestrial Life and Intelligence. *International Journal of Astrobiology*, **8**(02), 121.

2011. Spatio-Temporal Constraints on the Zoo Hypothesis, and the Breakdown of Total Hegemony. *International Journal of Astrobiology*, **10**(04), 341–347.

2013. On the Possibility of Detecting Class A Stellar Engines Using Exoplanet Transit Curves. *J. Br. Interplanet. Soc.*, **66**(jun), 144–154.

2014. Can Collimated Extraterrestrial Signals be Intercepted? *J. Br. Interplanet. Soc.*, **67**(oct), 232–236.

2017a. Exoplanet Transits as the Foundation of an Interstellar Communications Network. *IJA*, jul, in press.

2017b. The Galactic Club or Galactic Cliques? Exploring the Limits of Interstellar Hegemony and the Zoo Hypothesis. *International Journal of Astrobiology*, **16**(04), 349–354.

Forgan, Duncan H., and Elvis, Martin. 2011. Extrasolar Asteroid Mining as Forensic Evidence for Extraterrestrial Intelligence. *International Journal of Astrobiology*, **10**(04), 307–313.

Forgan, Duncan H., and Rice, Ken. 2010. Numerical Testing of the Rare Earth Hypothesis Using Monte Carlo Realization Techniques. *International Journal of Astrobiology*, **9**(02), 73.

Forgan, Duncan H., Papadogiannakis, Semeli, and Kitching, Thomas. 2013. The Effect of Probe Dynamics on Galactic Exploration Timescales. *J. Br. Interplanet. Soc.*, **66**, 171–177.

Forgan, Duncan H., Heller, René, and Hippke, Michael. 2018b. Photogravimagnetic Assists of Light Sails: A Mixed Blessing for Breakthrough Starshot? *Monthly Notices of the Royal Astronomical Society*, **474**(3), 3212–3220.

Forgan, Duncan H., Wright, Jason T., Tarter, Jill C., Korpela, Eric J., Siemion, Andrew P. V., Almár, Iván, and Piotelat, Elisabeth. 2018c. Rio 2.0 – Revising the Rio Scale for SETI Detections. *IJA*, in press.

Forshaw, Jason, Aglietti, Guglielmo, Navarathinam, Nimal, Kadhem, Haval, Salmon, Thierry, Joffre, Eric, Chabot, Thomas, Retat, Ingo, Axthelm, Robert, Barraclough, Simon, and Others. 2015. An In-Orbit Active Debris Removal Mission-REMOVEDEBRIS: Pre-Launch Update. In: *Int. Astronautical Congress, IAC'2015*.

Forward, R. 1989. Grey Solar Sails. In: *25th Joint Propulsion Conference*. Reston, Virigina: American Institute of Aeronautics and Astronautics.

1984. Roundtrip Interstellar Travel Using Laser-Pushed Lightsails. *Journal of Spacecraft and Rockets*, **21**(2), 187–195.

1985. Starwisp – An Ultra-Light Interstellar Probe. *Journal of Spacecraft and Rockets*, **22**(3), 345–350.

Foster, K., and Ratnieks, F. 2005. A New Eusocial Vertebrate? *Trends in Ecology & Evolution*, **20**(7), 363–364.

Fouchard, M., Froeschlé, Ch., Rickman, H., and Valsecchi, G. B. 2011. The Key Role of Massive Stars in Oort Cloud Comet Dynamics. *Icarus*, **214**(1), 334–347.

Fouts, R. S. 1973. Acquisition and Testing of Gestural Signs in Four Young Chimpanzees. *Science*, **180**(4089), 978–980.

Freeman, Chris, Ostle, Nick, and Kang, Hojeong. 2001. An Enzymic 'Latch' on a Global Carbon Store. *Nature*, **409**(6817), 149–149.

Freitas, Robert A. 1980. A Self-Reproducing Interstellar Probe. *J. Br. Interplanet. Soc.*, **33**, 251–264.

1983. The Search for Extraterrestrial Artifacts (SETA). *British Interplanetary Society*, **36**, 501–506.

2000. *Some Limits to Global Ecophagy by Biovorous Nanoreplicators, with Public Policy Recommendations*. Tech. rept. Foresight Institute.

Freudenthal, H. 1960. *Lincos: Design of a Language for Cosmic Intercourse*. Studies in Logic and the Foundations of Mathematics, no. v. 1. North-Holland.

Fripp, Deborah, Owen, Caryn, Quintana-Rizzo, Ester, Shapiro, Ari, Buckstaff, Kara, Jankowski, Kristine, Wells, Randall, and Tyack, Peter. 2005. Bottlenose Dolphin (Tursiops Truncatus) Calves Appear to Model Their Signature Whistles on the Signature Whistles of Community Members. *Animal Cognition*, **8**(1), 17–26.

Frisch, Priscilla C., and Mueller, Hans-Reinhard. 2013. Time-Variability in the Interstellar Boundary Conditions of the Heliosphere: Effect of the Solar Journey on the Galactic Cosmic Ray Flux at Earth. *Space Science Reviews*, **176**(1-4), 21–34.

Galama, T. J., Vreeswijk, P. M., van Paradijs, J., Kouveliotou, C., Augusteijn, T., Böhnhardt, H., Brewer, J. P., Doublier, V., Gonzalez, J.-F., Leibundgut, B., Lidman, C., Hainaut, O. R., Patat, F., Heise, J., in't Zand, J., Hurley, K., Groot, P. J., Strom, R. G., Mazzali, P. A., Iwamoto, K., Nomoto, K., Umeda, H., Nakamura, T., Young, T. R., Suzuki, T., Shigeyama, T., Koshut, T., Kippen, M., Robinson, C., de Wildt, P., Wijers, R. A. M. J., Tanvir, N., Greiner, J., Pian, E., Palazzi, E., Frontera, F., Masetti, N., Nicastro, L., Feroci, M., Costa, E., Piro, L., Peterson, B. A., Tinney, C., Boyle, B., Cannon, R., Stathakis, R., Sadler, E., Begam, M. C., and Ianna, P. 1998. Discovery of the Peculiar Supernova 1998bw in the Error Box of GRB980425. *Nature*, **395**(6703), 670–672.

Gale, Joseph. 2009. *The Astrobiology of Earth*. 1st edn. Oxford: Oxford University Press.

Gammie, C. 2001. Nonlinear Outcome of Gravitational Instability in Cooling, Gaseous Disks. *ApJ*, **553**(1), 174–183.

Garcia-Escartin, Juan Carlos, and Chamorro-Posada, Pedro. 2013. Scouting the Spectrum for Interstellar Travellers. *Acta Astronautica*, **85**(apr), 12–18.

Gardner, Chester S., Liu, Alan Z., Marsh, D. R., Feng, Wuhu, and Plane, J. M. C. 2014. Inferring the Global Cosmic Dust Influx to the Earth's Atmosphere from Lidar Observations of the Vertical Flux of Mesospheric Na. *Journal of Geophysical Research: Space Physics*, **119**(9), 7870–7879.

Gardner, R. Allen, and Gardner, Beatrice T. 1969. Teaching Sign Language to a Chimpanzee. *1Science*, **165**(3894), 664–672.

Garrett, M. A. 2015. Application of the Mid-IR Radio Correlation to the Ĝ Sample and the Search for Advanced Extraterrestrial Civilisations. *Astronomy & Astrophysics*, **581**(sep), L5.

Gerig, Austin, Olum, Ken D., and Vilenkin, Alexander. 2013. Universal Doomsday: Analyzing Our Prospects for Survival. *Journal of Cosmology and Astroparticle Physics*, **2013**(05), 013–013.

Gertz, J. 2016. Reviewing METI: A Critical Analysis of the Arguments. *JBIS*, **69**, 31–36.

Gillon, Michaël, Triaud, Amaury H. M. J., Demory, Brice-Olivier, Jehin, Emmanuël, Agol, Eric, Deck, Katherine M., Lederer, Susan M., de Wit, Julien, Burdanov, Artem, Ingalls, James G., Bolmont, Emeline, Leconte, Jeremy, Raymond, Sean N., Selsis, Franck, Turbet, Martin, Barkaoui, Khalid, Burgasser, Adam, Burleigh, Matthew R., Carey, Sean J., Chaushev, Aleksander, Copperwheat, Chris M., Delrez, Laetitia, Fernandes, Catarina S., Holdsworth, Daniel L., Kotze, Enrico J., Van Grootel, Valérie, Almleaky, Yaseen, Benkhaldoun, Zouhair, Magain, Pierre, and Queloz, Didier. 2017. Seven Temperate Terrestrial Planets Around the Nearby Ultracool Dwarf Star TRAPPIST-1. *Nature*, **542**(7642), 456–460.

Gisler, Galen, Weaver, Robert, and Gittings, Michael L. 2006. *Sage Calculations of the Tsunami Threat from La Palma*.

Glade, Nicolas, Ballet, Pascal, and Bastien, Olivier. 2012. A Stochastic Process Approach of the Drake Equation Parameters. *International Journal of Astrobiology*, **11**(02), 103–108.

Glansdorff, Nicolas, Xu, Ying, and Labedan, Bernard. 2008. The Last Universal Common Ancestor: Emergence, Constitution and Genetic Legacy of an Elusive Forerunner. *Biology Direct*, **3**(1), 29.

Glassmeier, Karl-Heinz, and Vogt, Joachim. 2010. Magnetic Polarity Transitions and Biospheric Effects. *Space Science Reviews*, **155**(1–4), 387–410.

Gleghorn, George, Asay, James, Atkinson, Dale, Flury, Walter, Johnson, Nicholas, Kessler, Donald, Knowles, Stephen, Rex, Dietrich, Toda, Susumu, Veniaminov, Stanislav, and Warren, Robert. 1995. *Orbital Debris: A Technical Assessment*. Tech. rept. National Academy Press, Washington, D.C.

Gonzalez, G., Brownlee, D., and Ward, P. 2001. The Galactic Habitable Zone: Galactic Chemical Evolution. *Icarus*, **152**(1), 185–200.

Good, Irving John. 1966. Speculations Concerning the First Ultraintelligent Machine. Pages 31–88 of: *Advances in Computers*. Elsevier.

Goodfellow, Ian J., Pouget-Abadie, Jean, Mirza, Mehdi, Xu, Bing, Warde-Farley, David, Ozair, Sherjil, Courville, Aaron, and Bengio, Yoshua. 2014. Generative Adversarial Networks. *ArXiv e-prints*, jun, ID 1406.2661.

Goodman, A. A., Benson, P. J., Fuller, G. A., and Myers, P. C. 1993. Dense Cores in Dark Clouds. VIII - Velocity Gradients. *ApJ*, **406**(apr), 528.

Gordon, J. P., Zeiger, H. J., and Townes, C. H. 1954. Molecular Microwave Oscillator and New Hyperfine Structure in the Microwave Spectrum of NH3. *Physical Review*, **95**(1), 282–284.

Gott, J. Richard. 1993. Implications of the Copernican Principle for Our Future Prospects. *Nature*, **363**(6427), 315–319.

Goudie, Andrew. 1992. *Environmental Change*. Oxford University Press.

Gould, S. J. 2000. *Wonderful Life: The Burgess Shale and the Nature of History*. Vintage.

Gowanlock, Michael G., Patton, David R., and McConnell, Sabine M. 2011. A Model of Habitability Within the Milky Way Galaxy. *Astrobiology*, **11**(9), 855–73.

Grace, Katja, Salvatier, John, Dafoe, Allan, Zhang, Baobao, and Evans, Owain. 2017. When Will AI Exceed Human Performance? Evidence from AI Experts. *Journal of Artificial Intelligence Research*, may, in press.

Gradie, Jonathan C., Chapman, Clark R., and Tedesco, Edward F. 1989. Distribution of Taxonomic Classes and the Compositional Structure of the Asteroid Belt. Pages 316–335 of: *Asteroids II*. Tucson: University of Arizona Press.

Gray, Robert H. 2015. The Fermi Paradox is Neither Fermi's Nor a Paradox. *Astrobiology*, **15**(3), 195–9.

Grimaldi, Claudio. 2017. Signal Coverage Approach to the Detection Probability of Hypothetical Extraterrestrial Emitters in the Milky Way. *Scientific Reports*, **7**(apr), 46273.

Gubbins, David, and Herrero-Bervera, Emilio. 2007. *Encyclopedia of Geomagnetism and Paleomagnetism*. Dordrecht: Springer Netherlands.

Gurzadyan, G. A. 1996. *Theory of Interplanetary Flights*. Amsterdam: Gordon and Breach.

Hacar, A., Tafalla, M., Kauffmann, J., and Kovács, A. 2013. Cores, Laments, and Bundles: Hierarchical Core Formation in the L1495/B213 Taurus Region. *Astronomy & Astrophysics*, **554**(jun), A55.

Hadjidemetriou, John D. 1963. Two-Body Problem with Variable Mass: A New Approach. *Icarus*, **2**(Jan), 440–451.

Hair, Thomas W. 2011. Temporal Dispersion of the Emergence of Intelligence: an Inter-Arrival Time Analysis. *International Journal of Astrobiology*, **10**(02), 131–135.

Hair, Thomas W., and Hedman, Andrew D. 2013. Spatial Dispersion of Interstellar Civilizations: A Probabilistic Site Percolation Model in Three Dimensions. *International Journal of Astrobiology*, **12**(01), 45–52.

Hall, Jeffrey C. 2008. Stellar Chromospheric Activity. *Living Reviews in Solar Physics*, **5**(2).

Hamkins, Joel David, and Lewis, Andy. 2000. Infinite Time Turing Machines. *The Journal of Symbolic Logic*, **65**(02), 567–604.

Hammond, Andrew, Galizi, Roberto, Kyrou, Kyros, Simoni, Alekos, Siniscalchi, Carla, Katsanos, Dimitris, Gribble, Matthew, Baker, Dean, Marois, Eric, Russell, Steven, Burt, Austin, Windbichler, Nikolai, Crisanti, Andrea, and Nolan, Tony. 2015. A CRISPR-Cas9 Gene Drive System Targeting Female Reproduction in the Malaria Mosquito Vector Anopheles Gambiae. *Nature Biotechnology*, **34**(1), 78–83.

Han, Zhidong, and Fina, Alberto. 2011. Thermal Conductivity of Carbon Nanotubes and Their Polymer Nanocomposites: A Review. *Progress in Polymer Science*, **36**(7), 914–944.

Hanson, Robin. 1998. *The Great Filter – Are We Almost Past It?*

Haqq-Misra, Jacob. 2018. Policy Options for the Radio Detectability of Earth. *Futures*, apr.

Haqq-Misra, Jacob, and Kopparapu, Ravi Kumar. 2012. On the Likelihood of Nonterrestrial Artifacts in the Solar System. *Acta Astronautica*, **72**(mar), 15–20.

Haqq-Misra, Jacob D., and Baum, Seth D. 2009. The Sustainability Solution to the Fermi Paradox. *J. Br. Interplanet. Soc.*, **62**, 47–51.

Harp, G. R., Richards, Jon, Shostak, Seth, Tarter, J. C., Vakoch, Douglas A., and Munson, Chris. 2016. Radio SETI Observations of the Anomalous Star KIC 8462852. *The Astrophysical Journal*, **825**(2), 155.

Hart, Michael H. 1975. Explanation for the Absence of Extraterrestrials on Earth. *QJRAS*, **16**, 128–135.

Hathaway, David H. 2010. The Solar Cycle. *Living Rev. Solar Phys.*, **7**, [Online Article]: cited 30/06/2016.

Hawkins, Jeff, and Ahmad, Subutai. 2016. Why Neurons Have Thousands of Synapses, a Theory of Sequence Memory in Neocortex. *Frontiers in Neural Circuits*, **10**(mar).

Hays, J. D., Imbrie, J., and Shackleton, N. J. 1976. Variations in the Earth's Orbit: Pacemaker of the Ice Ages. *Science (New York, N.Y.)*, **194**(4270), 1121–32.

Haywood, M., Lehnert, M. D., Di Matteo, P., Snaith, O., Schultheis, M., Katz, D., and Gómez, A. 2016. When the Milky Way Turned off the Lights: APOGEE Provides Evidence of Star Formation Quenching in our Galaxy. *Astronomy & Astrophysics*, **589**(apr), A66.

Haywood, Raphaëlle D. 2016. Stellar Activity as a Source of Radial-Velocity Variability. Pages 13–44 of: *Radial-velocity Searches for Planets Around Active Stars*. Springer.

Hein, Andreas M., Pak, M., Putz, D., and Reiss, P. 2012. World Ships – Architectures and Feasibility Revisited. *JBIS*, **65**, 119–133.

Heisler, J. 1986. The Influence of the Galactic Tidal Field on the Oort Comet Cloud. *Icarus*, **65**(1), 13–26.

Hejnol, A., and Martindale, M. Q. 2008. Acoel Development Supports a Simple Planula-like Urbilaterian. *Philosophical Transactions of the Royal Society B: Biological Sciences*, **363**(1496), 1493–1501.

Heller, René. 2017. Relativistic Generalization of the Incentive Trap of Interstellar Travel with Application to Breakthrough Starshot. *MNRAS*, **470**(3), 3664–3671.

Heller, René, and Barnes, Rory. 2013. Exomoon Habitability Constrained by Illumination and Tidal Heating. *Astrobiology*, **13**(1), 18–46.

Heller, Rene, and Hippke, Michael. 2017. Deceleration of High-Velocity Interstellar Photon Sails into Bound Orbits at Alpha Centauri. *ApJ*, **835**(2), L32.

Heller, René, and Pudritz, Ralph E. 2016. The Search for Extraterrestrial Intelligence in Earth's Solar Transit Zone. *Astrobiology*, **16**(4), 259–270.

Heller, René, Williams, Darren, Kipping, David, Limbach, Mary Anne, Turner, Edwin, Greenberg, Richard, Sasaki, Takanori, Bolmont, Emeline, Grasset, Olivier, Lewis, Karen, Barnes, Rory, and Zuluaga, Jorge I. 2014. Formation, Habitability, and Detection of Extrasolar Moons. *Astrobiology*, **14**(9), 798–835.

Heller, René, Hippke, Michael, and Kervella, Pierre. 2017. Optimized Trajectories to the Nearest Stars Using Lightweight High velocity Photon Sails. *The Astronomical Journal*, **154**(3), 115.

Helliwell, J. F., and Putnam, R. D. 2004. The Social Context of Well-Being. *Philosophical Transactions of the Royal Society B: Biological Sciences*, **359**(1449), 1435–1446.

Hennebelle, P., and Ciardi, A. 2009. Disk Formation During Collapse of Magnetized Protostellar Cores. *A&A*, **506**(2), L29–L32.

Herman, Louis M., Richards, Douglas G., and Wolz, James P. 1984. Comprehension of Sentences by Bottlenosed Dolphins. *Cognition*, **16**(2), 129–219.

Herzing, D L, and Johnson, C M. 2015. *Dolphin Communication and Cognition: Past, Present, and Future*. MIT Press.

Hilbert, Martin, and López, Priscila. 2011. The World's Technological Capacity to Store, Communicate, and Compute Information. *Science (New York, N.Y.)*, **332**(6025), 60–5.

Hippke, Michael. 2018. Spaceflight from Super-Earths Is Difficult. *IJA*, Apr, in press.

Hippke, Michael, and Forgan, Duncan H. 2017. Interstellar Communication. VI. Searching X-ray Spectra for Narrowband Communication. *eprint arXiv:1712.06639*, dec.

Hirata, K., Kajita, T., Koshiba, M., Nakahata, M., Oyama, Y., Sato, N., Suzuki, A., Takita, M., Totsuka, Y., Kifune, T., Suda, T., Takahashi, K., Tanimori, T., Miyano, K., Yamada, M., Beier, E. W., Feldscher, L. R., Kim, S. B., Mann, A. K., Newcomer, F. M., Van, R., Zhang, W., and Cortez, B. G. 1987. Observation of a Neutrino Burst from the Supernova SN1987A. *Physical Review Letters*, **58**(14), 1490–1493.

Hobaiter, Catherine, and Byrne, Richard W. 2011. The Gestural Repertoire of the Wild Chimpanzee. *Animal Cognition*, **14**(5), 745–767.

Horner, J., and Jones, B. W. 2008a. Jupiter – Friend or Foe? I: The Asteroids. *International Journal of Astrobiology*, **7**(3–4), 251.

Horner, J., and Jones, B. W. 2008b. Jupiter – Friend or Foe? II: The Centaurs. *International Journal of Astrobiology*, **8**(02), 75.

Horner, J., Jones, B. W., and Chambers, J. 2010. Jupiter – Friend or Foe? III: The Oort Cloud Comets. *International Journal of Astrobiology*, **9**(01), 1.

Horvath, P., and Barrangou, R. 2010. CRISPR/Cas, the Immune System of Bacteria and Archaea. *Science*, **327**(5962), 167–170.

Howard, Andrew, Horowitz, Paul, Mead, Curtis, Sreetharan, Pratheev, Gallicchio, Jason, Howard, Steve, Coldwell, Charles, Zajac, Joe, and Sliski, Alan. 2007. Initial Results from Harvard All-Sky Optical SETI. *Acta Astronautica*, **61**(1–6), 78–87.

Howard, Andrew W., Horowitz, Paul, Wilkinson, David T., Coldwell, Charles M., Groth, Edward J., Jarosik, Norm, Latham, David W., Stefanik, Robert P., Willman, Jr., Alexander J., Wolff, Jonathan, and Zajac, Joseph M. 2004. Search for Nanosecond Optical Pulses from Nearby Solar-Type Stars. *ApJ*, **613**(2), 1270–1284.

Hring, Nadine, and Rix, Hans-Walter. 2004. On the Black Hole Mass-Bulge Mass Relation. *The Astrophysical Journal*, **604**(2), L89–L92.

Huang, Su-Shu. 1959. The Problem of Life in the Universe and the Mode of Star Formation. *PASP*, **71**(oct), 421.

Hulse, R. A., 1994. The Discovery of the Binary Pulsar. *Bulletin of the Astronomical Society*, **26**, 971–972.

Hunt, B. G. 1979. The Effects of Past Variations of the Earth's Rotation Rate on Climate. *Nature*, **281**(5728), 188–191.

Huybers, Peter. 2007. Glacial Variability over the Last Two Million Years: an Extended Depth-Derived Agemodel, Continuous Obliquity Pacing, and the Pleistocene Progression. *Quaternary Science Reviews*, **26**(1–2), 37–55.

Huybers, Peter, and Wunsch, Carl. 2004. A Depth-Derived Pleistocene Age Model: Uncertainty Estimates, Sedimentation Variability, and Nonlinear Climate Change. *Paleoceanography*, **19**(1), n/a–n/a.

IceCube. 2017. Search for Astrophysical Sources of Neutrinos Using Cascade Events in IceCube. *The Astrophysical Journal*, **846**(2), 136.

Ida, S., Guillot, T., and Morbidelli, A. 2016. The Radial Dependence of Pebble Accretion Rates: A Source of Diversity in Planetary Systems. *Astronomy & Astrophysics*, **591**(jul), A72.

IEA. 2014. *2014 Key World Energy Statistics*. Tech. rept. International Energy Agency.

IPCC. 2013. *Climate Change 2013: The Physical Science Basis*. Tech. rept. IPCC.

Isaacson, Howard, Siemion, Andrew P. V., Marcy, Geoffrey W., Lebofsky, Matt, Price, Danny C., MacMahon, David, Croft, Steve, DeBoer, David, Hickish, Jack, Werthimer, Dan, Sheikh, Sofia, Hellbourg, Greg, and Enriquez, J. Emilio. 2017. The Breakthrough Listen Search for Intelligent Life: Target Selection of Nearby Stars and Galaxies. *eprint arXiv:1701.06227*, jan.

Ishino, Y., Shinagawa, H., Makino, K., Amemura, M., and Nakata, A. 1987. Nucleotide Sequence of the iap Gene, Responsible for Alkaline Phosphatase Isozyme Conversion in Escherichia Coli, and Identification of the Gene Product. *Journal of Bacteriology*, **169**(12), 5429–5433.

Jacobson, S. A., and Morbidelli, A. 2014. Lunar and Terrestrial Planet Formation in the Grand Tack Scenario. *Philosophical Transactions of the Royal Society A: Mathematical, Physical and Engineering Sciences*, **372**(2024), 20130174–20130174.

James, S.R. 1989. Hominid Use of Fire in the Lower and Middle Pleistocene: A Review of the Evidence. *Current Anthropology*, **30**, 1–26.

Janik, Vincent M. 2009. Chapter 4 Acoustic Communication in Delphinids. *Advances in the Study of Behavior*, **40**, 123–157.

Javaux, Emmanuelle J. 2007. Eukaryotic Membranes and Cytoskeleton: Origins and Evolution. Chap. The Early, pages 1–19 of: *The Early Eukaryotic Fossil Record*. New York, NY: Springer New York.

Jerison, Harry J. 1973. *Evolution of the Brain and Intelligence*. 1st edn. New York: Academic Press Inc.

Jewitt, David. 2000. Astronomy: Eyes Wide Shut. *Nature*, **403**(6766), 145–148.

Jiménez-Torres, Juan J., Pichardo, Bárbara, Lake, George, and Segura, Antígona. 2013. Habitability in Different Milky Way Stellar Environments: A Stellar Interaction Dynamical Approach. *Astrobiology*, **13**(5), 491–509.

Johansen, Anders, Oishi, Jeffrey S., Low, Mordecai-Mark Mac, Klahr, Hubert, Henning, Thomas, and Youdin, Andrew. 2007. Rapid Planetesimal Formation in Turbulent Circumstellar Disks. *Nature*, **448**(7157), 1022–1025.

Johnson, N. L., Office, O. D. P., and Cemter, N. J. S. 2008. *History of On-Orbit Satellite Fragmentations (14th Edition)*. Books Express Publishing.

Jones, Andrew, and Straub, Jeremy. 2017. Concepts for 3D Printing-Based Self-Replicating Robot Command and Coordination Techniques. *Machines*, **5**(2), 12.

Jones, Eric M. 1985. *'Where Is Everybody?' An Account of Fermi's Question*. Tech. rept. Los Alamos National Laboratory.

Jones, Rhys, Haufe, Patrick, Sells, Edward, Iravani, Pejman, Olliver, Vik, Palmer, Chris, and Bowyer, Adrian. 2011. RepRap – the Replicating Rapid Prototyper. *Robotica*, **29**(01), 177–191.

Joordens, Josephine C. A., D'Errico, Francesco, Wesselingh, Frank P., Munro, Stephen, de Vos, John, Wallinga, Jakob, Ankjærgaard, Christina, Reimann, Tony, Wijbrans,

Jan R., Kuiper, Klaudia F., Mücher, Herman J., Coqueugniot, Hélène, Prié, Vincent, Joosten, Ineke, van Os, Bertil, Schulp, Anne S., Panuel, Michel, van der Haas, Victoria, Lustenhouwer, Wim, Reijmer, John J. G., and Roebroeks, Wil. 2014. Homo Erectus at Trinil on Java Used Shells for Tool Production and Engraving. *Nature*, **518**(7538), 228–231.

Jorda, L., Gaskell, R., Capanna, C., Hviid, S., Lamy, P., Ďurech, J., Faury, G., Groussin, O., Gutiérrez, P., Jackman, C., Keihm, S.J., Keller, H. U., Knollenberg, J., Kührt, E., Marchi, S., Mottola, S., Palmer, E., Schloerb, F. P., Sierks, H., Vincent, J.-B., A'Hearn, M. F., Barbieri, C., Rodrigo, R., Koschny, D., Rickman, H., Barucci, M. A., Bertaux, J. L., Bertini, I., Cremonese, G., Da Deppo, V., Davidsson, B., Debei, S., De Cecco, M., Fornasier, S., Fulle, M., Güttler, C., Ip, W.-H., Kramm, J. R., Küppers, M., Lara, L. M., Lazzarin, M., Lopez Moreno, J. J., Marzari, F., Naletto, G., Oklay, N., Thomas, N., Tubiana, C., and Wenzel, K.-P. 2016. The Global Shape, Density and Rotation of Comet 67P/Churyumov-Gerasimenko from preperihelion Rosetta/OSIRIS Observations. *Icarus*, **277**(oct), 257–278.

Jortner, Joshua. 2006. Conditions for the Emergence of Life on the Early Earth: Summary and Reflections. *Philosophical Transactions of the Royal Society of London. Series B, Biological Sciences*, **361**(1474), 1877–91.

Kaiser, Brent N., Gridley, Kate L., Ngaire Brady, Joanne, Phillips, Thomas, and Tyerman, Stephen D. 2005. The Role of Molybdenum in Agricultural Plant Production. *Annals of Botany*, **96**(5), 745–754.

Kanas, Nick. 2015. *Humans in Space*. Cham: Springer International Publishing.

Kardashev, N. S. 1964. Transmission of Information by Extraterrestrial Civilizations. *Soviet Astronomy*, **8**, 217–221.

Kasting, James F. 1988. Runaway and Moist Greenhouse Atmospheres and the Evolution of Earth and Venus. *Icarus*, **74**(3), 472–494.

Kasting, J. F., Whitmire, D. P., and Reynolds, R. T. 1993. Habitable Zones around Main Sequence Stars. *Icarus*, **101**(1), 108–128.

Keeling, Patrick J., and Palmer, Jeffrey D. 2008. Horizontal Gene Transfer in Eukaryotic Evolution. *Nature Reviews Genetics*, **9**(8), 605–618.

Kennett, D. J., and Winterhalder, B. 2006. *Behavioral Ecology and the Transition to Agriculture*. Origins of Human Behavior and Culture. University of California Press.

Kerfeld, C. A. 2005. Protein Structures Forming the Shell of Primitive Bacterial Organelles. *Science*, **309**(5736), 936–938.

Kessler, Donald J., and Cour-Palais, Burton G. 1978. Collision Frequency of Artificial Satellites: The Creation of a Debris Belt. *Journal of Geophysical Research*, **83**(A6), 2637.

Khodachenko, Maxim L., Ribas, Ignasi, Lammer, Helmut, Grießmeier, Jean-Mathias, Leitner, Martin, Selsis, Franck, Eiroa, Carlos, Hanslmeier, Arnold, Biernat, Helfried K., Farrugia, Charles J., and Rucker, Helmut O. 2007. Coronal Mass Ejection (CME) Activity of Low Mass M Stars as An Important Factor for The Habitability of Terrestrial Exoplanets. I. CME Impact on Expected Magnetospheres of Earth-Like Exoplanets in Close-In Habitable Zones. *Astrobiology*, **7**(1), 167–184.

Kiefer, Flavien, des Etangs, Alain Lecavelier, Vidal-Madjar, Alfred, Hébrard, Guillaume, Bourrier, Vincent, and Wilson, Paul-Anthony. 2017. Detection of a Repeated Transit Signature in the Light Curve of the Enigma Star KIC 8462852: a 928-day period? *arXiv e-print 1709.01732*, sep.

King, S. L., Sayigh, L. S., Wells, R. S., Fellner, W., and Janik, V. M. 2013. Vocal Copying of Individually Distinctive Signature Whistles in Bottlenose Dolphins. *Proceedings of the Royal Society B: Biological Sciences*, **280**(1757), 20130053–20130053.

Kipping, D. M., Bakos, G. Á., Buchhave, L., Nesvorný, D., and Schmitt, A. 2012. The Hunt for Exomoons with Kepler (HEK) I. Description of a New Observational Project. *ApJ*, **750**(2), 115.

Kipping, David M. 2009a. Transit Timing Effects Due to an Exomoon. *MNRAS*, **392**(1), 181–189.

2009b. Transit Timing Effects Due to an Exomoon - II. *MNRAS*, **396**(3), 1797–1804.

Kipping, David M., and Teachey, Alex. 2016. A Cloaking Device for Transiting Planets. *MNRAS*, **459**(2), 1233–1241.

Kipping, David M., Forgan, Duncan, Hartman, Joel, Nesvorný, D., Bakos, Gáspár Á., Schmitt, A., and Buchhave, L. 2013. The Hunt for Exomoons with Kepler (HEK) III. The First Search for an Exomoon around a Habitable-Zone Planet. *ApJ*, **777**(2), 134.

Kislev, Mordechai E, Hartmann, Anat, and Bar-Yosef, Ofer. 2006. Early Domesticated Fig in the Jordan Valley. *Science*, **312**(5778), 1372–1374.

Kleiber, Max. 1932. Body Size and Metabolism. *Hilgardia*, **6**(11), 315–353.

Klein, R G. 2009. *The Human Career: Human Biological and Cultural Origins, Third Edition*. University of Chicago Press.

Kluyver, Thomas A., Jones, Glynis, Pujol, Benoît, Bennett, Christopher, Mockford, Emily J., Charles, Michael, Rees, Mark, and Osborne, Colin P. 2017. Unconscious Selection Drove Seed Enlargement in Vegetable Crops. *Evolution Letters*, **1**(2), 64–72.

Kohonen, Teuvo. 1982. Self-Organized Formation of Topologically Correct Feature Maps. *Biological Cybernetics*, **43**(1), 59–69.

Kokubo, Eiichiro, and Ida, Shigeru. 1998. Oligarchic Growth of Protoplanets. *Icarus*, **131**(1), 171–178.

Kolbert, Elizabeth. 2014. *The Sixth Extinction: an Unnatural History*. 1st edn. New York: Henry Holt and Company, 2014.

Kono, Masaru, and Roberts, Paul H. 2002. Recent Geodynamo Simulations and Observations of the Geomagnetic Field. *Reviews of Geophysics*, **40**(4), 1013.

Kopparapu, Ravi K., Ramirez, Ramses M., SchottelKotte, James, Kasting, James F., Domagal-Goldman, Shawn, and Eymet, Vincent. 2014. Habitable Zones around Main-Sequence Stars: Dependence on Planetary Mass. *ApJ*, **787**, L29.

Kopparapu, Ravi Kumar, Ramirez, Ramses, Kasting, James F., Eymet, Vincent, Robinson, Tyler D., Mahadevan, Suvrath, Terrien, Ryan C., Domagal-Goldman, Shawn, Meadows, Victoria, and Deshpande, Rohit. 2013. Habitable Zones around Main Sequence Stars: New Estimates. *ApJ*, **765**(2), 131.

Korenaga, Jun. 2010. On the Likelihood of Plate Tectonics on Super-Earths: Does Size Matter? *The Astrophysical Journal*, **725**(1), L43–L46.

Korpela, Eric J., Sallmen, Shauna M., and Greene, Diana Leystra. 2015. Modeling Indications of Technology in Planetary Transit Light Curves – Dark Side Illumination. *ApJ*, **809**(2), 139.

Korycansky, D. G., and Asphaug, Erik. 2006. Low-Speed Impacts Between Rubble Piles Modeled as Collections of Polyhedra. *Icarus*, **181**(2), 605–617.

Koshland Jr., D. E. 2002. Special Essay: The Seven Pillars of Life. *Science*, **295**(5563), 2215–2216.

Kozmik, Zbynek, Ruzickova, Jana, Jonasova, Kristyna, Matsumoto, Yoshifumi, Vopalensky, Pavel, Kozmikova, Iryna, Strnad, Hynek, Kawamura, Shoji, Piatigorsky, Joram, Paces, Vaclav, and Vlcek, Cestmir. 2008. Assembly of the Cnidarian Camera-Type Eye from Vertebrate-like Components. *Proceedings of the National Academy of Sciences*, **105**(26), 8989–8993.

Krissansen-Totton, Joshua, Olson, Stephanie, and Catling, David C. 2018. Disequilibrium Biosignatures over Earth History and Implications for Detecting Exoplanet Life. *Science Advances*, **4**(1), eaao5747.

Krolik, J. H. 1999. *Active Galactic Nuclei*. 3rd edn. Princeton: Princeton University Press.

Kump, Lee, Bralower, Timothy, and Ridgwell, Andy. 2009. Ocean Acidification in Deep Time. *Oceanography*, **22**(4), 94–107.

Laakso, T., Rantala, J., and Kaasalainen, M. 2006. Gravitational Scattering by Giant Planets. *Astronomy and Astrophysics*, **456**(1), 373–378.

Lacki, Brian C. 2015. SETI at Planck Energy: When Particle Physicists Become Cosmic Engineers. *ArXiv e-prints*, mar.

Lacy, J. H., Townes, C. H., Geballe, T. R., and Hollenbach, D. J. 1980. Observations of the Motion and Distribution of the Ionized Gas in the Central Parsec of the Galaxy. II. *The Astrophysical Journal*, **241**(oct), 132.

Lacy, J. H., Evans, Neal J., II, Achtermann, J. M., Bruce, D. E., Arens, J. F., and Carr, J. S. 1989. Discovery of Interstellar Acetylene. *The Astrophysical Journal*, **342**(Jul), L43.

Lambrechts, M., and Johansen, A. 2012. Rapid Growth of Gas-Giant Cores by Pebble Accretion. *A&A*, **544**(Jul), A32.

Landauer, R. 1961. Irreversibility and Heat Generation in the Computing Process. *IBM Journal of Research and Development*, **5**(3), 183–191.

Landenmark, Hanna K. E., Forgan, Duncan H., and Cockell, Charles S. 2015. An Estimate of the Total DNA in the Biosphere. *PLOS Biology*, **13**(6), e1002168.

Landis, G. A. 1998. The Fermi Paradox: An Approach Based on Percolation Theory. *Journal of the British Interplanetary Society*, **51**.

Larmor, J. 1919. Possible Rotational Origin of Magnetic Fields of Sun and Earth. *Electrical Review*, **85**, 412.

Laskar, J., and Robutel, P. 1993. The Chaotic Obliquity of the Planets. *Nature*, **361**(6413), 608–612.

Latham, Katherine J. 2013. Human Health and the Neolithic Revolution: An Overview of Impacts of the Agricultural Transition on Oral Health, Epidemiology, and the Human Body. *Nebraska Anthropologist*, **28**, 95–102.

Learned, J. G., Pakvasa, S., and Zee, A. 2009. Galactic Neutrino Communication. *Physics Letters B*, **671**(jan), 15–19.

Lederberg, J. 1965. Signs of Life. Criterion-System of Exobiology. *Nature*, **207**(992), 9–13.

Leinhardt, Z., and Richardson, D. 2002. N-Body Simulations of Planetesimal Evolution: Effect of Varying Impactor Mass Ratio. *Icarus*, **159**(2), 306–313.

Letunic, Ivica, and Bork, Peer. 2016. Interactive Tree of Life (iTOL) v3: An Online Tool for the Display and Annotation of Phylogenetic and Other Trees. *Nucleic Acids Research*, **44**(W1), W242–W245.

Li, Di, and Pan, Zhichen. 2016. The Five-Hundred-Meter Aperture Spherical Radio Telescope Project. *Radio Science*, **51**(7), 1060–1064.

Lilly, John C. 1962. Vocal Behavior of the Dolphin. *Proceedings of the American Philosophical Society*, **106**(6), 520–529.

Lin, Douglas N. C., and Pringle, John E. 1990. The Formation and Initial Evolution of Protostellar Disks. *ApJ*, **358**(aug), 515.

Lin, Henry W., Abad, Gonzalo Gonzalez, and Loeb, Abraham. 2014. Detecting Industrial Pollution in the Atmospheres of Earth-like Exoplanets. *ApJ*, **792**(1), L7.

Lineweaver, C. 2001. An Estimate of the Age Distribution of Terrestrial Planets in the Universe: Quantifying Metallicity as a Selection Effect. *Icarus*, **151**(2), 307–313.

Lineweaver, Charles H., Fenner, Yeshe, and Gibson, Brad K. 2004. The Galactic Habitable Zone and the Age Distribution of Complex Life in the Milky Way. *Science*, **303**(5654), 59–62.

Lingam, Manasvi, and Loeb, Abraham. 2017. Fast Radio Bursts from Extragalactic Light Sails. *The Astrophysical Journal*, **837**(2), L23.

Lissauer, Jack J., Barnes, Jason W., and Chambers, John E. 2012. Obliquity Variations of a Moonless Earth. *Icarus*, **217**(1), 77–87.

Lloyd, Seth. 2000. Ultimate Physical Limits to Computation. *Nature*, **406**(6799), 1047–1054.

Loeb, Abraham, and Turner, Edwin L. 2012. Detection Technique for Artificially Illuminated Objects in the Outer Solar System and Beyond. *Astrobiology*, **12**(4), 290–4.

Loeb, Abraham, and Zaldarriaga, Matias. 2007. Eavesdropping on Radio Broadcasts from Galactic Civilizations with Upcoming Observatories for Redshifted 21 cm Radiation. *Journal of Cosmology and Astroparticle Physics*, **1**, 020–020.

Lomb, N. R. 1976. Least-Squares Frequency Analysis of Unequally Spaced Data. *Astrophysics and Space Science*, **39**(2), 447–462.

López-Morales, Mercedes, Gómez-Pérez, Natalia, and Ruedas, Thomas. 2011. Magnetic Fields in Earth-like Exoplanets and Implications for Habitability around M-dwarfs. *Origins of Life and Evolution of Biospheres*, **41**(6), 533–537.

Lorimer, D. R., Bailes, M., McLaughlin, M. A., Narkevic, D. J., and Crawford, F. 2007. A Bright Millisecond Radio Burst of Extragalactic Origin. *Science*, **318**(5851), 777–780.

Lovelock, J. E. 1965. A Physical Basis for Life Detection Experiments. *Nature*, **207**(4997), 568–570.

Lovelock, J. E., and Kaplan, I. R. 1975. Thermodynamics and the Recognition of Alien Biospheres [and Discussion]. *Proceedings of the Royal Society B: Biological Sciences*, **189**(1095), 167–181.

Lubin, Philip. 2016. A Roadmap to Interstellar Flight. *JBIS*, **69**(Apr), 40–72.

Lubin, Philip, and Hughes, Gary B. 2015. Directed Energy for Planetary Defense. Pages 941–991 of: Pelton, Joseph N, and Allahdadi, Firooz (eds), *Handbook of Cosmic Hazards and Planetary Defense*. Cham: Springer International Publishing.

Lutz, Christian, and Giljum, Stefan. 2009. Global Resource Use in a Business-as-Usual World up to 2030: Updated Results from the GINFORS Model. Pages 30–41 of: *Sustainable Growth and Resource Productivity: Economic and Global Policy Issues*. Greenleaf Publishing Limited.

Lyra, W., Johansen, A., Zsom, A., Klahr, H., and Piskunov, N. 2009. Planet Formation Bursts at the Borders of the Dead Zone in 2D Numerical Simulations of Circumstellar Disks. *A&A*, **497**(3), 869–888.

MacArthur, Robert H., and Wilson, Edward O. 2016. *The Theory of Island Biogeography*. Princeton University Press.

Maccone, Claudio. 2010. The Statistical Drake Equation. *Acta Astronautica*, **67**(11-12), 1366–1383.

2012. *Mathematical SETI*. 2012 edn. Springer.

Maiman, T. 1960. Stimulated Emission in Ruby. *Nature*, **187**, 493–494.

Males, Jared R., Close, Laird M., Guyon, Olivier, Morzinski, Katie, Puglisi, Alfio, Hinz, Philip, Follette, Katherine B., Monnier, John D., Tolls, Volker, Rodigas, Timothy J., Weinberger, Alycia, Boss, Alan, Kopon, Derek, Wu, Ya-lin, Esposito, Simone, Riccardi, Armando, Xompero, Marco, Briguglio, Runa, and Pinna, Enrico. 2014 (Jul). Direct Imaging of Exoplanets in the Habitable Zone with Adaptive Optics. Page

914820 of: Marchetti, Enrico, Close, Laird M., and Véran, Jean-Pierre (eds.), *Society of Photo-Optical Instrumentation Engineers (SPIE) Conference Series*, vol. 9148.

Malmberg, Daniel, Davies, Melvyn B., and Heggie, Douglas C. 2011. The Effects of Fly-Bys on Planetary Systems. *MNRAS*, **411**(2), 859–877.

Mamajek, Eric E., Barenfeld, Scott A., Ivanov, Valentin D., Kniazev, Alexei Y., Väisänen, Petri, Beletsky, Yuri, and Boffin, Henri M. J. 2015. The Closest Known Flyby of a Star to the Solar System. *ApJ*, **800**(1), L17.

Manchester, Zachary, and Loeb, Abraham. 2017. Stability of a Light Sail Riding on a Laser Beam. *The Astrophysical Journal Letters*, **837**(2), L20.

Maoz, Dan, and Mannucci, Filippo. 2012. Type-Ia Supernova Rates and the Progenitor Problem: A Review. *PASA*, **29**(04), 447–465.

Marmur, Abraham. 2004. The Lotus Effect: Superhydrophobicity and Metastability. *Langmuir*, **20**(9), 3517–3519.

Marois, Christian, Macintosh, Bruce, Barman, Travis, Zuckerman, B., Song, Inseok, Patience, Jennifer, Lafrenière, David, and Doyon, René. 2008. Direct Imaging of Multiple Planets Orbiting the Star HR 8799. *Science*, **322**(5906), 1348–52.

Martinelli, D. 2010. *A Critical Companion to Zoosemiotics: People, Paths, Ideas*. Biosemiotics. Springer Netherlands.

Martínez-Barbosa, C. A., Jílková, L., Portegies Zwart, S., and Brown, A. G. A. 2017. The Rate of Stellar Encounters Along a Migrating Orbit of the Sun. *MNRAS*, **464**(2), 2290–2300.

Marx, G. 1966. Interstellar Vehicle Propelled by Terrestrial Laser Beam. *Nature*, **211**(5044), 22–23.

Matloff, G. L. 2006. *Deep Space Probes: To the Outer Solar System and Beyond*. 2nd edn. Springer Science & Business Media.

Maury, A. J., André, Ph., Hennebelle, P., Motte, F., Stamatellos, D., Bate, M., Belloche, A., Duchêne, G., and Whitworth, A. 2010. Toward Understanding the Formation of Multiple Systems. *A&A*, **512**(Mar), A40.

Mayr, E. 1995. The SETI Debate: A Critique of the Search for Extraterrestrial Intelligence. *Bioastronomy News*, **7**, 3.

Mayr, Ernst. 1993. The Search for Intelligence. *Science*, **259**(5101), 1522–1523.

McCarthy, John. 1960. *Programs with Common Sense*. RLE and MIT Computation Center.

McInnes, Colin R., and Brown, John C. 1990. The Dynamics of Solar Sails with a Non-Point Source of Radiation Pressure. *Celestial Mechanics and Dynamical Astronomy*, **49**(3), 249–264.

McKay, Chris P. 2004. What Is Life – and How Do We Search for It in Other Worlds? *PLoS Biology*, **2**(9), e302.

McKee, Christopher F., and Ostriker, Eve C. 2007. Theory of Star Formation. *ARA&A*, **45**(1), 565–687.

McNamara, K. J. 2010. *The Star-Crossed Stone: The Secret Life, Myths, and History of a Fascinating Fossil*. Chicago: University of Chicago Press.

Meadows, Victoria S., Reinhard, Christopher T., Arney, Giada N., Parenteau, Mary N., Schwieterman, Edward W., Domagal-Goldman, Shawn D., Lincowski, Andrew P., Stapelfeldt, Karl R., Rauer, Heike, DasSarma, Shiladitya, Hegde, Siddharth, Narita, Norio, Deitrick, Russell, Lyons, Timothy W., Siegler, Nicholas, and Lustig-Yaeger, Jacob. 2017. Exoplanet Biosignatures: Understanding Oxygen as a Biosignature in the Context of Its Environment. may.

Mehrabian, A. 2017. *Nonverbal Communication*. Taylor & Francis.

Mejia, Annie C., Durisen, R.H., Pickett, Megan K., and Cai, Kai. 2005. The Thermal Regulation of Gravitational Instabilities in Protoplanetary Disks. II. Extended Simulations with Varied Cooling Rates. *ApJ*, **619**(2), 1098–1113.

Melott, A. L., Thomas, B. C., Kachelrieß, M., Semikoz, D. V., and Overholt, A. C. 2017. A Supernova at 50 pc: Effects on the Earth's Atmosphere and Biota. *ApJ*, **840**(2), 105.

Melott, Adrian L., and Thomas, Brian C. 2009. Late Ordovician Geographic Patterns of Extinction Compared with Simulations of Astrophysical Ionizing Radiation Damage. *Paleobiology*, **35**(03), 311–320.

2011. Astrophysical Ionizing Radiation and Earth: A Brief Review and Census of Intermittent Intense Sources. *Astrobiology*, **11**(4), 343–361.

Merrill, R. T., McElhinny, M. W., and McFadden, P. L. 1996. *The Magnetic Field of the Earth: Paleomagnetism, the Core, and the Deep Mantle.* International geophysics series. Academic Press.

Messerschmitt, David G. 2013. End-to-End Interstellar Communication System Design for Power Efficiency. *Arxiv e-prints 1305.4684*, May.

2015. Design for Minimum Energy in Interstellar Communication. *Acta Astronautica*, **107**(Feb), 20–39.

Mikhail, Sami, and Heap, Michael J. 2017. Hot Climate Inhibits Volcanism on Venus: Constraints from Rock Deformation Experiments and Argon Isotope Geochemistry. *Physics of the Earth and Planetary Interiors*, **268**(Jul), 18–34.

Milankovitch, Milutin. 1941. *Canon of Insolation and the Ice Age Problem.* Belgrade: Royal Serbian Academy.

Miller, S. L. 1953. A Production of Amino Acids Under Possible Primitive Earth Conditions. *Science*, **117**(3046), 528–529.

Millman, Peter M. 1975. Seven Maxims of UFOs – A Scientific Approach. *Journal of the Royal Astronomical Society of Canada*, **69**, 175–188.

Mironova, Irina A., Aplin, Karen L., Arnold, Frank, Bazilevskaya, Galina A., Harrison, R. Giles, Krivolutsky, Alexei A., Nicoll, Keri A., Rozanov, Eugene V., Turunen, Esa, and Usoskin, Ilya G. 2015. Energetic Particle Influence on the Earth's Atmosphere. *Space Science Reviews*, **194**(1), 1–96.

Mithen, S. 2011. *The Singing Neanderthals: The Origins of Music, Language, Mind and Body.* Orion.

Mojzsis, S. J., Arrhenius, G., McKeegan, K. D., Harrison, T. M., Nutman, A. P., and Friend, C. R. L. 1996. Evidence for Life on Earth before 3,800 Million Years Ago. *Nature*, **384**(6604), 55–59.

Montet, Benjamin T., and Simon, Joshua D. 2016. KIC 8462852 Faded Throughout the Kepler Mission. *The Astrophysical Journal*, **830**(2), L39.

Morbidelli, Alessandro. 2005. Origin and Dynamical Evolution of Comets and their Reservoirs. *35th Saas-Fee Advanced Course, available as arXiv e-print 0512256*, Dec.

Morgan, D. L. 1982. Concepts for the Design of an Antimatter Annihilation Rocket. *JBIS*, **35**, 405–413.

Morrison, Philip, Billingham, John, and Wolfe, John. 1977. The Search for Extraterrestrial Intelligence, SETI. *The Search for Extraterrestrial Intelligence*.

Mortier, A., Faria, J. P., Correia, C. M., Santerne, A., and Santos, N. C. 2015. BGLS: A Bayesian Formalism for the Generalised Lomb-Scargle Periodogram. *A&A*, **573**(Jan), A101.

Mosqueira, Ignacio, and Estrada, Paul R. 2003a. Formation of the Regular Satellites of Giant Planets in an Extended Gaseous Nebula I: Subnebula Model and Accretion of Satellites. *Icarus*, **163**(1), 198–231.

2003b. Formation of the Regular Satellites of Giant Planets in an Extended Gaseous Nebula II: Satellite Migration and Survival. *Icarus*, **163**(1), 232–255.

Mozdzen, T. J., Bowman, J. D., Monsalve, R. A., and Rogers, A. E. E. 2017. Improved Measurement of the Spectral Index of the Diffuse Radio Background Between 90 and 190 MHz. *Monthly Notices of the Royal Astronomical Society*, **464**(4), 4995–5002.

Murray, James D. 2004. *Mathematical Biology*. Interdisciplinary Applied Mathematics, vol. 17. New York, NY: Springer New York.

Nadell, Carey D., Xavier, Joao B., and Foster, Kevin R. 2009. The Sociobiology of Biofilms. *FEMS Microbiology Reviews*, **33**(1), 206–224.

NASA. 2007. *Near Earth Object Survey and Deflection: Analysis of Alternatives – Report to Congress*. Tech. rept. NASA, Washington, D.C.

Nascimbeni, V., Mallonn, M., Scandariato, G., Pagano, I., Piotto, G., Micela, G., Messina, S., Leto, G., Strassmeier, K. G., Bisogni, S., and Speziali, R. 2015. An LBT View of the Atmosphere of GJ1214b. *eprint arXiv:1505.01488*.

Nayakshin, Sergei. 2010a. Formation of Planets by Tidal Downsizing of Giant Planet Embryos. *MNRAS*, **408**(1), L36–L40.

2010b. Grain Sedimentation Inside Giant Planet Embryos. *MNRAS*, **408**(4), 2381–2396.

2011. Formation of Terrestrial Planet Cores Inside Giant Planet Embryos. *MNRAS*, **413**(2), 1462–1478.

Neuenswander, Ben, and Melott, Adrian. 2015. Nitrate Deposition Following an Astrophysical Ionizing Radiation Event. *Advances in Space Research*, **55**(12), 2946–2949.

Newman, William I., and Sagan, Carl. 1981. Galactic Civilizations: Population Dynamics and Interstellar Diffusion. *Icarus*, **46**(3), 293–327.

Nicholson, Arwen, and Forgan, Duncan. 2013. Slingshot Dynamics for Self-Replicating Probes and the Effect on Exploration Timescales. *International Journal of Astrobiology*, **12**(04), 337–344.

NOAA, NASA, UNEP, WMO, and EC. 2010. *Scientific Assessment of Ozone Depletion: 2010*. Tech. rept. World Meteorological Organisation.

Noack, Lena, and Breuer, Doris. 2014. Plate Tectonics on Rocky Exoplanets: Influence of Initial Conditions and Mantle Rheology. *Planetary and Space Science*, **98**(aug), 41–49.

Norris, R.S., and Arkin, W.M. 1996. Known Nuclear Tests Worldwide, 1945-1995. *The Bulletin of the Atomic Scientists, Volume 52, Issue 3*, May, 61–63.

Nowak, Martin A., Tarnita, Corina E., and Wilson, Edward O. 2010. The Evolution of Eusociality. *Nature*, **466**(7310), 1057–1062.

O'Brien, David P., Walsh, Kevin J., Morbidelli, Alessandro, Raymond, Sean N., and Mandell, Avi M. 2014. Water Delivery and Giant Impacts in the 'Grand Tack' Scenario. *Icarus*, **239**(Sep), 74–84.

Ochman, Howard, Lawrence, Jeffrey G., and Groisman, Eduardo A. 2000. Lateral Gene Transfer and the Nature of Bacterial Innovation. *Nature*, **405**(6784), 299–304.

Odenwald, Sten F., and Green, James L. 2007. Forecasting the Impact of an 1859-Caliber Superstorm on Geosynchronous Earth-Orbiting Satellites: Transponder Resources. *Space Weather*, **5**(6), S06002.

Olum, Ken D. 2000. The Doomsday Argument and the Number of Possible Observers. *arXiv e-print 0009081*, Sep, 18.

O'Malley-James, Jack T., Greaves, Jane S., Raven, John A., and Cockell, Charles S. 2013. Swansong Biospheres: Refuges for Life and Novel Microbial Biospheres on Terrestrial Planets Near the End of Their Habitable Lifetimes. *International Journal of Astrobiology*, **12**(02), 99–112.

O'Malley-James, Jack T., Cockell, Charles S., Greaves, Jane S., and Raven, John A. 2014. Swansong Biospheres II: The Final Signs of Life on Terrestrial Planets Near the End of Their Habitable Lifetimes. *International Journal of Astrobiology*, **13**(03), 229–243.

Omarov, T. B. 1962. On Differential Equations for Oscillating Elements in the Theory of Variable Mass Movement [in Russian]. *Izv. Astrofiz. Inst. Acad. Nauk. KazSSR*, **14**, 66–71.

O'Neill, C., and Lenardic, A. 2007. Geological Consequences of Super-Sized Earths. *Geophysical Research Letters*, **34**(19), L19204.

O'Neill, Craig, Lenardic, Adrian, Weller, Matthew, Moresi, Louis, Quenette, Steve, and Zhang, Siqi. 2016. A Window for Plate Tectonics in Terrestrial Planet Evolution? *Physics of the Earth and Planetary Interiors*, **255**(Jun), 80–92.

O'Neill, Gerard K. 1974. The Colonization of Space. *Physics Today*, **27**(9), 32–40.

  1977. *The High Frontier: Human Colonies in Space*. William Morrow.

Oort, J. H. 1950. The Structure of the Cloud of Comets Surrounding the Solar System and a Hypothesis Concerning its Origin. *Bulletin of the Astronomical Institutes of the Netherlands*, **11**, 91–110.

O'Regan, Gerard. 2016. *Introduction to the History of Computing*. Undergraduate Topics in Computer Science. Cham: Springer International Publishing.

Orgel, Leslie E. 1973. *The Origins of Life: Molecules and Natural Selection*. Wiley.

Osmanov, Z. 2018. Are the Dyson Rings Around Pulsars Detectable? *International Journal of Astrobiology*, **17**(02), 112–115.

Paczynski, B. 1978. A Model of Self-Gravitating Accretion Disk. *Acta Astronomica*, **28**, 91–109.

Paluszek, Michael, and Thomas, Stephanie. 2017. *MATLAB Machine Learning*. Berkeley, CA: Apress.

Pan, Zhijian. 2007. *Artificial Evolution of Arbitrary Self-Replicating Cellular Automata*. Ph.D. thesis.

Papagiannis, Michael D. 1978. Are We All Alone, or Could They Be in the Asteroid Belt? *QJRAS*, **19**.

Parfrey, Laura Wegener, and Lahr, Daniel J. G. 2013. Multicellularity Arose Several Times in the Evolution of Eukaryotes (Response to DOI 10.1002/bies.201100187). *BioEssays*, **35**(4), 339–347.

Parker, Eugene N. 1955. Hydromagnetic Dynamo Models. *ApJ*, **122**(Sep), 293.

Patel, Bhavesh H., Percivalle, Claudia, Ritson, Dougal J., Duffy, Colm D., and Sutherland, John D. 2015. Common Origins of RNA, Protein and Lipid Precursors in a Cyanosulfidic Protometabolism. *Nature Chemistry*, **7**(4), 301–307.

Patruno, A., and Kama, M. 2017. Neutron Star Planets: Atmospheric Processes and Irradiation. *Astronomy & Astrophysics*, **608**(Dec), A147.

Patterson, F. G. 1981. Ape Language. *Science*, **211**(4477), 86–87.

  1978. The Gestures of a Gorilla: Language Acquisition in Another Pongid. *Brain and Language*, **5**(1), 72–97.

Pavlov, Alexander A. 2005a. Catastrophic Ozone Loss During Passage of the Solar System Through an Interstellar Cloud. *Geophysical Research Letters*, **32**(1), L01815.

  2005b. Passing Through a Giant Molecular Cloud: 'Snowball' Glaciations Produced by Interstellar Dust. *Geophysical Research Letters*, **32**(3), L03705.

Penny, Alan John. 2013. The SETI Episode in the 1967 Discovery of Pulsars. *The European Physical Journal H*, Feb.

Petroff, E., Barr, E. D., Jameson, A., Keane, E. F., Bailes, M., Kramer, M., Morello, V., Tabbara, D., and van Straten, W. 2016. FRBCAT: The Fast Radio Burst Catalogue. *Publications of the Astronomical Society of Australia*, **33**(Sep), e045.

Phoenix, C., and Drexler, E. 2004. Safe Exponential Manufacturing. *Nanotechnology*, **15**, 869–872.

Pinker, Steven. 1999. How the Mind Works. *Annals of the New York Academy of Sciences*, **882**(1), 119–127.

Pinker, Steven, and Jackendoff, Ray. 2005. The Faculty of Language: What's Special About It? *Cognition*, **95**(2), 201–236.

Piran, Tsvi, and Jimenez, Raul. 2014. Possible Role of Gamma Ray Bursts on Life Extinction in the Universe. *Physical Review Letters*, **113**(23), 231102.

Power, C. 1998. Old Wives' Tales: the Gossip Hypothesis and the Reliability of Cheap Signals. In: *Approaches to the Evolution of Language*. Cambridge: Cambridge University Press.

Powner, Matthew W., Gerland, Béatrice, and Sutherland, John D. 2009. Synthesis of Activated Pyrimidine Ribonucleotides in Prebiotically Plausible Conditions. *Nature*, **459**(7244), 239–242.

Prantzos, Nikos. 2007. On the 'Galactic Habitable Zone'. *Space Science Reviews*, **135**(1-4), 313–322.

2013. A Joint Analysis of the Drake Equation and the Fermi Paradox. *International Journal of Astrobiology*, **12**(03), 246–253.

Price, Daniel J., and Bate, Matthew R. 2007. The Impact of Magnetic Fields on Single and Binary Star Formation. *MNRAS*, **377**(1), 77–90.

Putnam, Robert D. 2000. *Bowling Alone: The Collapse and Revival of American Community*. Simon & Schuster.

2007. E Pluribus Unum: Diversity and Community in the Twenty-first Century The 2006 Johan Skytte Prize Lecture. *Scandinavian Political Studies*, **30**(2), 137–174.

Rabinowitz, David, Helin, Eleanor, Lawrence, Kenneth, and Pravdo, Steven. 2000. A Reduced Estimate of the Number of Kilometre-Sized Near-Earth Asteroids. *Nature*, **403**(6766), 165–166.

Rafikov, Roman R. 2005. Can Giant Planets Form by Direct Gravitational Instability? *ApJ*, **621**(1), L69–L72.

Raghavan, D., McMaster, H. A., Henry, T. J., Latham, D. W., Marcy, G. W., Mason, B. D., Gies, D. R., White, R. J., and ten Brummelaar, T. A. 2010. A Survey of Stellar Families: Multiplicity of Solar-Type Stars. *ApJSS*, **190**, 1–42.

Rahvar, Sohrab. 2016. Gravitational Microlensing Events as a Target for the SETI Project. *The Astrophysical Journal*, **828**(1), 19.

Ramankutty, Navin, and Foley, Jonathan A. 1999. Estimating Historical Changes in Global Land Cover: Croplands from 1700 to 1992. *Global Biogeochemical Cycles*, **13**(4), 997–1027.

Rampadarath, H., Morgan, J. S., Tingay, S. J., and Trott, C. M. 2012. The First Very Long Baseline Interferometric SETI Experiment. *The Astronomical Journal*, **144**(2), 38.

Rampino, M. 2002. Supereruptions as a Threat to Civilizations on Earth-like Planets. *Icarus*, **156**(2), 562–569.

Raup, D. M., and Valentine, J. W. 1983. Multiple Origins of Life. *PNAS*, **80**(10), 2981–2984.

Raup, David M., and Sepkoski, J. John. 1982. Mass Extinctions in the Marine Fossil Record. *Science*, **215**(4539), 1501–1503.

Raymo, M. E. 1997. The Timing of Major Climate Terminations. *Paleoceanography*, **12**(4), 577–585.

Raymond, Sean N., Kokubo, Eiichiro, Morbidelli, Alessandro, Morishima, Ryuji, and Walsh, Kevin J. 2013. Terrestrial Planet Formation at Home and Abroad. dec, 24.

Rice, W. K. M., Lodato, Giuseppe, Pringle, J. E., Armitage, P. J., and Bonnell, Ian A. 2004. Accelerated Planetesimal Growth in Self-Gravitating Protoplanetary Discs. *MNRAS*, **355**(2), 543–552.

Rice, W. K. M., Lodato, Giuseppe, and Armitage, P. J. 2005. Investigating Fragmentation Conditions in Self-Gravitating Accretion Discs. *MNRAS*, **364**(Oct), L56.

Rice, W. K. M., Armitage, P. J., Mamatsashvili, G. R., Lodato, G., and Clarke, C. J. 2011. Stability of Self-Gravitating Discs Under Irradiation. *MNRAS*, **418**(2), 1356–1362.

Rich, E. 1983. *Artificial Intelligence*. 2 edn. McGraw-Hill Series in Artificial Intelligence, no. Volume 1. McGraw-Hill.

Rickman, Hans, Fouchard, Marc, Froeschlé, Christiane, and Valsecchi, Giovanni B. 2008. Injection of Oort Cloud Comets: The Fundamental Role of Stellar Perturbations. *Celestial Mechanics and Dynamical Astronomy*, **102**(1–3), 111–132.

Riles, K. 2013. Gravitational Waves: Sources, Detectors and Searches. *Progress in Particle and Nuclear Physics*, **68**(jan), 1–54.

Robin, Eugene D. 1973. The Evolutionary Advantages of Being Stupid. *Perspectives in Biology and Medicine*, **16**(3), 369–380.

Robock, Alan, Ammann, Caspar M., Oman, Luke, Shindell, Drew, Levis, Samuel, and Stenchikov, Georgiy. 2009. Did the Toba Volcanic Eruption of $\sim$74 ka B.P. Produce Widespread Glaciation? *Journal of Geophysical Research*, **114**(D10), D10107.

Roebroeks, Wil, and Villa, Paola. 2011. On the Earliest Evidence for Habitual Use of Fire in Europe. *Proceedings of the National Academy of Sciences of the United States of America*, **108**(13), 5209–14.

Rogers, Patrick D., and Wadsley, James. 2012. The Fragmentation of Protostellar Discs: the Hill Criterion for Spiral Arms. *MNRAS*, **423**(2), 1896–1908.

Rose, Christopher, and Wright, Gregory. 2004. Inscribed Matter as an Energy-Efficient Means of Communication with an Extraterrestrial Civilization. *Nature*, **431**(7004), 47–49.

Rose, William I., and Chesner, Craig A. 1990. Worldwide Dispersal of Ash and Gases from Earth's Largest Known Eruption: Toba, Sumatra, 75 ka. *Palaeogeography, Palaeoclimatology, Palaeoecology*, **89**(3), 269–275.

Roush, Bradley Monton, and Sherri. 2001. Gott's Doomsday Argument. *Phil. Sci. Archive*, 23.

Rushby, Andrew J, Claire, Mark W, Osborn, Hugh, and Watson, Andrew J. 2013. Habitable Zone Lifetimes of Exoplanets Around Main Sequence Stars. *Astrobiology*, **13**(9), 833–49.

Ryabov, Vyacheslav A. 2016. The Study of Acoustic Signals and the Supposed Spoken Language of the Dolphins. *St. Petersburg Polytechnical University Journal: Physics and Mathematics*, **2**(3), 231–239.

Sacks, Oliver. 1993 (May). *To See and Not To See*.

Sagan, C., and Newman, W. I. 1983. The Solipsist Approach to Extraterrestrial Intelligence. *QJRAS*, **24**, 113–121.

Sagan, Carl. 1973. *Carl Sagan's Cosmic Connection: An Extraterrestrial Perspective*. Cambridge University Press.

Sampson, Robert. 2012. *Great American City: Chicago and the Enduring Neighborhood Effect*. Chicago: University of Chicago Press.

Sánchez, Joan-Pau, and McInnes, Colin R. 2015. Optimal Sunshade Configurations for Space-Based Geoengineering near the Sun-Earth L1 Point. *PLOS ONE*, **10**(8), e0136648.

Sandberg, A. 1999. The Physics of Information Processing Superobjects: Daily Life Among the Jupiter Brains. *Journal of Evolution and Technology*, **5**, 1.

Sandberg, Anders, Armstrong, Stuart, and Ćirković, Milan M. 2017. That Is Not Dead Which Can Eternal Lie: The Aestivation Hypothesis for Resolving Fermi's Paradox. *JBIS*, Apr, submitted.

Sapir, E. 1921. Language Defined. Chap. 1, page 3 of: *Language: An Introduction to the Study of Speech*.

Schawlow, A. L., and Townes, C. H. 1958. Infrared and Optical Masers. *Physical Review*, **112**(6), 1940–1949.

Schmidt, Gavin A., and Frank, Adam. 2018. The Silurian Hypothesis: Would It Be Possible to Detect an Industrial Civilization in the Geological Record? *International Journal of Astrobiology*, Apr, in press.

Schneider, Peter. 2005. *Weak Gravitational Lensing*. Berlin: Springer.

Schödel, R., Ott, T., Genzel, R., Hofmann, R., Lehnert, M., Eckart, A., Mouawad, N., Alexander, T., Reid, M. J., Lenzen, R., Hartung, M., Lacombe, F., Rouan, D., Gendron, E., Rousset, G., Lagrange, A.-M., Brandner, W., Ageorges, N., Lidman, C., Moorwood, A. F. M., Spyromilio, J., Hubin, N., and Menten, K. M. 2002. A Star in a 15.2-Year Orbit Around the Supermassive Black Hole at the Centre of the Milky Way. *Nature*, **419**(6908), 694–696.

Schröder, K.-P., and Connon Smith, Robert. 2008. Distant Future of the Sun and Earth Revisited. *MNRAS*, **386**(1), 155–163.

Schubert, G., and Soderlund, K. M. 2011. Planetary Magnetic Fields: Observations and Models. *Physics of the Earth and Planetary Interiors*, **187**(3–4), 92–108.

Schuetz, Marlin, Vakoch, Douglas A., Shostak, Seth, and Richards, Jon. 2016. Optical SETI Observations of the Anomalous Star KIC 8462852. *The Astrophysical Journal*, **825**(1), L5.

Schulkin, Jay, and Raglan, Greta B. 2014. The Evolution of Music and Human Social Capability. *Frontiers in Neuroscience*, **8**, 292.

Schwieterman, Edward W., Kiang, Nancy Y., Parenteau, Mary N., Harman, Chester E., DasSarma, Shiladitya, Fisher, Theresa M., Arney, Giada N., Hartnett, Hilairy E., Reinhard, Christopher T., Olson, Stephanie L., Meadows, Victoria S., Cockell, Charles S., Walker, Sara I., Grenfell, John Lee, Hegde, Siddharth, Rugheimer, Sarah, Hu, Renyu, and Lyons, Timothy W. 2017. Exoplanet Biosignatures: A Review of Remotely Detectable Signs of Life. may.

Seager, S., Bains, W., and Hu, R. 2013. A Biomass-Based Model To Estimate the Plausibility of Exoplanet Biosignature Gases. *ApJ*, **775**(2), 104.

Seager, S., Bains, W., and Petkowski, J.J. 2016. Toward a List of Molecules as Potential Biosignature Gases for the Search for Life on Exoplanets and Applications to Terrestrial Biochemistry. *Astrobiology*, **16**(6), 465–485.

Segura, Antígona, Walkowicz, Lucianne M., Meadows, Victoria, Kasting, James, and Hawley, Suzanne. 2010. The Effect of a Strong Stellar Flare on the Atmospheric Chemistry of an Earth-like Planet Orbiting an M Dwarf. *Astrobiology*, **10**(7), 751–771.

Seifried, D., Banerjee, R., Pudritz, R. E., and Klessen, R. S. 2013. Turbulence-Induced Disc Formation in Strongly Magnetized Cloud Cores. *MNRAS*, **432**(4), 3320–3331.

Shepherd, L. 1952. Interstellar Flight. *JBIS*, **11**, 149–167.

Shettleworth, S. J. 2010. *Cognition, Evolution, and Behavior*. Oxford University Press.

Shkadov, L. M. 1987. Possibility of Controlling Solar System Motion in the Galaxy. *IAF, International Astronautical Congress, 38th*, 10–17.

Shostak, Seth. 2010. What ET Will Look Like and Why Should We Care. *Acta Astronautica*, **67**(9–10), 1025–1029.

Shostak, Seth, and Almar, Ivan. 2002. The Rio Scale Applied to Fictional. Pages IAA–9–1–06 of: *34th COSPAR Scientific Assembly, Second World Space Congress*.

Shu, Frank H., Adams, Fred C., and Lizano, Susana. 1987. Star Formation in Molecular Clouds: Observation and Theory. *Annual Review of Astronomy and Astrophysics*, **25**(1), 23–81.

Shuch, H. Paul. 2011. *Searching for Extraterrestrial Intelligence*. The Frontiers Collection. Berlin, Heidelberg: Springer Berlin Heidelberg.

Siemion, Andrew P. V., Demorest, Paul, Korpela, Eric, Maddalena, Ron J., Werthimer, Dan, Cobb, Jeff, Howard, Andrew W., Langston, Glen, Lebofsky, Matt, Marcy, Geoffrey W., and Tarter, Jill. 2013. A 1.1-1.9 GHz SETI Survey of the Kepler Field. I. A Search for Narrow-Band Emission from Select Targets. *ApJ*, **767**(1), 94.

Silagadze, Z. K. 2008. SETI and Muon Collider. *Acta Physica Polonica B*, **39**(11), 2943.

Simpson, G. G. 1964. The Nonprevalence of Humanoids. *Science*, **143**(3608), 769–775.

Smartt, S. J., Chen, T.-W., Jerkstrand, A., Coughlin, M., Kankare, E., Sim, S. A., Fraser, M., Inserra, C., Maguire, K., Chambers, K. C., Huber, M. E., Krühler, T., Leloudas, G., Magee, M., Shingles, L. J., Smith, K. W., Young, D. R., Tonry, J., Kotak, R., Gal-Yam, A., Lyman, J. D., Homan, D. S., Agliozzo, C., Anderson, J. P., Angus, C. R., Ashall, C., Barbarino, C., Bauer, F. E., Berton, M., Botticella, M. T., Bulla, M., Bulger, J., Cannizzaro, G., Cano, Z., Cartier, R., Cikota, A., Clark, P., De Cia, A., Della Valle, M., Denneau, L., Dennefeld, M., Dessart, L., Dimitriadis, G., Elias-Rosa, N., Firth, R. E., Flewelling, H., Flörs, A., Franckowiak, A., Frohmaier, C., Galbany, L., González-Gaitán, S., Greiner, J., Gromadzki, M., Guelbenzu, A. Nicuesa, Gutiérrez, C. P., Hamanowicz, A., Hanlon, L., Harmanen, J., Heintz, K. E., Heinze, A., Hernandez, M.-S., Hodgkin, S. T., Hook, I. M., Izzo, L., James, P. A., Jonker, P. G., Kerzendorf, W. E., Klose, S., Kostrzewa-Rutkowska, Z., Kowalski, M., Kromer, M., Kuncarayakti, H., Lawrence, A., Lowe, T. B., Magnier, E. A., Manulis, I., Martin-Carrillo, A., Mattila, S., McBrien, O., Müller, A., Nordin, J., O'Neill, D., Onori, F., Palmerio, J. T., Pastorello, A., Patat, F., Pignata, G., Podsiadlowski, Ph., Pumo, M. L., Prentice, S. J., Rau, A., Razza, A., Rest, A., Reynolds, T., Roy, R., Ruiter, A. J., Rybicki, K. A., Salmon, L., Schady, P., Schultz, A. S. B., Schweyer, T., Seitenzahl, I. R., Smith, M., Sollerman, J., Stalder, B., Stubbs, C. W., Sullivan, M., Szegedi, H., Taddia, F., Taubenberger, S., Terreran, G., van Soelen, B., Vos, J., Wainscoat, R. J., Walton, N. A., Waters, C., Weiland, H., Willman, M., Wiseman, P., Wright, D. E., Wyrzykowski, Ł., and Yaron, O. 2017. A Kilonova as the Electromagnetic Counterpart to a Gravitational-Wave Source. *Nature*, Oct.

Smartt, Stephen J. 2009. Progenitors of Core-Collapse Supernovae. *Annual Review of Astronomy and Astrophysics*, **47**(1), 63–106.

Smith, David S., and Scalo, John M. 2009. Habitable Zones Exposed: Astrosphere Collapse Frequency as a Function of Stellar Mass. *Astrobiology*, **9**(7), 673–681.

Smith, J. M. 1982. *Evolution and the Theory of Games*. Cambridge University Press.

Smith, John Maynard, and Szathmary, Eors. 1997. *The Major Transitions in Evolution*. Oxford University Press, Oxford.

2000. *The Origins of Life: From the Birth of Life to the Origin of Language*. Oxford University Press, Oxford.

Smith, Peter. 2003. *An Introduction to Formal Logic*. 6th edn. Vol. 6. Cambridge: Cambridge University Press.

Smith, Reginald D. 2009. Broadcasting But Not Receiving: Density Dependence Considerations for SETI Signals. *International Journal of Astrobiology*, **8**(02), 101.

Snyder, Donald Lee, and Miller, Michael I. 1991. *Random Point Processes in Time and Space*. New York, NY, USA: Springer-Verlag.

Socas-Navarro, Hector. 2018. Possible Photometric Signatures of Moderately Advanced Civilizations: The Clarke Exobelt. *ApJ*, **855**(2), 110.

Society, Royal. 2004. *Nanoscience and Nanotechnologies: Opportunities and Uncertainties*. Tech. rept. The Royal Society.

Sockol, M. D., Raichlen, D. A., and Pontzer, H. 2007. Chimpanzee Locomotor Energetics and the Origin of Human Bipedalism. *Proceedings of the National Academy of Sciences*, **104**(30), 12265–12269.

Sojo, Victor, Herschy, Barry, Whicher, Alexandra, Camprubí, Eloi, and Lane, Nick. 2016. The Origin of Life in Alkaline Hydrothermal Vents. *Astrobiology*, **16**(2), 181–197.

Sotos, John G. 2017. Biotechnology and the Lifetime of Technical Civilizations. *arXiv e-print 1709.01149*, Sep.

Spiegel, D. 1995. Hypnosis and Suggestion. Chap. 4, pages 129–173 of: Schacter, Daniel L., Coyle, J. S., Fishbach, G. D., Mesulam, M. M., and Sullivan, L. E. (eds.), *Memory Distortion: How Minds, Brains, and Societies Reconstruct the Past*. Harvard University Press.

Spoor, F., Leakey, M. G., Gathogo, P. N., Brown, F. H., Antón, S. C., McDougall, I., Kiarie, C., Manthi, F. K., and Leakey, L. N. 2007. Implications of New Early Homo Fossils from Ileret, East of Lake Turkana, Kenya. *Nature*, **448**(7154), 688–91.

Stamenković, Vlada, Noack, Lena, Breuer, Doris, and Spohn, Tilman. 2012. The Influence of Pressure-Dependent Viscosity on the Thermal Evolution of Super-Earths. *The Astrophysical Journal*, **748**(1), 41.

Stanek, K. Z., Matheson, T., Garnavich, P. M., Martini, P., Berlind, P., Caldwell, N., Challis, P., Brown, W. R., Schild, R., Krisciunas, K., Calkins, M. L., Lee, J. C., Hathi, N., Jansen, R. A., Windhorst, R., Echevarria, L., Eisenstein, D. J., Pindor, B., Olszewski, E. W., Harding, P., Holland, S. T., and Bersier, D. 2003. Spectroscopic Discovery of the Supernova 2003dh Associated with GRB 030329. *The Astrophysical Journal*, **591**(1), L17–L20.

Steadman, D. 2003. The Late Quaternary Extinction and Future Resurrection of Birds on Pacific Islands. *Earth-Science Reviews*, **61**(1–2), 133–147.

Stefano, R. Di, and Ray, A. 2016. Globular Clusters as Cradles of Life and Advanced Civilizations. *ApJ*, **827**(1), 54.

Stephenson, D. G. 1977. Factors Limiting the Interaction Between Twentieth Century Man and Interstellar Cultures. *Journal of the British Interplanetary Society*, **30**.

Stern, Robert J. 2016. Is Plate Tectonics Needed to Evolve Technological Species on Exoplanets? *Geoscience Frontiers*, **7**(4), 573–580.

Stevens, Adam, Forgan, Duncan, and James, Jack O'Malley. 2016. Observational Signatures of Self-Destructive Civilizations. *International Journal of Astrobiology*, **15**(04), 333–344.

Stevens, Adam H. 2015. The Price of Air. Pages 51–61 of: *Human Governance Beyond Earth*.

Stewart, Sarah T., and Leinhardt, Zoë M. 2009. Velocity-Dependent Catastrophic Disruption Criteria for Planetesimals. *ApJ*, **691**(2), L133–L137.

Stone, R.P.S., Wright, S.A., Drake, F., Muñoz, M., Treffers, R., and Werthimer, D. 2005. Lick Observatory Optical SETI: Targeted Search and New Directions. *Astrobiology*, **5**(5), 604–611.

Stringer, Chris. 2012. Evolution: What Makes a Modern Human. *Nature*, **485**(7396), 33–5.

Strughold, Hubertus. 1953. *The Green and Red Planet: A Physiological Study of the Possibility of Life on Mars*. University of New Mexico Press.

Stuart, Joseph Scott, and Binzel, Richard P. 2004. Bias-Corrected Population, Size Distribution, and Impact Hazard for the Near-Earth Objects. *Icarus*, **170**(2), 295–311.

Sugihara, Kokichi. 2016. Anomalous Mirror Symmetry Generated by Optical Illusion. *Symmetry*, **8**(4), 21.

Surdin, V. G. 1986. Launch of a Galactic Probe using a Multiple Perturbation Maneuver. *Astronomicheskii Vestnik*, **19**, 354–358.

Swinney, R. W., Long, K. F., Hein, A., Galea, P., Mann, A., Crowl, A., and Obousy, R. 2012. Project Icarus: Exploring the Interstellar Roadmap Using the Icarus Pathfinder and Starfinder Probe Concepts. *Journal of the British Interplanetary Society,*, **65**, 244–254.

Tainter, J. 1988. *The Collapse of Complex Societies*. New Studies in Archaeology. Cambridge University Press.

Tainter, Joseph A. 2000. Problem Solving: Complexity, History, Sustainability. *Population and Environment*, **22**(1), 3–41.

Talbot Jr, R. J., and Newman, M. J. 1977. Encounters Between Stars and Dense Interstellar Clouds. *The Astrophysical Journal Supplement Series*, **34**(Jul), 295.

Tallis, Melisa, Maire, Jerome, Wright, Shelley, Drake, Frank D., Duenas, Andres, Marcy, Geoffrey W., Stone, Remington P. S., Treffers, Richard R., Werthimer, Dan, and NIROSETI. 2016. A Near-Infrared SETI Experiment: A Multi-time Resolution Data Analysis. *American Astronomical Society, AAS Meeting #228, id.120.03*, **228**.

Tarter, Jill. 2007. Searching for Extraterrestrial Intelligence. Chap. 26, pages 513–536 of: Sullivan III, Woodruff T., and Baross, Thomas (eds.), *Planets and Life – The Emerging Science of Astrobiology*. Cambridge: Cambridge University Press.

Taylor, Jim. 2014. *Deep Space Communications*. 1st edn. Pasadena: Jet Propulsion Laboratory.

Taylor, Joseph H. 1994. Binary Pulsars and Relativistic Gravity. *Reviews of Modern Physics*, **66**(3), 711–719.

Terrace, H., Petitto, L., Sanders, R., and Bever, T. 1979. Can an Ape Create a Sentence? *Science*, **206**(4421), 891–902.

Thomas, Brian C., Melott, Adrian L., Jackman, Charles H., Laird, Claude M., Medvedev, Mikhail V., Stolarski, Richard S., Gehrels, Neil, Cannizzo, John K., Hogan, Daniel P., and Ejzak, Larissa M. 2005a. Gamma-Ray Bursts and the Earth: Exploration of Atmospheric, Biological, Climatic, and Biogeochemical Effects. *ApJ*, **634**(1), 509–533.

Thomas, Brian C., Jackman, Charles H., Melott, Adrian L., Laird, Claude M., Stolarski, Richard S., Gehrels, Neil, Cannizzo, John K., and Hogan, Daniel P. 2005b. Terrestrial Ozone Depletion Due to a Milky Way Gamma-Ray Burst. *ApJ*, **622**(2), L153–L156.

Thomas, Brian C., Engler, E. E., Kachelrieß, M., Melott, A. L., Overholt, A. C., and Semikoz, D. V. 2016. Terrestrial Effects of Nearby Supernovae in the Early Pleistocene. *ApJ*, **826**(May).

Thommes, E. W., Duncan, M. J., and Levison, H. F. 2003. Oligarchic Growth of Giant Planets. *Icarus*, **161**(2), 431–455.

Thompson, R F. 2000. *The Brain: A Neuroscience Primer*. Worth Publishers.

Tian, Feng, France, Kevin, Linsky, Jeffrey L., Mauas, Pablo J. D., and Vieytes, Mariela C. 2014. High Stellar FUV/NUV Ratio and Oxygen Contents in the Atmospheres of Potentially Habitable Planets. *Earth and Planetary Science Letters*, **385**(jan), 22–27.

Tiec, Alexandre Le, and Novak, Jérôme. 2016. Theory of Gravitational Waves. In: Auger, G., and Plagnol, E. (eds.), *An Overview of Gravitational Waves: Theory and Detection*, 1st edn. World Scientific.

Timofeev, M. Y., Kardashev, N. S., and Promyslov, V. G. 2000. A Search of the IRAS Database for Evidence of Dyson Spheres. *Acta Astronautica*, 655–659.

Tipler, Frank J. 1993 (Aug). SETI – A Waste of Time! Pages 28–35 of: Kingsley, Stuart A. (ed), *SPIE 1867, The Search for Extraterrestrial Intelligence (SETI) in the Optical Spectrum*.

Tobin, John J., Looney, Leslie W., Wilner, David J., Kwon, Woojin, Chandler, Claire J., Bourke, Tyler L., Loinard, Laurent, Chiang, Hsin-Fang, Schnee, Scott, and Chen, Xuepeng. 2015. A Sub-Arcsecond Survey Toward Class 0 Protostars in Perseus: Searching for Signatures of Protostellar Disks. *ApJ*, **805**(2), 125.

Tomida, Kengo, Okuzumi, Satoshi, and Machida, Masahiro N. 2015. Radiation Magnetohydrodynamic Simulations of Protostellar Collapse: Nonideal Magnetohydrodynamic Effects and Early Formation of Circumstellar Disks. *ApJ*, **801**(2), 117.

Toynbee, A. J., and Somervell, D. C. 1987. *A Study of History*. A Study of History. Oxford University Press.

Tsander, F. 1961. *Problems of Flight by Jet Propulsion: Interplanetary Flights*. Moscow: Nawzhio – Technical Publishing House.

Tsiganis, K., Gomes, R., Morbidelli, A., and Levison, H. F. 2005. Origin of the Orbital Architecture of the Giant Planets of the Solar System. *Nature*, **435**(7041), 459–461.

UN. 2004. *World Population to 2300*. Tech. rept. New York.

United Nations Department of Economic and Social Affairs. 2015. *World Population Prospects: The 2015 Revision*. Tech. rept. United Nations, New York.

Valencia, Diana, O'Connell, Richard J., and Sasselov, Dimitar. 2006. Internal Structure of Massive Terrestrial Planets. *Icarus*, **181**(2), 545–554.

 2007. Inevitability of Plate Tectonics on Super-Earths. *The Astrophysical Journal*, **670**(1), L45–L48.

Valencia, Diana, Tan, Vivian Yun Yan, and Zajac, Zachary. 2018. Habitability from Tidally Induced Tectonics. *The Astrophysical Journal*, **857**(2), 106.

Vaughan, Naomi E., and Lenton, Timothy M. 2011. A Review of Climate Geoengineering Proposals. *Climatic Change*, **109**(3), 745–790.

Veras, Dimitri. 2016. Post-Main-Sequence Planetary System Evolution. *Royal Society Open Science*, **3**(2), id 150571.

Verendel, Vilhelm, and Häggström, Olle. 2015. Fermi's Paradox, Extraterrestrial Life and the Future of Humanity: a Bayesian Analysis. *arXiv e-print 1510.08684*, Oct, 16.

Verhulst, Pierre-Francois. 1845. Mathematical Researches into the Law of Population Growth Increase. *Nouveaux Mémoires de l'Académie Royale des Sciences et Belles-Lettres de Bruxelles*, **18**, 1–42.

Verpooten, J. 2015. *Art and Signaling in a Cultural Species: Proefschrift*. Proefschriften UA-WET.

Vida, K., Kővári, Zs., Pál, A., Oláh, K., and Kriskovics, L. 2017. Frequent Flaring in the TRAPPIST-1 System–Unsuited for Life? *The Astrophysical Journal*, **841**(2), 124.

Vidotto, A. A., Jardine, M., Morin, J., Donati, J.-F., Lang, P., and Russell, A. J. B. 2013. Planetary Protection in the Extreme Environments of Low-Mass Stars. *Proceedings of the International Astronomical Union*, **9**(S302), 237–238.

Villarroel, Beatriz, Imaz, Inigo, and Bergstedt, Josefine. 2016. Our Sky Now and Then: Searches for Lost Stars and Impossible Effects as Probes of Advanced Civilizations. *The Astronomical Journal*, **152**(3), 76.

von Kiedrowski, Günter. 1986. A Self-Replicating Hexadeoxynucleotide. *Angewandte Chemie International Edition in English*, **25**(10), 932–935.

Von Neumann, John. 1966. *Theory of Self-Reproducing Automata*. Urbana: University of Illinois Press.

Vukotic, B., and Ćirković, Milan M. 2007. On the Timescale Forcing in Astrobiology. *Serbian Astronomical Journal*, **175**(175), 45–50.

 2008. Neocatastrophism and the Milky Way Astrobiological Landscape. *Serbian Astronomical Journal*, **176**(176), 71–79.

2012. Astrobiological Complexity with Probabilistic Cellular Automata. *Origins of Life and Evolution of the Biosphere: The Journal of the International Society for the Study of the Origin of Life*, **42**(4), 347–71.

Wachtershauser, G. 1990. Evolution of the First Metabolic Cycles. *Proceedings of the National Academy of Sciences*, **87**(1), 200–204.

Wainscoat, Richard J., Chambers, Kenneth, Lilly, Eva, Weryk, Robert, Chastel, Serge, Denneau, Larry, and Micheli, Marco. 2015. The Pan-STARRS Search for Near Earth Objects. *American Astronomical Society, DPS Meeting #47*, id.301.07.

Wang, Bo, and Han, Zhanwen. 2012. Progenitors of Type Ia Supernovae. *New Astronomy Reviews*, **56**(4), 122–141.

Wang, Junfeng, Zhou, Zongzheng, Zhang, Wei, Garoni, Timothy M., and Deng, Youjin. 2013. Bond and Site Percolation in Three Dimensions. *Physical Review. E, Statistical, Nonlinear, and Soft Matter Physics*, **87**(5), 052107.

Ward, S. 2000. Asteroid Impact Tsunami: A Probabilistic Hazard Assessment. *Icarus*, **145**(1), 64–78.

Ward, Steven N., and Asphaug, Erik. 2003. Asteroid Impact Tsunami of 2880 March 16. *Geophysical Journal International*, **153**(3), F6–F10.

Ward, Steven N., and Day, Simon. 2001. Cumbre Vieja Volcano-Potential Collapse and Tsunami at La Palma, Canary Islands. *Geophysical Research Letters*, **28**(17), 3397–3400.

Ward, William R., and Canup, Robin M. 2010. Circumplanetary Disk Formation. *The Astronomical Journal*, **140**(5), 1168–1193.

Webb, Stephen. 2002. *If the Universe Is Teeming with Aliens – Where Is Everybody?: Fifty Solutions to the Fermi Paradox and the Problem of Extraterrestrial Life*. Vol. 1. Springer.

Weiss, Madeline C., Sousa, Filipa L., Mrnjavac, Natalia, Neukirchen, Sinje, Roettger, Mayo, Nelson-Sathi, Shijulal, and Martin, William F. 2016. The Physiology and Habitat of the Last Universal Common Ancestor. *Nature Microbiology*, **1**(9), 16116.

Wells, R., Poppenhaeger, K., Watson, C. A., and Heller, R. 2018. Transit Visibility Zones of the Solar System Planets. *MNRAS*, **473**(1), 345–354.

Wetherill, George W. 1994. Possible Consequences of Absence of Jupiters in Planetary Systems. *Astrophysics and Space Science*, **212**(1–2), 23–32.

White, C. R., and Seymour, R. S. 2003. Mammalian Basal Metabolic Rate Is Proportional to Body Mass 2/3. *Proceedings of the National Academy of Sciences*, **100**(7), 4046–4049.

Whitehead, A. N., and Russell, B. 1927. *Principia Mathematica*. Cambridge Mathematical Library, no. v. 1. Cambridge University Press.

Wiley, Keith B. 2011. The Fermi Paradox, Self-Replicating Probes, and the Interstellar Transportation Bandwidth. *arXiv e-prints 1111.6131*, Nov.

Willcox, George, Fornite, Sandra, and Herveux, Linda. 2008. Early Holocene Cultivation Before Domestication in Northern Syria. *Vegetation History and Archaeobotany*, **17**(3), 313–325.

Wilson, E. O., and Holldobler, B. 2005. Eusociality: Origin and Consequences. *Proceedings of the National Academy of Sciences*, **102**(38), 13367–13371.

Windmark, F., Birnstiel, T., Güttler, C., Blum, J., Dullemond, C. P., and Henning, Th. 2012. Planetesimal Formation by Sweep-up: How the Bouncing Barrier Can Be Beneficial to Growth. *A&A*, **540**(Mar), A73.

Winn, J. N. 2011. Exoplanet Transits and Occultations. Pages 55–78 of: Seager, S. (ed.), *Exoplanets*. Tucson: University of Arizona Press.

Wolszczan, A., and Frail, D. A. 1992. A Planetary System Around the Millisecond Pulsar PSR1257 + 12. *Nature*, **355**(6356), 145–147.

Wrangham, Richard W., Wilson, Michael L., and Muller, Martin N. 2006. Comparative Rates of Violence in Chimpanzees and Humans. *Primates*, **47**(1), 14–26.

Wright, J. T., Mullan, B., Sigurdsson, S., and Povich, M. S. 2014. The Ĝ Infrared Search for Extraterrestrial Civilizations with Large Energy Supplies. I. Background and Justification. *The Astrophysical Journal*, **792**(1), 26.

Wright, Jason T. 2017. Prior Indigenous Technological Species. *eprint arXiv: 1704.07263*, apr.

2018. SETI is Part of Astrobiology. jan.

Wright, Jason T., and Oman-Reagan, Michael P. 2017. Visions of Human Futures in Space and SETI. *International Journal of Astrobiology*, Aug, 1–12.

Wright, Jason T., and Sigurdsson, Steinn. 2016. Families of Plausible Solutions to the Puzzle of Boyajian's Star. *ApJ*, **829**(1), L3.

Wright, Jason T., Cartier, Kimberly M. S., Zhao, Ming, Jontof-Hutter, Daniel, and Ford, Eric B. 2015. The Ĝ Search for Extraterrestrial Civilizations with Large Energy Supplies. IV. The Signatures and Information Content of Transiting Megastructures. *ApJ*, **816**(1), 17.

Wünnemann, K., Collins, G. S., and Weiss, R. 2010. Impact of a Cosmic Body into Earth's Ocean and the Generation of Large Tsunami Waves: Insight From Numerical Modeling. *Reviews of Geophysics*, **48**(4), RG4006.

Wurster, James, Price, Daniel J., and Bate, Matthew R. 2016. Can Non-ideal Magnetohydrodynamics Solve the Magnetic Braking Catastrophe? *MNRAS*, **457**(1), 1037–1061.

WWF. 2016. *Living Planet Report 2016*. Tech. rept. World Wildlife Fund.

Xu, Yan, Newberg, Heidi Jo, Carlin, Jeffrey L., Liu, Chao, Deng, Licai, Li, Jing, Schönrich, Ralph, and Yanny, Brian. 2015. Rings and Radial Waves in the Disk of the Milky Way. *ApJ*, **801**(2), 105.

Youdin, Andrew N., and Goodman, Jeremy. 2005. Streaming Instabilities in Protoplanetary Disks. *The Astrophysical Journal*, **620**(1), 459–469.

Zackrisson, Erik, Calissendorff, Per, González, Juan, Benson, Andrew, Johansen, Anders, and Janson, Markus. 2016. Terrestrial Planets Across Space and Time. *The Astrophysical Journal*, **833**(2), 214.

Zappes, Camilah Antunes, Andriolo, Artur, Simões-Lopes, Paulo César, and Di Beneditto, Ana Paula Madeira. 2011. 'Human-Dolphin (Tursiops Truncatus Montagu, 1821) Cooperative Fishery' and Its Influence on Cast Net Fishing Activities in Barra de Imbé/Tramandaí, Southern Brazil. *Ocean & Coastal Management*, **54**(5), 427–432.

Zeki, Semir. 1999. *Inner Vision: An Exploration of Art and the Brain*. Oxford University Press.

Zerkle, Aubrey L., Claire, Mark W., Domagal-Goldman, Shawn D., Farquhar, James, and Poulton, Simon W. 2012. A Bistable Organic-Rich Atmosphere on the Neoarchaean Earth. *Nature Geoscience*, **5**(5), 359–363.

Zhang, Bing. 2013. A Possible Connection Between Fast Radio Bursts and Gamma-Ray Bursts. *The Astrophysical Journal*, **780**(2), L21.

Zhang, Zhihua, Moore, John C., Huisingh, Donald, and Zhao, Yongxin. 2015. Review of Geoengineering Approaches to Mitigating Climate Change. *Journal of Cleaner Production*, **103**(Sep), 898–907.

Zubrin, R. M. 1995. Detection of Extraterrestrial Civilizations via the Spectral Signature of Advanced Interstellar Spacecraft. Page 592 of: Shostak, Seth (ed.), *Progress in the Search for Extraterrestrial Life: 1993 Bioastronomy Symposium, Santa Cruz, California*. San Francisco: Astronomical Society of the Pacific.

Index

abiogenesis, 100
absolute magnitude, 185
adaptation, 117
afterglow, 210
air showers, 208
albedo, 71
allele, 117, 284
allele frequencies, 116
amino acid, 99
ancestor-simulation, 331
anthropic principle, 139, 262
anticipatory cognition, 141
apparent magnitude, 185
arc volcano, 170
artefact SETI, 35
artificial intelligence (AI), 234
astro-engineering, 35
asymptotic giant branch, AGB, 188
aurora, 181
aurorae, 174
automatic theorem provers, 237

backward propagation, 241
Bayes's Theorem, 340
Bekenstein bound, 334
biodiversity, 222
biosignatures/biomarkers, 64
Boussinesq approximation, 175
Bracewell Probe, 338
Bremermann's limit, 335
brightness temperature, 22
brown dwarf, 37, 189

Cambrian explosion, 105
carbon fixation, 109
chromosome, 108
chromosphere, 180
chron, 177
circumstellar habitable zone (CHZ), 71

Clustered Regularly Interspaced Short Palindromic Repeats (CRISPR), 228
cognitive bias, 238
colour, stellar, 185
connected component, 323
connectionism, 239
continuous circumstellar habitable zone (CCHZ), 73
convergent evolution, 128
coral bleaching, 223
corona, 180
cosmic microwave background, 22
CRISPR-Cas9, 228

data science, 241
decision tree, 237
Declination (Dec), 15
diploid, 108
Drake Equation, 7
Drake, Frank, 7
dualism, 331
Dyson sphere, 35

Earth, age of, 3
Earth-like, 71
effective temperature, 180
electron neutrino, 197
endocast, 119
endosymbiont theory, 111
energetically optimised communication, 291
entropic level analysis, 304
entropy, 99
epigenomics, 229
eukaryotes, 111
eusocial society, 124
evidence, 13
evolution, 115
evolutionarily stable strategy, ESS, 122
exaptation, 148
exomoon, 97

Fact A, 15
false negative, 66
false positive biosignature, 66
Fast Radio Burst, FRB, 364
Fermi, Enrico, 5
Fermi's Paradox, 3
Fisher's Fundamental Theorem, 117
fitness, 117
forward propagation, 241
fringe, 29

Galactic Cosmic Ray (GCR), 208
Galactic Technological Zone, 318
game theory, 121
gene drive, 229
gene flow, 116
gene transfer, horizontal, 115
gene transfer, vertical, 115
generation ship, 319
genetic drift, 116
genetics, 115
genomics, 229
genotype, 128
geomagnetic reversal, 177
geosynchronous orbit (GEO), 177
Glashow Resonance, 297
Global Electric Circuit, 208
Gödel's completeness theorem, 235
gradient descent, 241
Grand Tack, 96
graph theory, 322
Great Filter, The, 11

halting problem, 236
Hart, Michael, 6
heliosphere, 181
heliosphere,astrosphere, 208
High Earth Orbit (HEO), 224
Hill radius, 81
horizontal gene transfer, 118
hydrogen line (HI), 22
hydrothermal vents, 103

ideal magnetohydrodynamics (MHD), 77, 175
induction equation, 77
interstellar ark, 319
interstellar coherence hole, 293
interstellar medium, 75
interstellar ramjet, 274
island biogeography, 320
ISM, 75

knapping, 147
Konopinski, Emil, 5

Lagrange point, 95
Large Synoptic Survey Telescope, 342
lattice QCD, 337

learning rate, 241
lepton, 296
Life, definition of, 98
likelihood, 13
lithosphere, 169, 171
Low Earth Orbit (LEO), 224, 225
LUCA, 100
Lurker Probe, 338

machine learning, 241
macro-engineering, 35
magnetic braking, 76
magnetopause, 174
magnetosphere, 174
magnetotail, 174
magnitude, stellar, 185
main sequence, 185
Malthusian check, 212
mantle, 171
Mariner, 344
Medium Earth Orbit (MEO), 224
megafauna, 132
meme, 138
meteoroids, 225
Milky Way, 3
minimum spanning tree, MST, 324
molecular cloud, 75, 208
Monte Carlo Realisation, 361
Moore's Law, 330
multicellular organism, 112
Multilevel Keying (MLK), 294
muon, 296
Mutually Assured Destruction (MAD), 253

nearest neighbour heuristic (NNH), 269
negentropy, 99
neural network, 239
neuron, 119
neurotransmitter, 239
neutrino, 296
neutrinosphere, 197
non-ideal magnetohydrodynamics, 77
NRAO, Green Bank, 7
nucleic acid, 99

O'Neill colony/cylinder, 319
Occam's razor, 369
On-Off Keying (OOK), 294
opacity, 72
optimisation, 242

parasocial societies, 124
pebble accretion, 80
peptide, 99
perceptron, 240
percolation theory, 265
phenotype, 128
photosphere, 180

planet formation, core accretion, 78
planet formation, disc instability theory, 82
planetary protection, 326
plate tectonics, 169, 171
ploidy, 116
polarimetry, 18
post-detection protocol, 40
post-biological intelligence, 125
posterior, 13
predicate logic, first-order (PL1), 236
predicate logic, second-order (PL2), 238
prestellar core, 75
principle of assumed benevolence, 316
principle of precautionary malevolence, 316
principle of preliminary neutrality, 316
prior, 13
probabilistic cellular automata, PCA, 361
protein, 99
protolanguage, 140
push-down automaton, 244

radio frequency interference (RFI), 24
Radio SETI, 21
radiogenic heating, 173
Rayleigh Number, 172
recombination, 116
red giant branch, RGB, 188, 191
relativistic aberration, 280
reply protocol, 40
Right Ascension (RA), 15
Rio Scale, 38
Rossby number, 176
rubble pile, 80
RuBiSCo, 110

seed magnetic field, 176
semiotics, 137
SETA, 35
Shannon limit, 292
shield volcano, 170
shot noise, 22
sidelobe, 27
Singularity, the, 234
snowball glaciation, 209
social capital, 256

space debris, 225
spandrel, 148
stagnant-lid regime, 172
standoff distance, 177
streaming instability, 80
strong AI, 234
subduction, 171
subduction zone, 169
subgiant branch, 188
substrate independence, 330
superchron, 177
superintelligence, 235
supernova, Type I, 194
supernova, Type II, 196
supervolcano, 170
synapse, 119, 239

tachocline, 180
tau lepton, 296
technosignature, 68
Teller, Edward, 5
theory of mind, 141
Tipler, Frank J., 6
training set, 242
Turing machine, 235
Turing-Church thesis (see also: halting problem), 236

Universal Grammar (UG), 139
Universal Turing Machine, 235
urban heat islands, 222

Van Allen radiation belts, 174
vegetative red edge, 68
Venera, 344
Volcanic Explosivity Index (VEI), 170

weak AI, 234
working fluid, 273
Wright's Adaptive Landscape, 117

York, Herbert, 5

Zipf's law, 306
zooxanthellae, 223